D0782291

Neurochemistry

a practical approach

TITLES PUBLISHED IN

THE PRACTICAL APPROACH SERIES

Gel electrophoresis of proteins
Gel electrophoresis of nucleic acids
Iodinated density-gradient media
Centrifugation (2nd Edition)
Transcription and translation
Oligonucleotide synthesis
Microcomputers in biology
Mutagenicity testing
Affinity chromatography
Virology
DNA cloning
Plant cell culture
Immobilised cells and enzymes
Nucleic acid hybridisation
Animal cell culture
Human cytogenetics
Photosynthesis: energy transduction
Drosophila
Human genetic diseases
H.p.l.c. of small molecules
Carbohydrate analysis
Biochemical toxicology
Electron microscopy in molecular biology
Nucleic acid and protein sequence analysis
Teratocarcinomas and embryonic stem cells
Spectrophotometry and Spectrofluorimetry
Plasmids
Biological membranes

Neurochemistry

a practical approach

Edited by

A J Turner

Department of Biochemistry, University of Leeds, Leeds LS2 9JT, UK

H S Bachelard

Department of Biochemistry, St. Thomas's Hospital Medical School, London SE1 7EH, UK

IRL PRESS

Oxford · Washington DC

IRL Press Limited
PO Box 1,
Eynsham,
Oxford OX8 1JJ,
England

British Library Cataloguing in Publication Data

Neurochemistry : a practical approach.—(Practical
approach series)
1. Neurochemistry 2. Neurophysiology
3. Mammals—Physiology
I. Turner,A.J. (Anthony John) II. Bachelard,H.S.
III. Series
599.01′88 QL739.2

ISBN 1-85221-028-1 (hardbound)
ISBN 1-85221-027-3 (softbound)

Cover illustration: Diagram of a nerve ending.
Inset: growth cones from rat brain (Chapter 1, figure 4).

Printed by Information Printing Ltd, Oxford, England.

Preface

In selecting the topics for this book we have tried to emphasize the diversity of techniques currently available to neurochemists. Our aim has also been to provide newcomers to the field, whether postgraduates or seasoned research workers, with sufficient background material and practical information on basic methods. Of course, many of the techniques described are not the exclusive preserve of neurochemists but the very complexity of the nervous system engenders special problems not encountered in other areas of biochemistry. Wherever possible, we have tried to organize the practical procedures in a step-by-step format with tabulation of key materials and methods. We trust that this will make the techniques easy to reproduce in the investigator's own laboratory.

It is now 25 years since Victor Whittaker and Eduardo de Robertis and their respective colleagues independently reported that homogenization and fractionation of brain tissue allowed the isolation of intact, re-sealed nerve-ending particles, termed synaptosomes by Whittaker. In the intervening period, this single technique has probably contributed more than any other to the rapid advances we have seen in neurochemistry. Today it is the techniques of molecular biology that are already producing dramatic new increases in our understanding of neural systems. This book begins and ends with these themes. The intervening chapters reflect some of the many other methodologies currently being employed by neurochemists.

Thus, Chapter 1 deals with the isolation of synaptosomes and their subcellular components. A number of different isolation procedures are described in detail and the limitations and advantages of each are discussed. Chapter 2 considers procedures for the production and maintenance of neuronal and glial cells in culture and for identifying the various cell types. Chapter 3 aims to provide a basic practical guide to the use of immunocytochemistry in chemical neuroanatomy. The various applications of bioluminescence techniques to neurobiology are featured in Chapter 4 with particular emphasis on the cholinergic system. The techniques and pitfalls of assaying and isolating neurotransmitter receptors are discussed critically in Chapter 5 and the two subsequent chapters deal with aspects of receptor transduction systems: protein phosphorylation and phosphoinositide turnover. Chapter 8 turns to the cell nucleus and describes methods for the isolation of nuclei from both neuronal and glial cells and describes their applications, for example in studies of transcriptional activity. In the final chapter, the special requirements for the functional expression of mRNAs for cell-surface receptors and ion channels are described in detail and successful applications of this technique are highlighted. Finally, we would like to thank all our contributors who participated in producing this text, particularly for bearing with our editorial interference.

A.J.Turner
H.S.Bachelard

Contributors

E.A.Barnard
MRC Molecular Neurobiology Unit, MRC Centre, Hills Road, Cambridge CB2 2QH, UK

I.H.Batty
Department of Pharmacology and Therapeutics, Medical Sciences Building, University of Leicester, Leicester LE1 7RH, UK

G.Bilbe
Centre for Molecular Biology, University of Heidelberg, Postfach 106249, Heidelberg 1, FRG

J.de Vellis
Departments of Anatomy and Psychiatry, Mental Retardation Research Center, University of California, Los Angeles, CA 90024, USA

P.R.Gordon-Weeks
Department of Anatomy, King's College (K.Q.C.) London, The Strand, London WC2R 2LS, UK

J.N.Hawthorne
Department of Biochemistry, University Hospital and Medical School, Clifton Boulevard, Nottingham NG7 2UH, UK

M.Israel
Département de Neurochimie, Laboratoire de Neurobiologie Cellulaire et Moléculaire, CNRS, 91190 Gif-sur-Yvette, France

B.Lesbats
Département de Neurochimie, Laboratoire de Neurobiologie Cellulaire et Moléculaire, CNRS, 91190 Gif-sur-Yvette, France

G.G.Lunt
Department of Biochemistry, 4 West, University of Bath, Claverton Down, Bath BA2 7AY, UK

P.Mobley
Department of Pharmacology, University of Texas Health Science Center, 7703 Floyd Curl Drive, San Antonio, TX 78284, USA

H.C.Palfrey
Department of Pharmacological and Physiological Sciences, University of Chicago, 947 East 58th Street, Chicago, IL 60637, USA

J.V.Priestley
United Medical and Dental Schools, Departments of Physiology and Biochemistry, St. Thomas's Campus, Lambeth Palace Road, London SE1 7EH, UK

R.P.Saneto
Departments of Anatomy and Psychiatry, Mental Retardation Research Center, University of California, Los Angeles, CA 90024, USA

C.M.F.Simpson
Howard Hughes Medical Institute at Seattle, University of Washington, Mail Stop SL-15, Seattle, WA 98195, USA

R.J.Thompson
Department of Clinical Biochemistry, School of Clinical Medicine, Addenbrooke's Hospital, Hills Road, Cambridge CB2 2QR, UK

Contents

Abbreviations

ABC	avidin–biotin–peroxidase complex
ACh	acetylcholine
AD	adrenaline (epinephrine)
ADM	astrocyte-defined medium
ANSA	1-anilino-8-naphthalene sulphonic acid
bp	base pairs
BSA	bovine serum albumin
CaM	calmodulin
cDNA	complementary DNA
CHAPS	3-(3-cholamidopropyl)-dimethylammoniopropane sulphonate
ChAT	choline acetyl transferase
CMC	critical micelle concentration
CNPase	2′,3′-cyclic nucleotide 3′-phosphodiesterase
CNS	central nervous system
DA	dopamine
DAB	3′,3-diaminobenzidine
dbcAMP	N^6,O^2-dibutyryl cyclic adenosine monophosphate
DBH	dopamine β-hydroxylase
DEAE	diethyl-aminoethyl
DMSO	dimethyl sulphoxide
DPC	diethylpyrocarbonate
DTT	dithiothreitol
EDTA	ethylenediamine tetra-acetic acid
EGTA	ethyleneglycobis(β-aminoethyl)ether tetra-acetic acid
EM	electron microscopic
FITC	fluorescein isothiocyanate
GABA	γ-aminobutyric acid
GABAR	GABA receptor complex
GAD	glutamic acid decarboxylase
GFAP	glial fibrillary acidic protein
GPDH	glycerol phosphate dehydrogenase
GroPIns	glycerophosphoinositol
GS	glutamine synthetase
GTC	guanidinium thiocyanate
Hepes	*N*-2-hydroxyethylpiperazine-*N*′-2-ethanesulphonic acid
HRP	horseradish peroxidase
5-HT	5-hydroxytryptamine (serotonin)
IEF	isoelectric focusing
Ins	inositol
$InsP_1$	inositol monophosphate
$InsP_2$	inositol bisphosphate
INT	*p*-iodonitrotetrazolium violet
KRB	Krebs–Ringer bicarbonate buffer
LDH	lactate dehydrogenase
LM	light microscopic
LP	lysate pellet
LS	lysate supernatant

MAO	monoamine oxidase
MAP	microtubule-associated protein
MBP	myelin basic protein
mRNA	messenger RNA
NA	noradrenaline (norepinephrine)
nAChR	nicotinic acetylcholine receptor
NDM	neuronally-defined medium
NSE	neuron-specific enolase
ODAB	oxidised 3′,3-diaminobenzidine
ODM	oligodendrocyte-defined medium
PAP	peroxidase–antiperoxidase
PBS	phosphate-buffered saline
PCA	perchloric acid
PDH	pyruvate dehydrogenase
PEG	polyethylene glycol
PHA-L	phytohaemagglutinin-L
PK	protein kinase
PMSF	phenylmethylsulphonyl fluoride
PNS	peripheral nervous system
PSD	post-synaptic density
PSM	post-synaptic membranes
PtdA	phosphatidic acid
PtdCho	phosphatidylcholine
PtdEtn	phosphatidylethanolamine
PtdIns	phosphatidylinositol
PtdSer	phosphatidylserine
RIA	radio-immunoassay
RMP	resting membrane potential
SDS	sodium dodecyl sulphate
SDS-PAGE	polyacrylamide gel electrophoresis in the presence of SDS
SFM	serum-free basal medium
Sp. g.	specific gravity
SPM	synaptic plasma membrane
SSM	serum-supplemented medium
SV	synaptic vesicle
TBPS	*t*-butylbicyclophosphorothionate
TCA	trichloroacetic acid
TEMED	*N,N,N′,N′*-tetramethylethylenediamine
Tes	*N*-tris(hydroxymethyl)methyl-2-aminoethane sulphonic acid
TMB	tetramethylbenzidine
TH	tyrosine hydroxylase
TRITC	tetramethylrhodamine isothiocyanate
TTX	tetrodotoxin
TX	Triton X-100
VIP	vasoactive intestinal peptide
WGA	wheatgerm agglutinin

CHAPTER 1

Isolation of synaptosomes, growth cones and their subcellular components

PHILLIP R.GORDON-WEEKS

1. INTRODUCTION

Neural tissue is composed of neurones and their support cells, the glia. Neurones are notable for their highly elaborate shapes and the possession of long and often branching processes which intermingle extensively with each other and with glia. Such cells do not survive homogenization intact; the cell bodies are sheared from their processes which break up into discrete fragments. The plasma membranes of cell fragments may reseal to form osmotically active particles and when such particles contain the organelles of the synapse they are known as synaptosomes (*Figure 1*). This chapter describes methods for the isolation of synaptosomes and their subcellular components from mammalian brain. Also included are two methods for the isolation of growth cones, the highly motile tips of growing neuronal processes that become, during the development of the nervous system, pre- or post-synaptic elements of the synapse.

Subcellular fractions enriched in synaptosomes are sufficiently pure to permit the study of certain physiological and pharmacological aspects of synaptic function and in this respect they have been enormously important in contributing to our ideas about synaptic mechanisms. These fractions are not, however, completely free of contaminating particles particularly axonal fragments, myelinated and unmyelinated, and mitochondria, nor are they homogeneous in one type of synapse (1). This has important consequences for the interpretation of experiments using synaptosomes for it is extremely difficult to assign a particular property of the fraction to a constituent, except under certain conditions such as when investigating the release of a radiolabelled neurotransmitter that is known to accumulate only in that constituent.

Unlike synaptosomal fractions, subcellular components derived from synaptosomes such as post-synaptic densities and synaptic vesicles (*Figure 1*) are sufficiently pure for biochemical analysis. These fractions have also been used for antibody production.

2. ISOLATION OF SYNAPTOSOMES

2.1 Introduction

In the original description of the isolation of synaptosomes from rat brain, sucrose density gradients were used (2,3). Although sucrose is ideal in many respects as an isolation medium, the concentration used to isolate synaptosomes is hypertonic and consequently synaptosomes lose water and are very shrunken when recovered from the gradient. This fact and therefore the necessity to return synaptosomes isolated on sucrose gra-

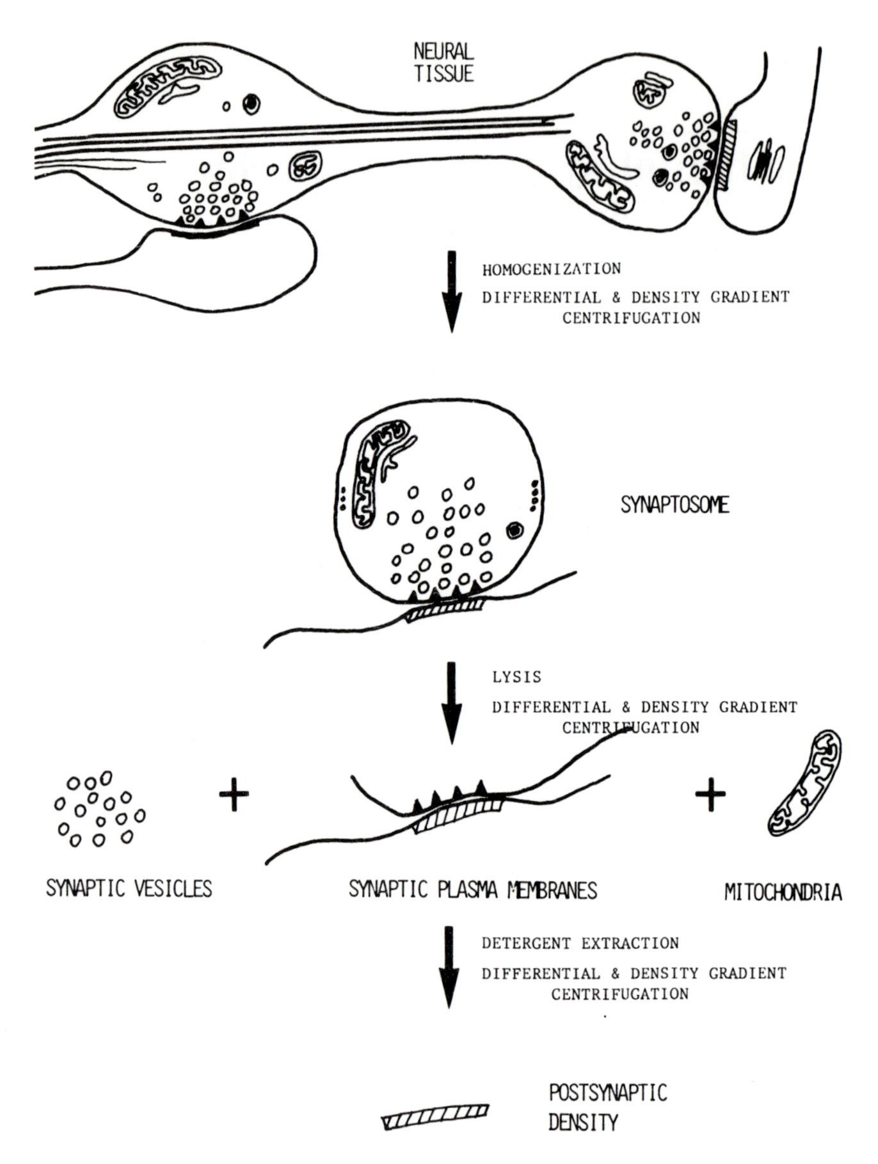

Figure 1. Diagram illustrating the derivation of synaptosomes and their subcellular fractions from a homogenate of neural tissue. Note that the glia, accounting for about half of the mass of the homogenate, have been left out.

dients to a warmed, physiological medium has not been generally appreciated. This is also important when these preparations are to be used as starting material for synaptic plasma membranes, etc., (see Section 4 below) because re-swollen synaptosomes are more susceptible to osmotic shock. For these reasons and because of the relatively long centrifugation times required for sucrose gradients I have included, in addition to a sucrose gradient method, protocols which are based on published methods using Ficoll (4) and Percoll (5); centrifugation media which do not suffer from these disad-

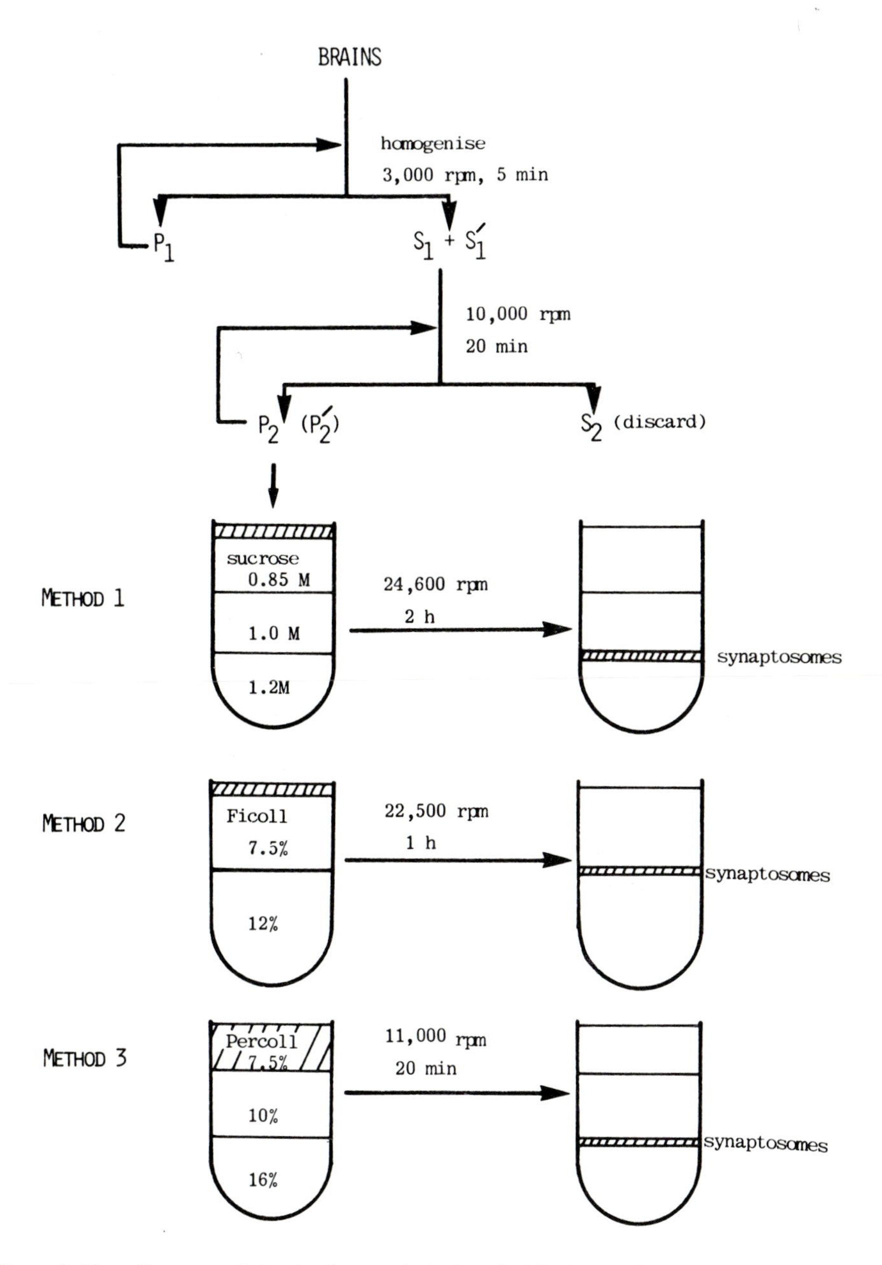

Figure 2. Flow diagram outlining the three methods described in the text for preparing synaptosomes.

vantages. The disadvantage of these media is that they are expensive.

The three methods described in this chapter for the isolation of synaptosomes have been standardized for a wet weight of starting material of 2–4 g (equivalent to two or three rat brains). A flow diagram of the three methods is shown in *Figure 2*. The methods have been used successfully with human material and with most laboratory mammals including rats, mice, guinea-pigs, gerbils, cats, rabbits and monkeys. A different procedure has been published for chickens (6) and there is also a method in the literature for the isolation of synaptosomes from post-mortem human brain (7).

2.2 **Equipment, solutions and gradients**

2.2.1 *Centrifuges and rotors*

For an account of the theory and practice of centrifugation see the book in this series called *'Centrifugation—A Practical Approach'* (8).

(i) A high-speed centrifuge (ultracentrifuge) capable of speeds up to 50 000 r.p.m.
(ii) A swing-out rotor for the ultracentrifuge with tube capacity of about 38 ml.
(iii) A medium-speed refrigerated centrifuge capable of speeds up to 15 000 r.p.m.
(iv) A fixed-angle rotor for the medium-speed centrifuge with tube capacity of about 50 ml.
(v) A bench centrifuge (microcentrifuge) capable of speeds up to 10 000 r.p.m.

2.2.2 *Homogenizers*

(i) A large homogenizer of the Potter–Elvehjem type having a glass mortar of capacity about 55 ml and a Teflon pestle (Jencons). The clearance should be about 0.2 mm. The pestle should be motor driven with continuously variable speed and high torque.
(ii) A small homogenizer of the Potter–Elvehjem type with a capacity of about 15 ml and a clearance of about 0.2 mm. This is hand-held.

2.2.3 *Solutions*

Use glass-distilled water and Analar reagents.

(i) Solution A (buffered isotonic sucrose). Dissolve 0.65 g of Hepes (sodium salt, 5 mM, Sigma) and 54.75 g of sucrose (320 mM) in 500 ml of water. Adjust to pH 7.4 by adding NaOH (1 M) or HCl (5 M).
(ii) Krebs' solution. Modified by Fried and Blaustein (9), see *Table 1*. The Ca^{2+} will precipitate out unless it is added, diluted with water, to the final solution made up to nearly the final volume. Adjust the pH to 7.4 by adding NaOH (1 M) or HCl (5 M).

Table 1. Krebs' solution.

Compounds	*Weight (g)/500 ml water*	*Concentration (mM)*
NaCl	4.25	145
KCl	0.185	5
$CaCl_2$ [a]	600 μl	1.2
$MgCl_2.6H_2O$	0.13	1.3
$NaH_2PO_4.2H_2O$ [b]	0.095	1.2
Glucose	0.9	10
Hepes (sodium salt)	2.6	20

[a]Use analytical volumetric solution (1 M), e.g. BDH Chemicals Ltd, UK.
[b]Other hydrates are unstable.

(iii) Ficoll solutions (for Method 2). Weight out 7.5 g and 12 g of Ficoll (mol. wt 400 000; Pharmacia Fine Chemicals) into two 100 ml volumetric flasks and add solution A [see (i) above] dropwise while mixing. Leave to dissolve (2–3 h or overnight at 4°C) and adjust to correct volume. Final concentrations of Ficoll are 7.5% (w/v) and 12% (w/v).

(iv) Percoll solutions (for Method 3). Percoll (Pharmacia Fine Chemicals) is a solution of colloidal silica coated in polyvinyl pyrollidone. It has a pH of about 9.0 and therefore, when made up in buffered solutions, it may be necessary to readjust the pH. Prepare Percoll solutions just before use. For more details about the characteristics of Percoll see the technical booklet 'Percoll: Methodology and Applications' and the literature list 'Percoll Reference List, January 1985' published by Pharmacia. Prepare a stock solution of 90% Percoll by adding nine volumes of Percoll to one volume of 3.2 M sucrose. Prepare lower density media by dilution with solution A [see (i) above].

2.2.4 *Gradients*

For all gradients use transparent, rigid-walled polycarbonate tubes. Discontinuous gradients such as the ones used here can be made by layering the individual steps on top of one another using a graduated pipette.

(i) Sucrose (for Method 1). Overlay 10 ml of 1.2 M sucrose (41.1% w/v) with 10 ml of 1.0 M sucrose (34.2% w/v) followed by 10 ml of 0.85 M sucrose (29.1% w/v). Make the gradient a few hours before use and leave at 4°C to allow the interfaces to 'smear'.

(ii) Ficoll (for Method 2). Overlay 15 ml of 12% (w/v) Ficoll (see Section 2.2.3) with 15 ml of 7.5% (w/v) Ficoll. These volumes assume a tube volume of 38 ml. Make the gradient a few hours before use and leave at 4°C to allow the interface to 'smear'.

(iii) Percoll (for Method 3). Overlay 4 ml of 16% (v/v) Percoll with 4 ml of 10% (v/v) Percoll. These volumes assume a tube volume of about 14 ml.

2.3 **Method 1 (sucrose gradient)**

This method is based on the original description by Gray and Whittaker (2) and employs a discontinuous sucrose gradient. Carry out all operations at 0–4°C except where stated otherwise.

(i) Kill animals by decapitation or cervical fracture.

(ii) Rapidly dissect the brain tissue from the animal, place it on a sheet of black Perspex and cover with a drop of solution A (see Section 2.2.3). Dissect and discard as much of the meninges and white matter as possible.

(iii) Chop the tissue into small pieces with one half of a double-edged razor blade or a scalpel.

(iv) Add the chopped tissue to 40 ml of solution A in the mortar of the large homogenizer (see Section 2.2.2). Note that the tissue to medium ratio is 5–10%.

(v) Homogenize at 800 r.p.m. with 12 up-and-down passes taking care not to withdraw the pestle so fast as to create suction forces.

(vi) Pour the homogenate into a centrifuge tube (50 ml, rigid-walled polycarbonate), top up with solution A and centrifuge at 3000 r.p.m. ($\sim$1000 g_{max}) for 5 min in the medium-speed centrifuge using the fixed-angle rotor (see Section 2.2.1) to produce a pellet (P_1) and a supernatant (S_1).

(vii) Aspirate the supernatant with a wide-tip Pasteur pipette into a flat bottomed flask with a side arm by connecting the side arm to a water pump. Place on ice.

(viii) Add a few millilitres of solution A to P_1 and resuspend thoroughly by 'whirlimixing', that is by placing the bottom of the tube onto a vortex mixer. Dilute to 50 ml with solution A.

(ix) Repeat steps (vi) and (vii) above to produce P_1' and S_1'.

(x) Combine S_1 and S_1' and put into two centrifuge tubes and top up to 50 ml with solution A.

(xi) Centrifuge at 10 000 r.p.m. (12 000 g_{max}) for 20 min in the medium-speed centrifuge using the fixed-angle rotor to produce P_2 and S_2. P_2 is the so-called 'crude mitochondrial pellet'.

(xii) Aspirate S_2 with a wide-tip Pasteur pipette attached to a pressure line and discard. Resuspend P_2 in a few millilitres of solution A by gentle whirlimixing. Re-homogenize in the small, hand-held homogenizer.

Note: P_2 has a buff-coloured portion at the bottom of the centrifuge tube that is mainly composed of mitochondria, unlike the rest of the pellet which is whiter and contains the majority of the synaptosomes. By careful whirlimixing it is possible to leave the bulk of the buff portion of the pellet behind.

(xiii) Resuspend P_2 in solution A to 50 ml and repeat step (xi) to produce P_2'.

(xiv) Resuspend P_2' in solution A to 8 ml and layer carefully onto the sucrose gradient (see Section 2.2.4) with a wide-tip Pasteur pipette.

(xv) Centrifuge the gradient at 24 600 r.p.m. (85 000 g_{av}) for 2 h in the high-speed centrifuge using the swing-out rotor (see Section 2.2.1).

(xvi) After centrifugation, recover the synaptosomes (visible as a cream coloured band) at the 1.0 M/1.2 M sucrose interface with a wide-tip Pasteur pipette after aspirating the material above. This operation is aided by illuminating the gradient at right angles to the line of view.

(xvii) Place the synaptosomes in a beaker, and while stirring slowly with a Teflon-coated 'flea', add Krebs' solution (see *Table 1*) dropwise at room temperature to a final volume of 50 ml.

(xviii) Repeat step (xi) above.

(xix) Lift the pellet off the bottom of the tube with a jet of Krebs' solution from a Pasteur pipette and then re-homogenize in the small homogenizer in a few millilitres of Krebs' until dispersed.

(xx) Incubate the synaptosome suspension at 37°C for 10 min before use.

Expect about 5–10 mg protein/g wet weight of starting material. The preparation time is about 4–5 h. To save time, but at the expense of yield, omit step (ix). To improve purity and in particular to eliminate most of the microsomal contamination (10) repeat step (xiii).

2.4 Method 2 (Ficoll gradient)

(i) Proceed as in Method 1 (see Section 2.3) as far as and including step (xiii).

(ii) Resuspend P_2' in solution A (see Section 2.2.3) to 8 ml and layer carefully onto the Ficoll gradient (see Section 2.2.4) with a wide-tip Pasteur pipette.
(iii) Centrifuge the gradient at 22 500 r.p.m. (~68 000 g_{av}) for 1 h in the high-speed centrifuge using the swing-out rotor (see Section 2.2.1).
(iv) Recover the synaptosomes (visible as a cream-coloured band) with a Pasteur pipette at the 7.5%/12% Ficoll interface after aspirating the material above. This operation is aided by illuminating the gradient at right angles to the line of view.
(v) Continue as in Method 1 beginning at step (xvii).

Note: a variation of this method, introduced by Booth and Clark (11) involves dispersing the P_2' pellet in the 12% Ficoll step of the gradient. During centrifugation the mitochondria sediment to form a pellet, the synaptosomes float up to the 7.5%/12% Ficoll interface and the myelin fragments float to the top of the 7.5% Ficoll step. Morphological analysis of the synaptosome fraction indicates that approximately 70% of the particles are synaptosomes (12).

2.5 Method 3 (Percoll gradient)

(i) Proceed as in Method 1 (see Section 2.3) as far as and including step (xiii).
(ii) Resuspend P_2' in solution A (see Section 2.2.3) to 1 ml and add 8 ml of 8.5% Percoll (see Section 2.2.3). Layer 6 ml onto the Percoll gradient (see Section 2.2.4) with a wide-tip Pasteur pipette.
(iii) Centrifuge the gradient at 11 000 r.p.m. (15 000 g_{av}) for 20 min in the high-speed centrifuge using the swing-out rotor (see Section 2.2.1).
(iv) Recover the synaptosomes (visible as a cream coloured band) at the 10%/16% Percoll interface with a wide-tip Pasteur pipette after aspirating the material above. This operation is aided by illuminating the gradient at right angles to the line of view.
(v) Continue as in Method 1 beginning at step (xvii).
(vi) If necessary the Percoll can be completely removed by repeating step (xi) in Method 1 (see Section 2.3) twice.

2.6 Electron microscopy

Before preparing synaptosomes for electron microscopy it is essential to incubate them in a physiological salt solution, for example Krebs' (*Table 1*) for about 10 min at 37°C. Polypropylene microcentrifuge tubes (1.5 ml; MSE, Eppendorf) are suitable and can be used throughout the fixation schedule.

Table 2. Fixative solutions.

Primary fixative:	Glutaraldehyde (3% v/v) and formaldehyde (2% w/v) in sodium cacodylate[a] (0.1 M) pH 7.4. Prepare formaldehyde fresh by depolymerizing paraformaldehyde at 80°C in water to which a few drops of NaOH (1 M) have been added.
Solution A:	Sucrose (5% w/v) in sodium cacodylate (0.1 M) pH 7.4.
Solution B:	OsO_4 (2% w/v) in sodium cacodylate (0.1 M)
Solution C:	A filtered, saturated solution of uranyl acetate (~8% w/v) in water.

[a]Sodium cacodylate can be made up as a 0.2 M stock solution and stored at 4°C for several months. The trihydrate is stable.

(i) Put the synaptosomes (200–500 μg protein) into a microcentrifuge tube and centrifuge at 10 000 g_{max} for 15 sec in a bench centrifuge.
(ii) Remove the supernatant with a Pasteur pipette and loosen the pellet by brief, gentle whirlimixing.
(iii) For fixation, fill the tube with primary fixative (*Table 2*) at room temperature and disperse the pellet by overturning. Fix for 1 h or overnight at 4°C.
(iv) Pellet the synaptosomes as in (i) above and remove the supernatant.
(v) Fill the tube with solution A (*Table 2*) at room temperature, leave for 5 min and repeat.
(vi) Remove solution A and add solution B (*Table 2*) at room temperature. To aid penetration of osmium tetroxide, free the pellet from the bottom of the tube by carefully inserting a suitably whittled down wooden applicator stick between the pellet and the wall of the tube. If the pellet is large, that is greater than 1 mm in any dimension, break it up into smaller pieces with the stick. Leave in solution B for 40 min. Operations with osmium tetroxide should be conducted in a fume cupboard.
(vii) Remove solution B and add solution C at room temperature for 30 min.
(viii) Dehydrate the sample with ethanol (50%, 75%, 90% and 100%, v/v) in water and then with ethanol:propylene oxide mixtures (1:1, 1:2) and finally pure propylene for 3 min at each step.
(ix) Prepare fresh Araldite or similar (Polaron, EMscope, Agar Aids) resin according to manufacturers' instructions. These resins can be prepared in bulk and stored for several months at −20°C. The physical properties of the cured resin, such as hardness, are determined by the proportions of the various components in the resin (see manufacturer's instructions). Infiltrate the sample with resin by immersing it successively in the following mixtures of resin and propylene oxide (1:2, 1:1 and 2:1 by vol) for 30 min at each step at room temperature. Finally, put the sample into pure resin in a suitable container and cure at 70°C for 24 h. The compartments of plastic ice trays with a thin layer of previously polymerized resin in the bottom, are ideal for this purpose. Alternatively, capsules are available commercially (e.g. Polaron, EMscope, Agar Aids).

2.7 Assessment of purity

Ideally a synaptosome preparation should contain only synaptosomes. In practice these subcellular fractions are contaminated to varying extents by free mitochondria, segments of axons and dendrites, myelin and glial cell fragments (*Figure 3*) and it is necessary, therefore, to assess the degree of purity. This can be done by the use of an enzyme or other biochemical marker whose association with a particular structure has been independently determined (see *Table 3*) by ultrastructural analysis or by utilizing some physiological property that is known to be restricted to certain structures, for example neurotransmitter uptake. The synaptosomes derived from complex neural tissues such as vertebrate brain are not homogeneous in neurotransmitter type. This can compromise further analysis and, consequently, there have been attempts to purify synaptosomes of one transmitter type from mixed synaptosome preparations. The most successful attempts have used an immunological approach. For instance, antibodies to the

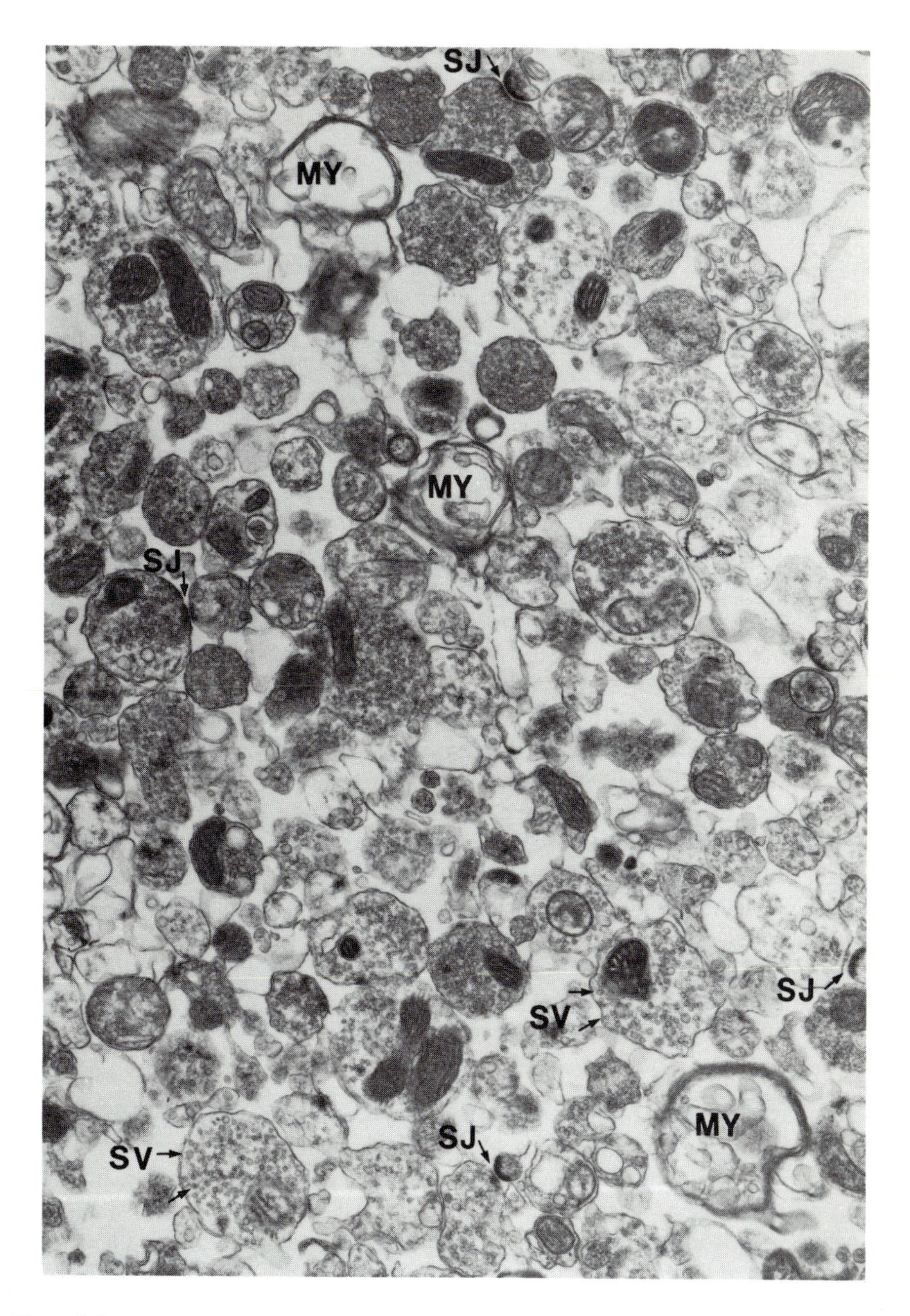

Figure 3. Low power electron micrograph of the synaptosome fraction prepared by Method 2 (Ficoll gradient) from rat forebrain. A large proportion (~ 70%) of the membrane bounded structures are synaptosomes, identified by their content of synaptic vesicles (SV) and, in favourable sections, presence of synaptic junctions (SJ). The fraction also contains fragments of myelin (MY) and small, unidentifiable structures. Free mitochondrial contamination is low. Prepared for electron microscopy as described in Section 2.6. The section was stained on the grid with lead. Magnification ×19 750.

Table 3. Markers for the subcellular fractions described in the text.

Fraction	*Marker*	*Reference*
Mitochondria	cytochrome *c* oxidase	25, 31
Myelin	2′,3′-cyclic nucleotide 3-phosphohydrolase	30
Plasma membrane	ouabain-sensitive (Na^+, K^+)-ATPase	32
Growth cones	electron microscopy	15
Synaptic vesicles	Synapsin I	47
Synaptosomes	electron microscopy	see Section 2.6
Post-synaptic densities	51 000-dalton phosphoprotein[a]	38
	180 000-dalton glycoprotein	39, 40, 42

[a]Does not apply to cerebellar post-synaptic densities

cholinergic terminal specific antigen Chol-1(50) have been used to affinity purify cholinergic terminals from a P_2 fraction of rat brain (50,51). With mammalian brain even those fractions of high purity such as synaptic vesicles and post-synaptic densities (see Section 4 below) are heterogeneous in that they are derived from a wide range of synapse types. The proportion of synaptosomes is probably best determined by ultrastructural means by, for instance, overlaying low power micrographs with a transparent sheet onto which a grid of suitably spaced lines has been drawn and then determining the number of line intersections covering synaptosomes as a proportion of all other intersections covering the fraction (13). It may be necessary to sample several levels within the pellet along the direction of centrifugal force if layering effects due to the latter are suspected, for instance when using long tubes. There is no differential sedimentation of the components of the fraction under the conditions described in Section 2.6.

To determine the extent of contamination by myelin and mitochondria, assay for their marker enzymes as indicated in Section 4.2.2 Method 1 step (xii), see also *Table 3*.

3. ISOLATION OF GROWTH CONES

3.1 Introduction

During the development of the central nervous system, zones of cell proliferation are established out of which neurones migrate to reach their final relative positions. At the end of their migration most neurones begin to grow an axon and, usually later, dendrites. At the tips of these growing processes are highly motile enlargements called growth cones. Growth cones are crucial structures in the development of the nervous system because they guide the growing axon (or dendrite) through the developing brain matrix, they recognize the target site and they transform into pre- or post-synaptic elements.

Two groups have reported methods for the isolation of subcellular fractions enriched in neuronal growth cones (14,15). These procedures differ considerably from each other and both are included in this chapter. Unlike the situation with synaptosomes, a substantial literature on the properties of these fractions has not yet built up.

3.2 Equipment, solutions, gradients and column chromatography

3.2.1 *Centrifuges and rotors*

As for Section 2.2.1 except that a vertical rotor, such as the MSE MVT 50.35 (35 ml tubes) or the Beckman VTi50 (39 ml tubes), is used instead of a swing-out rotor for Method 2.

3.2.2 *Homogenizers*

As for Section 2.2.2.

3.2.3 *Solutions*

Use glass-distilled water and Analar reagents.

(i) Solution A. Same as solution A in Section 2.2.3.

(ii) Solution B (buffered, isotonic sucrose). Dissolve 114.6 mg of N-Tris(hydroxymethyl)methyl-2-aminoethane-sulphonic acid (Tes) (1 mM, Sigma), 0.1 g of $MgCl_2 .6H_2O$ (1 mM) and 54.75 g of sucrose (320 mM) in 500 ml of water. Adjust to pH 7.3 by adding NaOH (1 M).

(iii) Ficoll solutions (for Method 1). Weight out 7 g and 14 g of Ficoll (mol. wt 400 000; Pharmacia Fine Chemicals) into two 100 ml volumetric flasks and add solution A [see (i) above] dropwise while mixing. Leave to dissolve (2–3 h or overnight at 4°C) and adjust to the correct volume. Final concentrations of Ficoll are 7% (w/v) and 14% (w/v).

(iv) Sucrose solutions for gradient (Method 2). As for (ii) above except use 128.4 g of sucrose (0.75 M) and 455.4 g of sucrose (2.66 M). Make up 30 ml of each solution.

3.2.4 *Gradients*

(i) Ficoll gradients (for Method 1). There are two gradients in this method only one of which is prepared beforehand. Overlay 18 ml of glycerol with 12 ml of 7% Ficoll (see Section 3.2.3) in a 38 ml rigid-walled polycarbonate centrifuge tube.

(ii) Sucrose (for Method 2). Overlay 10 ml of 2.66 M sucrose with 10 ml of 0.75 M sucrose (see Section 3.2.3) in a 35 ml (MSE) or 39 ml (Beckman) centrifuge tube.

3.2.5 *Column chromatography*

Pack a column (80 × 1 cm) with controlled-pore glass beads (No. CPG03000, 80–120 mesh, 300 nm mean pore diameter, Electro-Nucleonics, Inc., Fairfield, NJ, USA) and vibrate the column strongly to facilitate compaction. The beads should be previously coated with polyethylene glycol (PEG). Add 500 ml of glass beads to 2 l of a de-gassed solution of 1% (w/v) PEG 20 M (BDH Chemicals Ltd, UK). After the beads have settled, pour off the supernatant and repeat the process until no air bubbles appear in the supernatant (about three times). Wash the beads five times in water.

Equilibrate the column with 0.65 M sucrose containing 1 mM $MgCl_2$ and 1 mM Tes, pH 7.3. The column can be used continuously for 1 year during which time it should be re-coated once with PEG *in situ*.

3.3 Method 1

This method is that of Gordon-Weeks and Lockerbie (15) and the starting material is 5 day old (day of birth is day 1) rat forebrain. Carry out all operations at 0–4°C except where stated otherwise.

(i) Kill the animals from one litter by decapitation.
(ii) Rapidly dissect out the forebrain, place on a sheet of black Perspex and cover with a drop of solution A (see Section 3.2.3).
(iii) Chop the tissue into small pieces with one half of a double-edged razor blade or a scalpel.
(iv) Add half of the chopped tissue to 40 ml of solution A in the mortar of the large homogenizer (see Section 3.2.2).
(v) Homogenize at 800 r.p.m. with eight up-and-down passes taking care not to withdraw the pestle so fast as to create suction forces.
(vi) Repeat steps (iv) and (v) with the other half of the tissue.
(vii) Pour homogenate into 2 × 50 ml rigid-walled polycarbonate centrifuge tubes, top up with solution A and centrifuge at 3000 r.p.m. (~1000 g_{max}) for 10 min in the medium-speed centrifuge using the fixed-angle rotor (see Section 3.2.1) to produce a pellet P_1 and a supernatant S_1.
(viii) Aspirate the supernatants into a flat-bottomed flask with a side arm by connecting the side arm to a water pump. Place on ice.
(ix) Add a few millilitres of solution A to the pellets and resuspend thoroughly by 'whirlimixing', that is by placing the bottom of the tube onto a vortex mixer. Dilute to 100 ml with solution A.
(x) Repeat steps (vii) and (viii) above to produce P_1' and S_1'.
(xi) Combine S_1 and S_1' and distribute into 4 × 50 ml centrifuge tubes and top up the tubes with solution A.
(xii) Centrifuge at 10 500 r.p.m. (13 300 g_{max}) for 15 min in the fixed-angle rotor to produce P_2 and S_2.
(xiii) Aspirate S_2 with a wide-tip Pasteur pipette attached to a pressure line and discard. Read Note in Section 2.3, step (xii) above. Resuspend P_2 in a few millilitres of solution A by gentle whirlimixing. Re-homogenize in the small hand-held homogenizer (see Section 3.2.2).
(xiv) Dilute to 100 ml with solution A and repeat twice steps (xii) and (xiii) above to produce P_3.
(xv) Resuspend P_3 to a final volume of 4 ml of solution A per six to eight forebrains.
(xvi) With a wide-tip Pasteur pipette carefully layer 2 ml of P_3 onto the Ficoll gradient (see Section 3.2.4).
(xvii) Centrifuge at 14 750 r.p.m. (30 400 g_{av}) for 20 min in the high-speed centrifuge using the swing-out rotor (see Section 3.2.1).
(xviii) After centrifugation carefully remove the material banding at the interface between the sample and the 7% Ficoll with a wide-tip Pasteur pipette. This operation is aided by illuminating the gradient at right angles to the line of view.
(xix) Dilute to 8 ml in solution A and mix thoroughly with 8 ml of 14% Ficoll (see Section 3.2.3). Layer carefully with a graduated pipette onto 16 ml of glycerol in a 38 ml polycarbonate centrifuge tube.

(xx) Centrifuge at 11 500 r.p.m. (18 000 g_{av}) for 60 min in the high-speed centrifuge using the swing-out rotor.
(xxi) After centrifugation, remove the top two-thirds of the Ficoll step with a wide-tip Pasteur pipette.
(xxii) Place the growth cone fraction in a beaker and while stirring slowly with a Teflon-coated 'flea' add Krebs' solution (see *Table 1*) dropwise at room temperature to a final volume of 50 ml and put into a polycarbonate centrifuge tube.
(xxiii) Repeat step (xii) above.
(xxiv) Aspirate and discard the supernatant. Lift the pellet off the bottom of the tube with a jet of Krebs' solution from a Pasteur pipette and then re-homogenize in the small homogenizer in a few millilitres of Krebs' solution until dispersed.
(xxv) Incubate the growth cone fraction 37°C for 10 min before use. Expect about 500 μg protein/g wet weight of starting material.

To prepare isolated growth cones for electron microscopy follow the procedure in Section 2.6 above except use 0.12 M sodium cacodylate throughout.

3.4 Method 2

This method is that of Pfenninger *et al.* (14) and the starting material is fetal rat brain of 16–18 days (ideally 17 days) gestation. Carry out all operations at 0–4°C except where stated otherwise.

(i) Kill the animals by decapitation.
(ii) Rapidly dissect out the brain and put into the mortar of the pre-weighed large homogenizer (see Section 3.2.2) containing ice-cold solution B (see Section 3.2.3). Re-weigh and calculate the tissue wet weight.
(iii) Adjust the volume of solution B in the homogenizer so that the ratio of tissue wet weight to volume of solution B is 6. One litter (12 fetuses) yields about 1 g wet weight of brain material.
(iv) Homogenize at 100–150 r.p.m. with five up-and-down passes taking care not to withdraw the pestle so fast as to create suction forces.
(v) Pour the homogenate into 13 ml rigid-walled polycarbonate centrifuge tubes and centrifuge at 1660 g_{max} for 15 min in the bench centrifuge or a medium-speed centrifuge using the fixed-angle rotor (see Section 3.2.1) to produce a pellet P_1 and a supernatant S_1.
(vi) Remove S_1 with a wide-tip Pasteur pipette and carefully layer onto the sucrose gradient (see Section 3.2.4). Note: when using the MSE vertical rotor which has 35 ml tubes, add 15 ml of S_1, equivalent to about 2 litters. For the Beckman vertical rotor use 19 ml. Centrifuge at 50 000 r.p.m. (242 000 g_{max}) for 40 min in the high-speed centrifuge using the vertical rotor (see Section 3.2.1).
(vii) After centrifugation, remove the material at the interface between the sample and the 0.75 M sucrose with a wide-tip Pasteur pipette. This operation is aided by illuminating the gradient at right angles to the line of view.
(viii) Chromatograph the interface material on the controlled-pore glass bead column (see Section 3.2.5) at a flow-rate of about 1.7 ml/min at room temperature under light pressure. Monitor the optical density of the eluate continuously at 280 nm and collect the first peak (void peak).

Expect about 1 mg protein/g wet weight of starting material. Electron microscopy is as for Section 3.3, step (xxvii).

3.5 Assessment of purity

There are no identified biochemical markers for neuronal growth cones and therefore the purity of the fractions isolated by these two methods has been assessed by electron microcopy only (14, 15). Examples of electron micrographs of growth cones prepared by these two methods are shown in *Figure 4*. The fraction isolated by Method 1 (see Section 3.3) contains neuronal growth cones by several criteria, including the presence of filopodia (15); however there are also considerable quantities of large membrane bounded sacs (probably derived from the plasma membranes of lysed growth cones during the separation on the final, flotation gradient (Gordon-Weeks, unpublished observation) and unidentified cell fragments (*Figure 4a*). The fraction isolated by Method 2 (see Section 3.4) is more uniform in composition but the particles do not bear filopodia (*Figure 4b*) (14).

Information about the pharmacology of the fraction isolated by Method 1 (16,17) and the biochemistry of the fraction isolated by Method 2 (18,19) is beginning to appear. appear.

4. ISOLATION OF THE SUBCELLULAR COMPONENTS OF SYNAPTOSOMES

4.1 Introduction

Synaptosomes are osmotically active particles and will undergo lysis when subjected to hypotonic solutions (20). Lysis of a synaptosome fraction or the parent crude mitochondrial fraction [see Section 2.3 (xi)] releases the organelles within the synaptosome, such as synaptic vesicles and intra-terminal mitochondria, and these may then be fractionated and purified separately from the plasma membrane of the synaptosome (*Figure 1*). This membrane is referred to as the synaptic plasma membrane since it retains part of the synapse.

A synaptic plasma membrane preparation is one starting point for the isolation of post-synaptic densities. These are most commonly derived by detergent extraction of the parent membrane, a step which solubilizes membrane lipids and leaves the post-synaptic density as an insoluble residue which can be harvested by differential or gradient centrifugation (21).

4.2 Isolation of synaptic plasma membranes

Synaptosome fractions are contaminated chiefly by mitochondria and fragments of glia and myelin. These contaminants and the intrasynaptosomal organelles released during hypotonic lysis (mainly mitochondria and synaptic vesicles) must be separated from synaptic plasma membranes (SPM) if these are to be purified. Unfortunately, mitochondria and SPMs have similar buoyant densities and it has proved difficult to fractionate them on density gradients. Two strategies have proved useful in circumventing this problem. The first relies on the observation that if the lysis buffer is alkaline, mitochondria and SPMs are subsequently more effectively resolved on sucrose density gradients (22). The second strategy involves artificially increasing the buoyant density of mitochondria by filling them with an insoluble formazan precipitate (23).

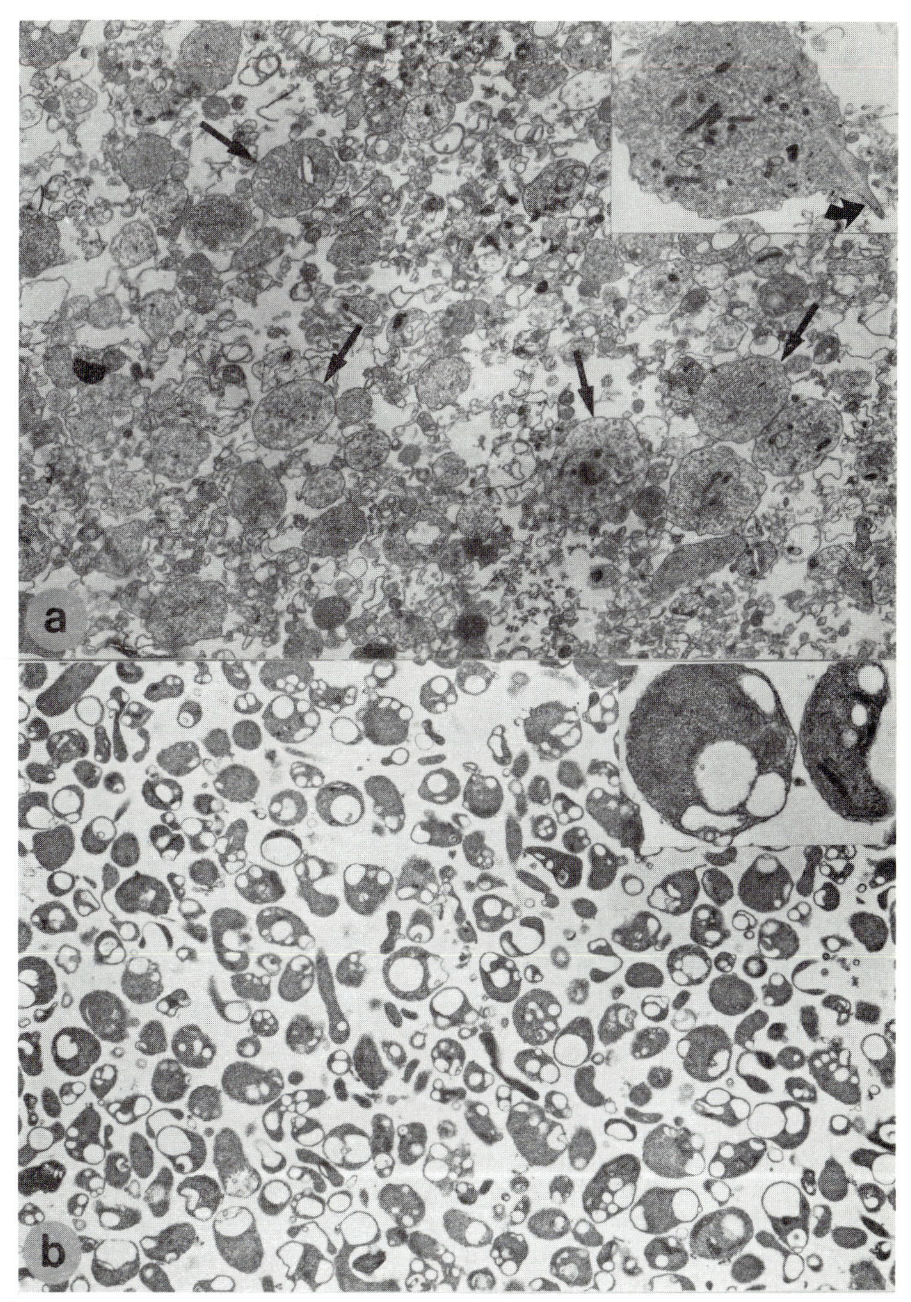

Figure 4. Low power electron micrographs of subcellular fractions enriched in growth cones prepared by the two methods described in the text. (**a**) Growth cones prepared by the method of Gordon-Weeks and Lockerbie (15) from rat forebrain. Most of the large, membrane-bound structures are growth cones (arrows). The contamination consists mainly of small, unidentifiable objects and empty, membrane sacs. The inset shows a high power view of an individual growth cone bearing a filopodium (arrow). Magnification ×6000 (Inset ×10 000). (**b**) Growth cones prepared by the method of Pfenninger *et al.* (14) from rat brain. Note the fairly uniform appearance of the particles. The inset shows a high power view of two growth cones. These do not bear filopodia. Magnification ×6500. (Inset ×19 800).

Two methods for isolating SPMs are described in this chapter. The first method is the more commonly employed one and involves fractionation of the synaptosomal lysate on a sucrose density gradient (22,24). The purity of the fraction is high as assessed by marker enzyme activities (>80%, see below and *Table 3*) but the procedure is lengthy and the yield is low. The second method is simpler and quicker (25).

4.2.1 *Equipment, solutions and gradients*

(i) Centrifuges and rotors. As for Section 2.2.1.
(ii) Homogenizers. As for Section 2.2.2 except that in addition a tight-fitting (0.1 mm clearance) all-glass homogenizer of the Dounce type (Jencons) is needed.
(iii) Solutions. Use glass-distilled water and Analar reagents.
  (a) Solution A. Lysis buffer. Dissolve 79 mg of Tris-HCl (5 mM, Sigma), in 100 ml of water, add Ca^{2+} to 50 μM (see footnote a in *Table 1*) and adjust to pH 8.1 by adding solid Tris-base or NaOH (1 M).
  (b) Solution B. Dissolve 25.3 mg of *p*-iodonitrotetrazolium violet (INT, Sigma) *followed by* 0.65 g of disodium succinate hexahydrate in 10 ml of 0.2 M sodium phosphate buffer, pH 7.4.
(iv) Gradient (for Method 1). Overlay 10 ml of 1.2 M sucrose (41.1% w/v) with 10 ml of 1.0 M sucrose (34.2% w/v) followed by 10 ml of 0.8 M sucrose (27.4% w/v) in a 38 ml rigid-walled polycarbonate centrifuge tube. Make the gradient a few hours before use and leave at 4°C to allow interfaces to 'smear'.

4.2.2 *Method 1*

This method is essentially that described by Cotman and Taylor (24).

(i) Prepare synaptosomes as described in Section 2.4, up to and including step (xx), Section 2.3. Ensure that the wet weight of the starting material is determined.
(ii) Osmotic lysis. After incubating the synaptosomes at 37°C for 10 min, centrifuge at 10 000 g_{max} for 15 sec in a bench centrifuge.
(iii) Resuspend the pellet in 30 ml of solution A [see Section 4.2.1 (iii)] by homogenization with six up-and-down strokes of the Dounce homogenizer (see Section 4.2.1). Leave at 4°C for 30 min.
(iv) Homogenize the lysate with six up-and-down strokes of the tight-fitting Dounce homogenizer.
(v) Centrifuge the lysate at 23 300 r.p.m. (100 000 g_{max}) for 30 min at 4°C in the high-speed centrifuge using the swing-out rotor (see Section 2.2).
(vi) Resuspend the pellet by homogenization in the small homogenizer [see Section 4.2.1 (ii)] to 26 ml in lysis buffer [solution A, see Section 4.2.1 (iii)] per 9 g wet weight of starting material. Add 3.6 ml of solution B followed by 0.2 M phosphate buffer to a final volume of 38 ml. Incubate at 30°C for 25 min. Note: This ratio of INT to starting material is optimal for separating mitochondria from SPMs (26) and is lower than that originally recommended by Davis and Bloom (23).
(vii) Repeat step (v) above.
(viii) Wash the pellet by resuspending in the small homogenizer in 0.16 M sucrose and repeating step (v) above.

(ix) Resuspend the pellet in the small homogenizer in 4 ml of 0.32 M sucrose in solution A per gram of wet weight of starting material and carefully layer 8 ml onto the sucrose gradient [see Section 4.2.1 (iv)] with a wide-tip Pasteur pipette.
(x) Centrifuge the gradient at 18 500 r.p.m. (63 580 g_{max}) for 105 min in the high-speed centrifuge using the swing-out rotor (see Section 2.2.1).
(xi) After centrifugation, recover the SPM fraction at the 1.0 M/1.2 M sucrose interface with a wide-tip Pasteur pipette after removing the material above by aspiration. This operation is aided by illuminating the gradient at right angles to the line of view.
(xii) Wash the SPM fraction by diluting with lysis buffer [solution A, see Section 4.2.1 (iii)] and repeating step (v) above.

Expect about 1–2 mg/g wet weight of starting material.

In addition to the production of formazan within mitochondria, the INT incubation step [see (vi)] also causes the formation of disulphide bonds from free sulphydryl groups (27). This may have adverse effects, for instance INT treatment quantitatively alters monoclonal antibody binding (28) and dramatically reduces the binding of [^{3}H]quinuclidinyl benzilate, a specific muscarinic receptor ligand (Murakami and Gordon-Weeks, unpublished observations). The activities of the marker enzymes, cytochrome *c* oxidase (mitochondria) and 2′,3′- cyclic nucleotide 3′-phosphohydrolase (myelin) are, however, unaffected (Murakami and Gordon-Weeks, unpublished observations). The INT step is not essential and if adverse effects are suspected omit steps (vi)–(viii) inclusive. This will, however, result in a less pure SPM fraction. For the cognoscenti the SPM fraction produced by the method of Mena *et al.* (29) is probably the purest, as assessed by marker enzymes. The method does not involve an INT step but is extremely laborious (14–16 h) and the yield is low (150–250 μg/g wet weight starting material).

Contamination of the SPM fraction by myelin can be determined by assaying for the myelin-specific enzyme, 2′,3′-cyclic nucleotide 3′-phosphohydrolase (30) and for mitochondrial contamination by assaying for cytochrome *c* oxidase (31, see modification in reference 25). The enrichment in plasma membranes can be measured by assaying for the plasma membrane marker enzyme, ($Na^+ + K^+$)-activated ouabain-inhibited ATPase (32) (see *Table 3*). Protein can be assayed for by the method of Bradford (33) as modified by Read and Northcote (34) and Gogstad and Krutnes (33). Note that INT interferes with the Lowry protein assay but not the Bradford method.

Preparation time is about 8–10 h.

4.2.3 *Method 2*

This method is that described by Jones and Matus (25) except that it includes an INT incubation step [see Section 4.2.2 step (xii) and above].

(i) Proceed as in Method 1 (Section 4.2.2) as far as and including step (viii).
(ii) Resuspend the pellet in lysis buffer [solution A, see Section 4.2.1 (iii)] and add 48% (w/w) sucrose so that the final volume is 18 ml and the sucrose concentration is 34% (w/w). Put into a 38 ml polycarbonate centrifuge tube. Use one gradient for 2–3 rat brains.
(iii) Overlay the 34% sucrose with 18 ml of 28.5% (w/w) sucrose and overlay this with a few millilitres of 10% (w/w) sucrose.

(iv) Centrifuge the gradient at 21 000 r.p.m. (60 000 g_{av}) for 110 min in the high-speed centrifuge using the swing-out rotor (see Section 2.2.1).
(v) After centrifugation, recover the SPM fraction at the 34%/28.5% sucrose interface with a wide-tip Pasteur pipette after aspirating the material above. This operation is aided by illuminating the gradient at right angles to the line of view.
(vi) Wash the SPM fraction by diluting with lysis buffer [solution A, see Section 4.2.1 (iii)] and repeating step (v) in Section 4.2.2 Method 1.

Expect 2–2.5 mg/g wet weight starting material. The preparation time is about 8–10 h.

The preparation time can be reduced by using Method 3 (see Section 2.5) to prepare the synaptosomes. The preparation time can be further reduced, but at the expense of purity, by lysing a crude mitochondrial fraction. Assess the purity of the fraction using the marker enzymes indicated in Section 4.2.2 Method 1 (xiv) and in *Table 3*.

4.3 **Isolation of post-synaptic densities**

4.3.1 *Introduction*

The post-synaptic density (PSD) is the densely staining material found subjacent to the post-synaptic plasma membrane at the synapse (*Figure 1*) (36). It has been suggested that the function of the PSD is to anchor plasma membrane proteins such as neurotransmitter receptors in the post-synaptic membrane (21). This idea is supported by the observation of high affinity neurotransmitter receptor binding in PSD fractions (37).

The PSD can be regarded as a specialized region of the neuronal membrane cytoskeleton. They are usually isolated by detergent extraction of synaptosomes or SPM fractions; a method analogous to those used for the isolation of cytoskeletons in general. The strength and type of detergent used (neutral or ionic) determines the protein composition of the PSD (21). The stronger the detergent the simpler the protein composition. There is, however, a characteristic set of polypeptides that are always present regardless of the type of detergent used. Among these are actin, tubulins and the major PSD protein (mol. wt. ~51 000 daltons) which has recently been shown to be a subunit of a calmodulin-dependent protein kinase (38).

Two methods for the isolation of PSDs are included in this chapter. The first method, which uses the detergent Triton X-100, is the more commonly used one and the fraction has been well characterized (21). The second method uses the milder detergent, *n*-octyl glucoside, and has only recently been introduced (39). Reports of the biochemical properties of these PSDs are beginning to appear (40–42) and a slightly modified method has been used to isolate chicken PSDs (43). This method has the advantage of a higher yield and a faster preparation time and the PSDs are not contaminated by myelin basic protein as are the PSDs isolated with Triton X-100 (44). By morphological criteria the preparations are similar and greater than 90% of PSDs (*Figure 5*) (39,46).

4.3.2 *Equipment, solutions and gradients*

(i) Centrifuges and rotors. As for Section 2.2.1.
(ii) Homogenizer. As for Section 4.2.1.
(iii) Solutions. Use glass-distilled water and Analar reagents.

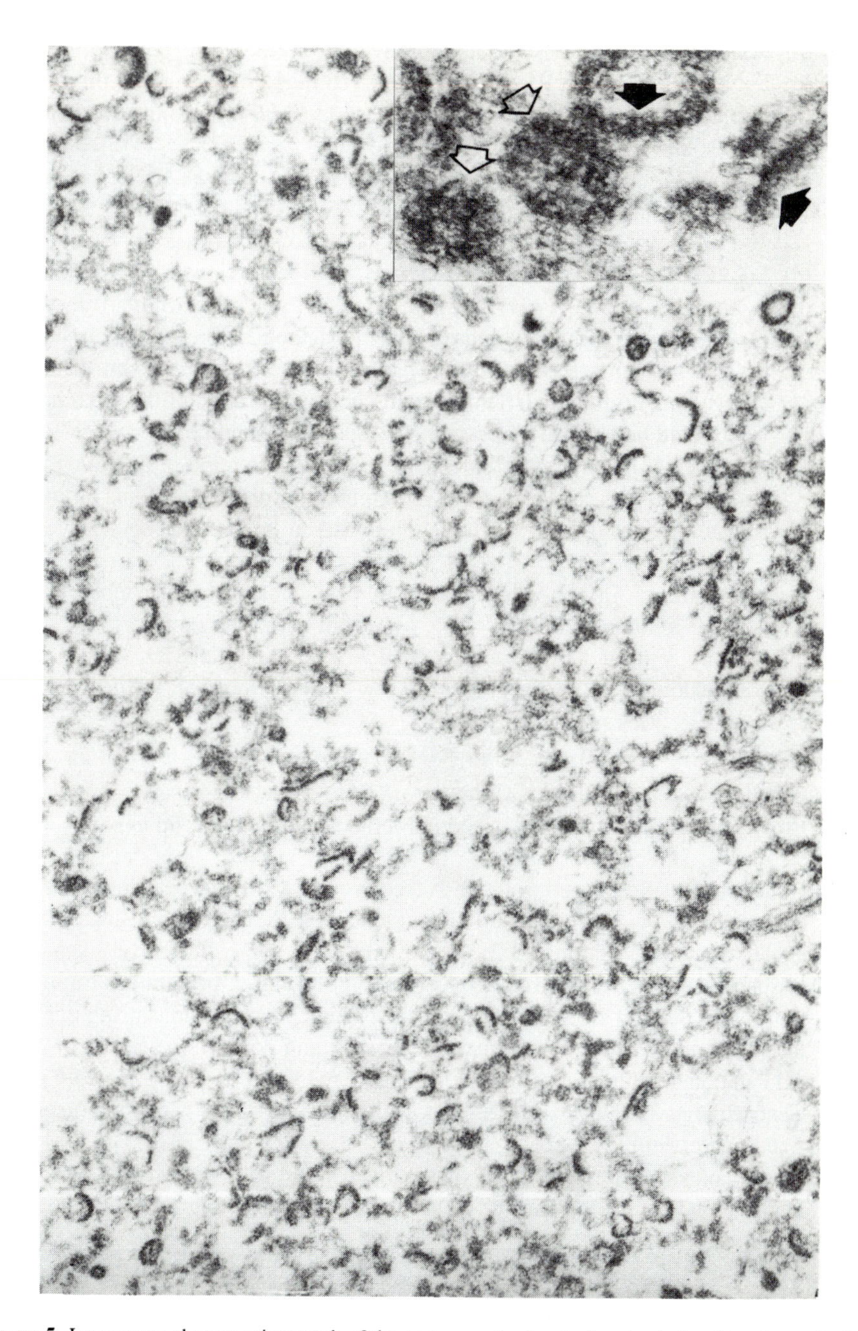

Figure 5. Low power electron micrograph of the post-synaptic density fraction isolated by the method of Gurd *et al.* (39) from rat forebrain. The majority of objects in the field are post-synaptic densities. The inset shows a high power view of post-synaptic densities. These appear as curved bars when cut transversely (closed arrows) and as discs *en face* (open arrows). Magnification ×19 750. (Inset ×70 800).

(a) Solution A. Same as solution A in Section 2.2.3.
(b) Solution B. Dissolve 94.8 g of Tris-HCl (6 mM, Sigma) in 100 ml of water and add Ca^{2+} to 50 μM (see footnote a in *Table 1*). Adjust the pH to 8.1 with solid Tris-base or NaOH (1 M).
(c) Solution C. Dissolve 189.6 g of Tris-HCl (12 mM, Sigma), 10.95 g of sucrose (320 mM) and 1 ml of Triton X-100 (1% v/v, BDH Chemicals Ltd, UK) in 100 ml of water. Adjust to pH 8.1 by adding solid Tris-base or NaOH (1 M).
(d) Solution D. Add Triton X-100 (1%, v/v) to a solution of KCl (150 mM).
(e) Solution E. A 20% (w/v) solution of *n*-octyl glucoside (Sigma) in water.
(f) Stock solutions for the two-phase system (Method 2). Prepare stock solutions of 20% (w/w) of dextran T500 (Pharmacia Fine Chemicals) and PEG 6000 (BDH Chemicals Ltd, UK) in water. The concentration of the dextran solution should be verified by polarimetry (specific rotation $[\alpha]_D = + 199°$, see reference 43) or by refractometry [a 20% (w/v) solution of dextran T500 in water has a refractive index of 1.3630] because dextran contains an indeterminate amount of water. These solutions can be kept for several years at −20°C.

(iv) Sucrose gradient. (for Method 1). Use 12 ml polyallomer centrifuge tubes. Overlay 4 ml of 2.0 M sucrose (68.5% w/v) with 3 ml of 1.5 M sucrose (51.3% w/v) followed by 3 ml of 1.0 M sucrose (34.2% w/v). Add $NaHCO_3$ (1 mM) to all sucrose solutions. Make the gradient a few hours before use and leave at 4°C to allow interfaces to 'smear'.

4.3.3 *Method 1*

This method is essentially that of Cohen *et al.* (46, see also 44).

(i) Prepare synaptosomes as described in Method 1 Section 2.3, up to and including step (xvi).
(ii) Dilute the synaptosomes in solution A (see Section 2.2.3) and centrifuge at 10 000 r.p.m. (12 000 g_{max}) for 20 min in the medium-speed centrifuge using the fixed-angle rotor (see Section 4.3.2).
(iii) Resuspend the pellet to a final concentration of 4 mg of protein per ml in solution B (see Section 4.3.2). Determine the protein concentrations using the Bradford (33) assay as modified in references 34 and 35.
(iv) Add an equal volume of solution C and stir for 15 min on ice or in a cold room.
(v) Centrifuge at 32 800 g_{av} for 20 min in the medium-speed centrifuge using the fixed-angle rotor.
(vi) Resuspend the pellet in 2.5 ml of solution B per 10 g wet weight of starting material and layer 2 ml carefully onto the sucrose density gradient (see Section 4.3.2).
(vii) Centrifuge the gradient at 201 800 g_{av} for 2 h in the high-speed centrifuge using the swing-out rotor (see Section 4.3.2).
(viii) After centrifugation, recover the PSD fraction at the 1.5 M/2.0 M sucrose interface with a plastic-tipped pipette after removing the material above by aspiration. This operation is aided by illuminating the gradient at right angles to the line of view.
(ix) Dilute the pipetted band to 6 ml with solution B and add an equal volume of solution D.

(x) Centrifuge the suspension at 201 800 g_{av} for 20 min in the high-speed centrifuge using 12 ml polyallomer tubes in the swing-out rotor.
(xi) Resuspend the pellet in solution B by vigorous pipetting through a plastic-tipped pipette.

Expect about 200–300 μg/g wet weight starting material. The preparation time is about 10 h.

Prepare the PSDs for electron microscopy exactly as for synaptosomes (see Section 2.6).

The PSD fraction is extremely adherent to glass and cellulose nitrate tubes and therefore polyallomer tubes, plastic beakers and plastic pipette tips should be used throughout. Homogenization in Teflon–glass homogenizers to resuspend pellets should be avoided.

4.3.4 *Method 2*

This method is that of Gurd *et al.* (39).

(i) Prepare SPM as described in Method 1 Section 4.2.2, up to and including step (xi).
(ii) Dilute the SPMs in three volumes of water containing 50 μM Ca^{2+}.
(iii) Centrifuge the SPMs at 100 000 g_{max} for 20 min in the high-speed centrifuge using the swing-out rotor (see Section 4.3.2).
(iv) Resuspend the SPMs in water to a final concentration of 8–12 mg/ml using a plastic-tipped pipette. Determine the protein concentration using the Bradford (33) method as modified in references 34 and 35. This is equivalent to 5–7 g wet weight of starting material per ml.
(v) Weigh out 1.25 g of the stock solution of PEG (see Section 4.3.2) and 1.5 g of the stock solution of dextran and add 0.5 ml of 0.5 M $NaHCO_3$ (pH 9.6), 0.2 ml of solution E (see Section 4.3.2) and 0.5 ml of the SPM suspension. Add sufficient water so that the final weight is 5 g.
(vi) Homogenize with five to six up-and-down strokes in the Dounce homogenizer (see Section 4.3.2).
(vii) Centrifuge the homogenate at 2250 g_{av} for 7.5 min in the bench centrifuge.
(viii) After centrifugation, recover the PSDs as a pink band at the interface between the PEG (top phase) and dextran (bottom phase) with a plastic-tipped pipette.
(ix) Dilute the PSDs with three or four volumes of solution B (see Section 4.3.2) and pellet in the bench centrifuge at 10 000 g_{max} for 5 min. Resuspend the pellet in solution B by vigorous triturition through a plastic-tipped pipette.

Expect about 80 μg of protein/g wet weight starting material. The preparation time is about 5–6 h.

Electron microscopy should be as for Section 4.3.3 and similar precautions should be taken.

4.4 **Isolation of synaptic vesicles**

Synaptic vesicles are the storage sites of neurotransmitters in the pre-synaptic nerve terminal. Although they can be isolated from the mammalian brain in a highly purified fraction, it should be remembered that they are heterogeneous with respect to the

Figure 6. Low power electron micrograph of the synaptic vesicle preparation isolated by the method of Huttner *et al*. (47) from rat cerebral cortex. Most of the objects in the field are synaptic vesicles. There is a small level of contamination from membrane fragments (arrows). Magnification ×18 200. This micrograph is reproduced from ref. 47 with permission.

neurotransmitters they contain. In contradistinction, synaptic vesicles isolated from the electric organs of fish are purely cholinergic (48,49). For an example of the electron microscopical appearance of the synaptic vesicle fraction prepared as described here, see *Figure 6*.

4.4.1 *Equipment, solutions, gradients and column chromatography*

(i) Centrifuges and rotors. As for Section 2.2.1.
(ii) Homogenizers. As for Section 2.2.2.
(iii) Solutions.
 (a) Solution A. Same as solution A in Section 2.2.3.
 (b) Solution B. Dissolve 11.3 g of glycine (300 mM), 0.65 g of Hepes (5 mM, sodium salt, Sigma) and 0.1 g of sodium azide (0.02% w/v) in 500 ml of water. Adjust the pH to 7.2 with NaOH (1 M).
(iv) A linear continuous sucrose density gradient is used. This gradient is formed in 70 ml cellulose nitrate centrifuge tubes from 25 ml of 800 mM sucrose and 29 ml of 50 mM sucrose. Use a two-chamber gradient former (BioRad) to make the gradient.
(v) For column chromatography, suspend 420 ml of dry controlled-pore glass beads (No GLY 03000B, 120–200 mesh, 300 nm mean pore diameter, 74–125 μm, Electro-Nucleonics Inc., Fairfield, NJ, USA) in water and de-gas the beads. Coat the beads with PEG as described in Section 3.2.5. Pour the beads into a column (2 cm internal diameter × 150 cm) and vibrate the column strongly to facilitate compaction. Equilibrate the column with solution B [see (iii) above] before use.

4.4.2 *Isolation of synaptic vesicles*

This method is that of Huttner *et al.* (47) and is based on methods developed by Whittaker's group (48,49). The starting material is rat cerebral cortex. Fourteen adult rats are needed. Carry out all operations at 0–4°C.

(i) Kill seven animals by decapitation or cervical fracture.
(ii) Rapidly remove the brain from the skull, place on a sheet of black Perspex and cover with a drop of solution A (see Section 2.2.3). Dissect the cerebral cortices from the rest of the brain and remove the meninges.
(iii) Chop the tissue into small pieces with one half of a double-edged razor blade or scalpel.
(iv) Pool the cerebral cortices of seven rats and homogenize in 50 ml of solution A in the large homogenizer (see Section 2.2.2) with 12 up-and-down strokes at 900 r.p.m. taking care not to withdraw the pestle so fast as to create suction forces.
(v) Repeat steps (i)–(iv) above with the remaining seven rats.
(vi) Pool the two homogenates and add sufficient volume of solution A to take the final volume to 150 ml.
(vii) Distribute the homogenate into three rigid-walled 50 ml polycarbonate centrifuge tubes.
(viii) Centrifuge at 3000 r.p.m. ($\sim 1000\ g_{max}$) for 10 min in the medium-speed centrifuge using the fixed-angle rotor (see Section 2.2.1) to produce a pellet (P_1) and a supernatant (S_1).
(ix) Remove S_1 and place it into three 50 ml centrifuge tubes, top up with solution A and centrifuge at 10 000 r.p.m. (12 000 g_{max}) for 15 min in the medium-speed centrifuge using the fixed-angle rotor (see Section 2.2.1) to produce S_2 and P_2.
(x) Aspirate S_2 with a wide-tip Pasteur pipette attached to a pressure line and discard.

(xi) Add a few millilitres of solution A to P_2 and resuspend thoroughly by 'whirlimixing', that is by placing the bottom of the tube onto a vortex mixer. Dilute each P_2 to 50 ml with solution A and repeat step (ix) above except at 10 500 r.p.m. (13 000 g_{max}) to produce P_2'.

(xii) Resuspend each P_2' in 13 ml of solution A. Transfer 6.5 ml to the large homogenizer and add 58.5 ml of ice-cold water. Homogenize with three up-and-down strokes at 3000 r.p.m. taking care not to withdraw the pestle so fast as to create suction forces.

(xiii) Immediately after homogenization pour the P_2' lysate rapidly into a beaker containing 500 ml of 1 M Hepes–NaOH buffer (pH 7.4).

(xiv) Repeat steps (x) and (xi) with the rest of P_2' and leave on ice for 30 min.

(xv) Place the lysate in 50 ml centrifuge tubes, top up with water and centrifuge at 16 500 r.p.m. (25 000 g_{av}) for 20 min in the medium-speed centrifuge using the fixed-angle rotor to yield a lysate pellet (LP_1) and a lysate supernatant (LS_1).

(xvi) Aspirate the LS_1s with a wide-tip Pasteur pipette and transfer into 6 × 38 ml polycarbonate tubes, top up with solution A and centrifuge at 34 500 r.p.m. (165 000 g_{av}) for 2 h in the high-speed centrifuge using the swing-out rotor.

(xvii) After centrifugation, aspirate the supernatants with a wide-tip Pasteur pipette attached to a pressure line and discard.

(xviii) Resuspend the pellets in a total of 6 ml of 40 mM sucrose.

(xix) Homogenize the resuspended pellets with 10 up-and-down strokes at 1200 r.p.m. in the large homogenizer.

(xx) Force the homogenate five times back and forth through a 25-gauge needle attached to a 10 ml disposable syringe.

(xxi) Layer the suspension onto the sucrose density gradient (see Section 4.4.1), top up with 40 mM sucrose and centrifuge the gradient at 22 600 r.p.m. (65 000 g_{av}) for 5 h in the high-speed centrifuge using the swing-out rotor.

(xxii) After centrifugation, recover the synaptic vesicles (SV) in the broad band of high turbidity in the 200–400 mM sucrose region by introducing a 25-gauge needle attached to a disposable syringe through the side of the tube. During this operation, illuminate the tube at right angles to the line of view to aid visualization of the banded material.

(xxiii) Chromatograph about 15 ml of the SV fraction overlayered with solution B on the controlled-pore glass bead column (see Section 4.4.1) at a flow-rate of about 40 ml/h and collect 5 ml fractions. Monitor the optical density of the eluate continuously at 280 nm and collect the first peak after the void peak.

(xxiv) Pool the peak fractions, put into 38 ml cellulose nitrate tubes and top up with solution B. Centrifuge the eluate at 41 500 r.p.m. (234 000 g_{av}) for 1 h in the high-speed centrifuge using the swing-out rotor to obtain a pellet of synaptic vesicles.

Preparation takes 1 day with step (xxiii) run overnight.

For electron microscopy, prepare the pelleted synaptic vesicles [see step (xxiv) above] as described in Section 2.6 except use 2% (v/v) glutaraldehyde in 10 mM sodium cacodylate buffer as the primary fixative.

5. ACKNOWLEDGEMENTS

I would like to thank Helen Appleton for typing the manuscript and Philip Batten for photographic work. I am particularly indebted to Professor Karl Pfenninger for providing details of his method for preparing growth cones and for providing *Figure 4b*, to Dr P.De Camilli and The Rockefeller University Press for permission to use *Figure 6* and to Dr Andrew Matus for helpful suggestions.

6. REFERENCES

1. Cotman,C.W. (1974) In *Methods in Enzymology*. Fleischer,S. and Packer,L. (eds), Academic Press Inc., London and New York, Vol. 31, p. 445.
2. Gray,E.G. and Whittaker,V.P. (1962) *J. Anat.*, **96**, 79.
3. De Robertis,E., Pellegrino De Iraldi,A., Rodriguez de Lores Arnaiz,G. and Salganicoff,L. (1962) *J. Neurochem.*, **9**, 23.
4. Autilio,L.A., Appel,S.H., Pettis,P. and Gambetti,P.L. (1968) *Biochemistry*, **7**, 2615.
5. Nagy,A. and Delgado-Escueta,A.V. (1984) *J. Neurochem.*, **43**, 1114.
6. Babitch,J.A., Breithaupt,T.B., Chiu,T.-C., Garadi,R. and Helseth,D.L. (1976) *Biochim. Biophys. Acta*, **433**, 75.
7. Dodd,P.R., Hardy,J.A., Bradford,H.F., Bennett,G.W., Edwardson,J.A. and Harding,B.N. (1979) *Neurosci. Lett.*, **11**, 87.
8. Rickwood,D. (1984) In *Centrifugation—A Practical Approach*. Rickwood,D. (ed.), IRL Press, Oxford and Washington DC, 2nd edn, p. 1.
9. Fried,R.C. and Blaustein,M.P. (1978) *J. Cell Biol.*, **78**, 685.
10. Morgan,I.G., Wolfe,L.S., Mandel,P. and Gombos,G. (1971) *Biochim. Biophys. Acta*, **241**, 737.
11. Booth,R.F.G. and Clark,J.B. (1978) *Biochem.J.*, **176**, 365.
12. Deutsch,C., Drown,C., Rafalowska,U. and Silver,I.A. (1981) *J. Neurochem.*, **36**, 2063.
13. Weibel,E.R. and Bolender,R.P. (1973) In *Principles and Techniques of Electron Microscopy: Biological Applications*. Hayat,M.A. (ed.), Van Nostrand-Reinhold, New York, Vol. 3, p. 239.
14. Pfenninger,K.H., Ellis,L., Johnson,M.P., Friedman,L.B. and Somlo,S. (1983) *Cell*, **35**, 573.
15. Gordon-Weeks,P.R. and Lockerbie,R.O. (1984) *Neuroscience*, **13**, 119.
16. Gordon-Weeks,P.R., Lockerbie,R.O. and Pearce,B.R. (1984) *Neurosci. Lett.*, **52**, 205.
17. Lockerbie,R.O., Gordon-Weeks,P.R. and Pearce,B.R. (1985) *Dev. Brain Res.*, **21**, 265.
18. Ellis,L., Katz,F. and Pfenninger,K.H. (1985) *J. Neurosci.*, **5**, 1393.
19. Ellis,L., Wallis,I., Abreu,E. and Pfenninger,K.H. (1985) *J. Cell Biol.*, **101**, 1977.
20. Johnson,M.K. and Whittaker,V.P. (1963) *Biochem. J.*, **88**, 404.
21. Gurd,J.W. (1982) In *Molecular Approaches to Neurobiology*. Brown,I.R. (ed.), Academic Press Inc., London and New York, p. 99.
22. Cotman,C.W. and Matthews,D.A. (1971) *Biochim. Biophys. Acta*, **249**, 380.
23. Davis,G.A. and Bloom,F.E. (1973) *Brain Res.*, **62**, 135.
24. Cotman,C.W. and Taylor,D. (1972) *J. Cell Biol.*, **55**, 696.
25. Jones,D.H. and Matus,A.I. (1974) *Biochim. Biophys. Acta*, **356**, 276.
26. Nieto-Sampedro,M., Bussineau,C.M. and Cotman,C.W. (1981) *Neurochem. Res.*, **6**, 307.
27. Cotman,C.W. and Kelly,P.T. (1980) In *The Cell Surface and Neuronal Function*. Cotman,C.W., Poste,G. and Nicholson,G.I. (eds), Elsevier, Amsterdam, p. 503.
28. Stoughton,R.L., Kelly,P.T. and Akeson,R. (1983) *Neurosci Lett.*, **35**, 215.
29. Mena,E.E., Hoeser,C.A. and Moore,B.W. (1980) *Brain Res.*, **188**, 207.
30. Agrawal,H.C., Trotter,J.L., Burton,R.M. and Mitchell,R.F. (1974) *Biochem. J.*, **140**, 99.
31. Duncan,H.M. and Mackler,B. (1966) *J. Biol. Chem.*, **241**, 1694.
32. Cotman,C., Herschman,H. and Taylor,D. (1971) *J. Neurobiol.*, **2**, 169.
33. Bradford,M.M. (1976) *Anal. Biochem.*, **72**, 248.
34. Read,S.M. and Northcote,D.M. (1981) *Anal. Biochem.*, **116**, 53.
35. Gogstad,G.O. and Krutnes,M.-B. (1982) *Anal. Biochem.*, **126**, 355.
36. Gray,E.G. (1952) *J. Anat.*, **93**, 420.
37. Matus,A. and Pehling,G. (1981) *J. Neurobiol.*, **12**, 67.
38. Kennedy,M.B., Bennett,M.K. and Erondu,N.E. (1983) *Proc. Natl. Acad. Sci. USA*, **80**, 7357.
39. Gurd,J.W., Gordon-Weeks,P.R. and Evans,W.H. (1982) *J. Neurochem.*, **39**, 1117.

40. Gurd,J.W., Gordon-Weeks,P.R. and Evans,W.H. (1983) *Brain Res.*, **276**, 141.
41. Gurd,J.W. (1985) *Brain Res.*, **333**, 385.
42. Gordon-Weeks,P.R. and Harding,S. (1983) *Brain Res.*, **277**, 380.
43. Murakami,K., Gordon-Weeks,P.R. and Rose,S.P.R. (1986) *J. Neurochem.*, **46**, 340.
44. Siekevitz,P. (1981) In *Research Methods in Neurochemistry*. Marks,N. and Rodnight,R. (eds), Plenum Press, London and New York, Vol. 5, p. 75.
45. Albertsson,P.A. (1971) *Partition of Cell Particles and Macromolecules*. 2nd edn, John Wiley and Sons, New York.
46. Cohen,R.S., Blomberg,F., Berzins,K. and Siekevitz,P. (1977) *J. Cell Biol.*, **74**, 181.
47. Huttner,W.B., Schiebler,W., Greengard,P. and De Camilli,P. (1983) *J. Cell Biol.*, **96**, 1374.
48. Nagy,A., Baker,R.R., Morris,S.J. and Whittaker,V.P. (1976) *Brain Res.*, **109**, 285.
49. Whittaker,V.P., Michaelson,I.A. and Kirkland,J.A. (1964) *Biochem J. (Tokyo)*, **90**, 293.
50. Jones,R.T., Walker,J.H., Richardson,P.J., Fox,G.Q. and Whittaker,V.P. (1981) *Cell Tissue Res.*, **218**, 355.
51. Richardson,P.J., Siddle,K. and Luzio,J.P. (1984) *Biochem. J.*, **219**, 647.

CHAPTER 2

Neuronal and glial cells: cell culture of the central nervous system

RUSSELL P.SANETO and JEAN DE VELLIS

1. INTRODUCTION

A critical question in neurobiology centres on the elucidation of cellular and molecular events controlling the development and function of the central nervous system (CNS). Based on anatomical observations Cajal (1) formulated with insight his 'principle of connectional specificity'. One tenet of this principle was that each cell forms specific and precise connections. Further research has shown Cajal's hypothesis to be correct. Today, neurobiologists think that proliferation, survival and differentiation is a highly ordered and controlled process. However, the complex interactions between genome and environment giving rise to the functional mature nervous system remain virtually unknown.

The limited elucidation of the cellular and molecular mechanisms underlying CNS structure and physiology resides in its extreme complexity. The hierarchy of cytoarchitecture and network of cellular processes combined with isolation (blood–brain barrier) makes studies *in vivo* difficult to perform and interpret. To circumvent these problems, early investigations utilized bulk isolation techniques (2,3) and dissociated culture of whole brain samples (4). However, these early methods could not provide live cell preparations of the various CNS cell types. Only recently have primary culture techniques evolved to the point where pure identifiable cultures of the major CNS cell types are attainable.

1.1 **Historical perspective of neural cell culture**

The usefulness of cell culture techniques in answering neurobiological questions was provided by Harrison, who developed the first real neural cultures (5). Prior to Harrison's observations, the origin of long axonal processes was controversial. By placing explants of frog embryo nervous tissue in drops of clotted lymph, Harrison provided the first evidence that the neurone produced elongated axonal processes. Subsequently, Ingebrigtsen (6) demonstrated that the mammalian brain could be grown in explant culture. Since these early reports, many different types of neuronal culture have been developed (for review, see 7,8). However, all of these culture systems have been heterogeneous in cellular make-up. For single cell analysis, electrical recordings, immunocytochemistry and receptor autoradiography, such mixed neuronal–glial cultures are useful. Nevertheless, the cellular heterogeneity of these cultures precludes their use for biochemical and molecular studies. Only within the last few years have techniques made pure neuronal cultures possible.

Although glial cells have historically contaminated neuronal cultures, pure glial cultures are of recent vintage. In 1966, Shein (9) observed that dissociated human fetal brain cell cultures yielded spongioblasts (immature oligodendrocytes) resting on a bed layer of astrocytes. Early attempts at glial culture yielded cultures consisting predominantly of astrocytes (for review, see 10). Only since 1980 has the production of reproducible mixed glial cultures been possible. McCarthy and de Vellis (11) found that mixed glial cultures could be obtained only if the initial seeding density was high. With the exception of a few studies which facilitate the separation and enrichment of specific cell types by immunological methods (12) and physical methods (13), the most widely used method of cell separation is the shaking technique developed in our laboratory (11).

1.2 **Culture types**

Fundamental to obtaining qualitative and quantitative data on cellular structure and function is the isolation and long-term culture of homogeneous populations of CNS cell types. To circumvent cellular heterogeneity and address the issue of cell function, bulk isolation techniques were developed. Methodologies such as velocity sedimentation (14), density gradient (3) and differential centrifugation (2) allowed the elucidation of cell-specific biochemical parameters. These procedures, however, have the shortcomings of low yields, questionable purity and variable viability. To address regulation of cell-specific properties, researchers turned to the use of clonal cell lines of neural origin. Clonal cell lines represent a useful model system since they provide a genetically homogeneous cell population that can be easily grown and maintained. Unfortunately, their tumorigenic state obscures many meaningful correlations of findings *in vitro* to responses of normal cells *in vivo*. Nevertheless, upon rigorous characterization of a specific cellular function with its cell counterpart *in vivo*, normal processes can be elucidated and well characterized.

The recent development of primary culture techniques have made these systems suitable to study CNS function. The primary systems of explant, re-aggregate and dissociated cultures have select inherent advantages. Explants retain much of their three-dimensional tissue organization; re-aggregate cultures recreate histotypic organization; and dissociated cultures form a simplified two-dimensional network. The importance of tissue organization and cell contact for the expression of cellular function is illustrated by the hydrocortisone-induced increase of retinal glutamine synthetase activity. Moscona and his colleagues (1971) found induction only in systems where three-dimensional organization was maintained (15). Thus the selection of the culture system is critical for proper examination of CNS phenomena *in vitro*.

Explant, as with whole organ culture, is the least disruptive of the primary culture techniques. In place of the whole brain, pieces of brain between 0.5 and 1.0 mm are placed in culture medium. Survival is dependent on exchange of nutrients and metabolites via simple diffusion, which makes cell death within the tissue variable from culture to culture. Re-aggregate cultures are more reproducible from experiment to experiment. By dissociation of specific brain areas, which is followed by aggregation in rotation suspension, model systems of specific areas to study histogenesis and cellular differentiation are created (16). Dissociated cultures produce model systems which are readily accessible to manipulation, a feature absent from re-aggregating cultures.

However, all of these culture systems are complicated by cellular heterogeneity which make elucidation of cell-specific function ambiguous. Better model systems to circumvent the unknowns of using transformed cell lines and heterogeneity of mixed cell cultures are homogeneous cultures of specific CNS cell types.

1.3 Theoretical considerations

There are several conceptual aspects of culture model systems as correlates of function *in vivo* that are important. Much biological research is aimed at discovery and analysis of molecular agents which regulate specific cellular function. For simplicity and uniformity, the definitions of what function these agents perform are important. Without proper and exact definitions of function, the scientific literature becomes very confused. Secondly, as model systems, cell types within the culture need to be fully characterized to reduce uncertainty of extrapolation to normal phenomena. Finally, the model systems developed are strictly that and should not become an end in themselves.

1.3.1 *Nomenclature of extrinsic regulators or factors*

Cell survival, proliferation and differentiation in the CNS appear to be regulated by macromolecular agents. The notions of 'molecular agent' or 'trophic factor' have been extensively reviewed and are not discussed here (for reviews, see 17–19). Much confusion has been created by misunderstanding of some of these definitions, especially in relation to primary cell culture, and will therefore be discussed here.

The terms 'trophic factor' and 'growth factor' have had different meanings in the literature. In one reference, 'trophic factor' may represent a mitogen, while in another it may mean a survival factor. Originally the term 'trophic factor' was used in reference to nerve growth factor and its effect on the survival, stimulation of cellular metabolism and hypertrophy of peripheral neurones (20). Although the term 'growth factor' has occasionally been used for survival and growth, it should be reserved for proliferation-promoting molecules.

Stimulation of cell-specific functions, such as neurotransmitter synthesis in neurones, induction of glycerol phosphate dehydrogenase (GPDH) activity in oligodendrocytes and glial fibrillary acidic protein (GFAP) expression in astrocytes, require different nomenclature. These specific 'differentiation-promoting' factors control the expression of a particular cellular phenotype and not survival, proliferation and general metabolism. The 'differentiation-promoting' factors have also been referred to in the literature as 'specifying' or 'instructive' factors. It should be noted that some factors can induce a multiplicity of effects. The colony-stimulating factors cannot only induce cell proliferation but also cell differentiation of macrophage precursor cells (21).

1.3.2 *Identification of cell types*

Until recently, the identification of neurones, astrocytes and oligodendrocytes relied on metallic staining techniques developed in the early 1900s (22,23). The development of electron microscopy has helped to characterize cell types further and this is still the most reliable method for distinguishing glial cell types. Recently, biochemical, molecular and immunological techniques have been developed to identify specific cell types. However, the small number of definitive markers and lack of markers for specific situations (e.g. stem cell detection) can make absolute definition of cell type a problem. Therefore, cell differentiation *in vitro* should not be based on the detection of only one cell marker. No single cell property should be used, by itself, as an exclusive and absolute standard for a cell type. Ideally, multiple identification methods should be used to characterize a specific cell type.

1.3.3 *In vitro assessment of in vivo function*

The culturing of specific neural cell types inherently precludes homeostatic controls found in the animal. In addition, the reduction of cellular heterogeneity and interactions produces losses in pre- and post-synaptic connections for neuronal elements and blood-borne molecules in all cell culture. Hence, directed effort to integrate culture data with findings in the whole animal is paramount. Since the aim of primary culture is the elucidation of physiological processes *in vivo*, normality *in vitro* must be judged relative to reality *in vivo*.

2. CELL CULTURE TECHNIQUE

2.1 **Enhancement of CNS primary culture**

2.1.1 *Source of tissue for neural culture*

The developmental age of the animal and the area of the brain cultured are important determinants of both the cellular milieu and the developmental stage of cells within the culture. Generally, in any CNS area, neurones become post-mitotic before glial cell proliferation takes place (24). Furthermore, areas of the brain mature at different times: for example the cerebellum is a later developing structure than the cerebrum. These conditions can be manipulated to produce the desired model system of cell culture. The ease of controlled breeding, quantity of available tissue and developmental time schedule made newborn rodents the choice of researchers for the culture of brain cells.

In our development of CNS culture techniques, we utilize the same brain area for the source of culture material. This allows for reproducibility and establishes credibility for extrapolating culture data to results found *in vivo*. For neuronal cultures we use two different developmental ages of the rat. Our rat colony is the Wistar strain but we have also used the Sprague–Dawley strain with equivocal results. For embryonic neuronal cultures, we use 16 day embryonic pups, the day where the vaginal plug is found is designated as day 0. Post-natal neuronal cultures are initiated from 0–1 day neonatal rat pups. The area of brain used for both of these cultures is cerebral cortex. Mixed glial cultures are initiated from 0–2 day neonates. If younger than newborn pups are used, neurones tend to survive the culture period of 7–10 days, and older pups increase the percentage of 'fibroblast-like' cells contaminating the culture. Donor age of rat pups has been used up to 3 days post-natal with similar results.

Care is taken to use only the cerebral cortical area for glial cultures. This allows the cells grown in culture to be more homogeneous with respect to both proliferation and differentiation potential.

2.1.2 *Sterile technique*

All procedures are performed to prevent bacterial and fungal contamination. Any area of the hood that is touched is assumed to be non-sterile and should not touch any material

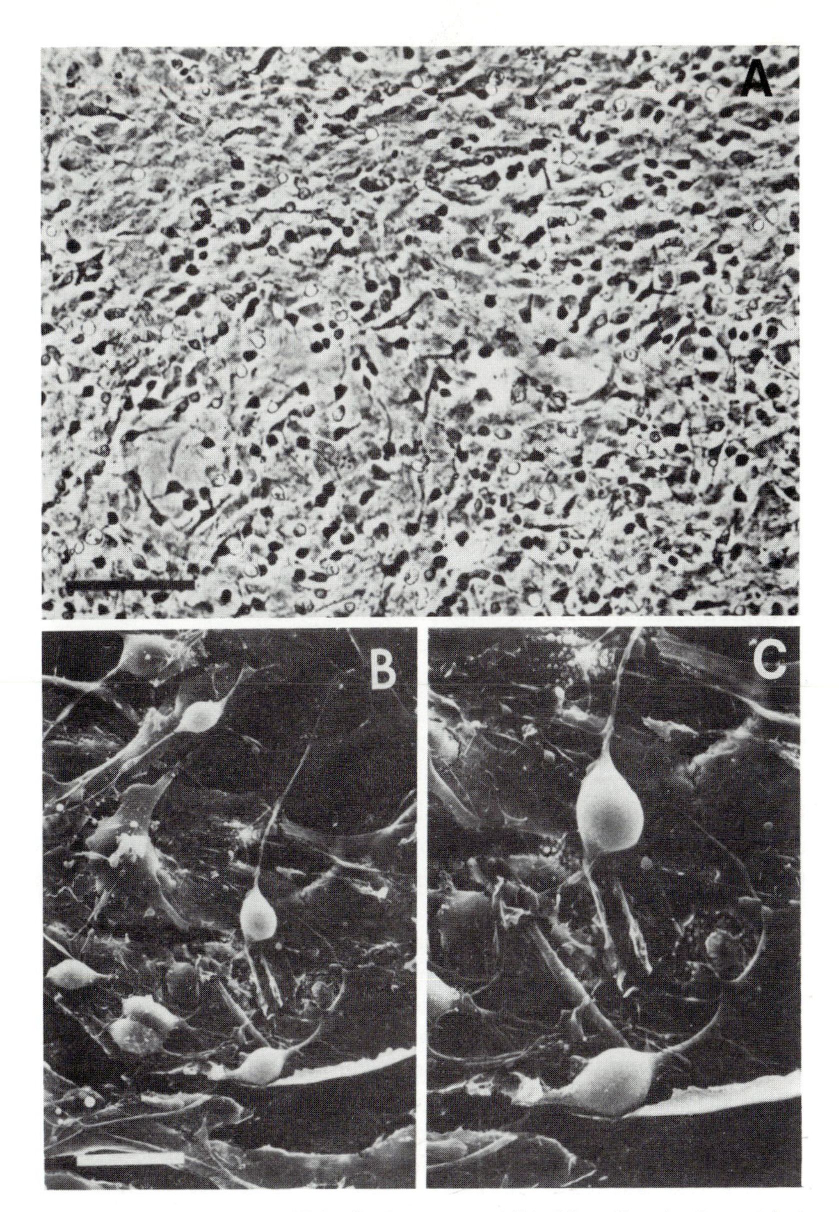

Figure 1. Mixed glial cell cultures. Glial cell cultures are established from dissociated neonatal rat cerebral cells. (**A**) This panel represents a phase-contrast micrograph of a 7-day primary culture. The phase-dark process-bearing cells, oligodendrocyte progenitor cells, the bed layer of phase-grey polygonal cells, astrocytes. (**B**) This represents a scanning electron micrograph of a culture similar to **A**. The stratification of cell types is clearly represented (**A**, ×240; **B**, ×965; **C**, ×1930).

to be cultured. Furthermore, any unnecessary movement around the dissection area should be avoided. It is best to have a separate, completely enclosed room to do dissections. Routine sterilization protocols and culture techniques may be found in several reviews (25,26).

2.1.3 *Dissection precautions needed for primary culture*

Tissue should not be allowed to dry during the dissection procedure. We use a balanced salt solution to avoid ionic losses and cell death throughout dissection. Solutions are used at 37°C; in addition to increased cell survival, this temperature helps remove blood from the dissected material. Mechanical trauma needs to be kept to a minimum. Direct cuts with a sharp instrument are better than teasing tissue apart as teasing disrupts more cells than cutting. Transported tissue should be carried and not squeezed by forceps in transfer.

2.2 Establishment of glial primary cultures

2.2.1 *Mixed glial culture*

With the exception of a few studies which utilize immunological methods (12) and physical methods (13) the most widely used method to generate separate populations of glia is the use of mixed glial culture. We have found that a critical developmental time window exists for optimal mixed glial cultures from cerebral tissue. Using 0–2 day neonatal rat cerebrum, neurones fail to survive and are eliminated before the second medium change. During this time, cells of two distinct morphologies can be seen undergoing proliferation. There are small phase-dark process-bearing cells residing on top of a phase-light confluent bed layer of cells (*Figure 1*). We have characterized both cell types in both mixed and isolated cell cultures (see Section 2.2.4). The cells that abut the bed layer of cells are identified by biochemical, ultrastructural and immunocytochemical criteria and judged to be oligodendrocytes. Those cells comprising the bed layer are identified as astrocytes by the same criteria.

(i) Dip newborn pups into 95% ethanol, allowing the thumb and forefinger also to be immersed. Allowing the thumb and forefinger to be immersed in alcohol while dipping the animals keeps the finger and thumb clean of blood and possible contamination.
(ii) Using a pair of sterilized scissors decapitate the pup and place the head on either a piece of sterile gauze or sterile paper towel.
(iii) Use another pair of sterile scissors to cut the skin at the midline of the head, cutting from the base of the head to the mid-eye area.
(iv) Fold back the skin by pulling the loose skin forward and hold open using the thumb and forefinger being careful not to touch the exposed skull.
(v) Use a third pair of scissors to cut the skull at the midline fissure. Care should be taken not to cut the brain tissue. To reduce the possibilities of contamination use each pair of scissors only for its specific function. After each use, dip the scissors in 95% ethanol (separate from the alcohol used to dip the pups) and place on a piece of sterile gauze to drain.
(vi) Holding and applying 'slight' pressure with the thumb and forefinger, remove the raised portion of the skull by sterile forceps by pulling the piece of skull away from the midline.
(vii) Subsequently excise the brain by carefully running a sterile spatula along the bottom and sides of the brain calvarium from the olfactory lobes to the posterior of the midbrain.
(viii) Once freed, place the brain in a 35 mm plastic Petri dish.

(ix) Dissect cerebral cortices from the rest of the brain using two microforceps. Steadying the brain with one pair of forceps at the median fissure, gently pull away one cerebral hemisphere from the rest of the brain at the median fissure with the second pair of forceps. The cortex is now lying with its superior surface folded over onto the surface of the Petri dish. With the second pair of forceps, remove the cortex by pinching the cortex against the Petri dish surface. Remove the other cortex in a similar manner.

(x) Once removed, immediately immerse the cortices in sterile saline solution I (138 mM NaCl, 5.4 mM KCl, 1.1 mM Na_2HPO_4, 1.1 mM KH_2PO_4, 22 mM glucose and 0.9 mM $CaCl_2$) or serum-free basal medium (SFM; Section 3.1) contained in another sterile Petri dish.

(xi) After collecting cortices from 15–20 pups, carefully remove the meninges. We use two microdissecting forceps (the two forceps used to remove cortices can be re-used), one to steady the cortex immersed in saline solution I and the other to remove the meninges. Use another Petri dish containing saline solution I for this procedure. Tease away the meninges from the isolated cortex by beginning with an exposed corner of the membrane on the outer surface of the cortex and gently pulling the meninges off. The meninges should pull off as an intact sheet.

(xii) Blot the removed meninges onto a piece of sterilized gauze.

(xiii) After removal of the meninges and blood vessels, transfer the cortex to another 35 mm Petri dish containing saline solution I. The use of different Petri dishes in this procedure allows the removal of blood and pieces of pulled off meninges that have been left behind. Dip both forceps in 95% ethanol and allow to drain.

Dissociation of cerebral tissue can be accomplished either by mechanical sieving or trypsinization, both techniques yield similar cellular compositions of mixed glial cultures.

2.2.2 *Trypsinization technique*

(i) After removal of the meninges, gently aspirate the cortices through a 10 ml pipette 2–3 times and subsequently transfer to a sterile, 50 ml screw-capped Erlenmeyer flask. The volume of tissue suspension should be recorded at this step so that the correct dilution of trypsin solution I (Section 3.3.1) can be added.

(ii) Place the partially disrupted tissue onto a rotary shaker and agitate for 15 min (80 r.p.m., 37°C), to loosen the tissue components.

(iii) Following this period, add trypsin solution I to a final concentration of 10% and replace the tissue suspension onto the rotary shaker for an additional 25 min.

(iv) Stop trypsinization by the addition of 10 ml of SSM medium (SFM containing 10% fetal calf serum; v/v). We do not use antibiotics but their addition while learning the technique is probably beneficial.

(v) Gently aspirate the tissue suspension through a 10 ml pipette three times and allow the tissue pieces to settle. This can be facilitated by tipping the Erlenmeyer flask at an angle so that the untrypsinized tissue pieces settle into a corner of the flask. Placing the flask on edge via use of paper towels accomplishes this well.

(vi) Taking care not to disturb the settled tissue, remove the cell suspension and filter either through Nitex 130 (mesh size, 130 μm; Tetko, Inc.) or tissue sieve 100 (mesh size, 140 μm).

(vii) Add 7 ml of culture medium to undissociated tissue and again aspirate the tissue pieces through a 10 ml pipette three times.
(viii) Allow the tissue pieces to settle again, remove the cell suspension, filter through Nitex 130 and combine it with the initial filtrate.
(ix) Add culture medium (3 ml) to the remaining undissociated tissue and after three forceful aspirations through a 10 ml pipette, pass it through Nitex 130 and combine it with the initial filtrate.
(x) Centrifuge the total filtrate at 40 *g* for 5 min using a clinical table top centrifuge at room temperature.
(xi) Resuspend the cell pellet in culture medium and re-filter through Nitex 33 (mesh size: 33 μm).
(xii) Determine the number of cells in the filtrate using a haemocytometer (AO instruments), and dilute the cell suspension such that the plating density is 15×10^6 cells for each 75 cm^2 plastic tissue culture flask (Section 3.2).
(xiii) After taking cell aliquots incubate the flasks at 37°C in a water-saturated 5% CO_2:95% air atmosphere incubator. Care should be taken so as to not disturb the flasks for 3 days. Movement of the flasks before day three reduces cell attachment and therefore should be avoided. Change the culture medium after day three and every other day thereafter.

2.2.3 *Mechanical sieving technique*

(i) After removal of the meninges, transfer the cortices to a 35 mm Petri dish containing 5 ml of serum-supplemented medium (SSM).
(ii) Make a sterile monofilament mesh bag from Nitex 210 (mesh size: 210 μm) with the measurements of approximately 1.5×3.0 inches. Sew three sides of this bag using a very fine stitch.
(iii) Place the Nitex bag in a 100 mm Petri dish containing 15–20 ml of culture medium.
(iv) Hold open the mouth (open end) of the bag with sterile forceps and pipette the tissue suspension through the opening. Close the opening of the bag with forceps, and immerse the entire bag in the Petri dish.
(v) Use light strokes of a sterile glass rod to tease the tissue through the mesh.
(vi) When approximately one-half of the tissue has been disrupted, remove the cell suspension and filter it either through Nitex 230 (mesh size: 230 μm) or a collector tissue sieve with screen 60 (mesh size: 240 μm; Bellco Glass, Inc., Vineland, NJ). Repeat the teasing procedure for the undissociated tissue.
(vii) Elevate the bag above the Petri dish and wash it with a 5 ml stream of culture medium so the medium and cells can drain into the Petri dish.
(viii) Filter this solution through Nitex 230 and add to the initial filtrate.
(ix) Filter the combined cell suspension through Nitex 130 or tissue sieve 100 (mesh size: 140 μm).
(x) Centrifuge the resulting filtrate at 40 *g* for 5 min in a clinical table top centrifuge at room temperature.
(xi) Resuspend the cell pellet in culture medium and count using a haemocytometer. Cell counts can also be made using an automatic cell counter, we use a Royco 927 TC cell tissue counter.

(xii) Dilute the cell suspension, such that the plating density is 15×10^6 cells for each 75 cm^2 plastic culture flask.
(xiii) Incubate the flasks as previously described for the trypsinization method.

2.2.4 *Isolation of purified astrocytes and oligodendrocytes*

The length of the initial culture period should be limited to 7–9 days. Periods shorter than this are not sufficient for stratification of astrocytes and oligodendrocytes. Periods longer than this can result in clustering of astrocytes above the bed layer. Upon subsequent isolation of oligodendrocytes the clustered astrocytes break up and contaminate the oligodendrocyte preparation. We have found that the oligodendrocytes tend to adhere more strongly the longer they remain in mixed culture, making the shaking procedure less efficient.

(i) At the end of the initial culture period, a large number of phase-dark process-bearing cells are overlaying a confluent bed layer of cells (*Figure 1*). Change the culture medium and tighten shut the plastic culture flask lids.
(ii) Place the flasks on a rotary shaker (Labline Junior Orbit Shaker, Scientific Products) and secure with fibre tape (Grayarc, Garden Grove, CA). Shake the flasks (250 r.p.m., 1.5 inch stroke diameter, 37°C) for 1 h, discard the medium, add fresh medium, re-fasten the flasks to the rotary shaker and shake them for 6 h (250 r.p.m.). We have found that insulating the top of the shaker with packing foam sheets helps dissipate heat generated from the shaker. The 1 h shake removes dividing astrocytes and other contaminating cells and therefore improves the purity of isolated oligodendrocyte progenitor cell cultures. Rotary shakers vary in their actual speed and hence need to be accurately calibrated before initial use.
(iii) After the 6 h shake, filter the cell suspension through Nitex 33 (mesh size: 33 μm).
(iv) Centrifuge the filtrate at 40 *g* for 3 min at room temperature.
(v) Resuspend the resulting cell pellet in fresh culture medium and filter through Nitex 17 (mesh size: 17 μm). The second filtration step increases the purity of oligodendrocyte progenitor cell cultures by removing the contaminating astrocytes which are present almost exclusively in clumps.
(vi) Re-plate the cell suspension into the appropriate culture vessel.
(vii) Replenish the same flasks used for the 6 h shake with fresh culture medium and shake them overnight (250 r.p.m., 1.5 inch stroke diameter, 37°C) for an additional 16–18 h.
(viii) Process the resulting cell suspension as described for the 6 h shake.

We have found the optimal plating density of oligodendrocytes to be 5×10^6 cells per 75 cm^2 culture flask; 1×10^5 cells per well in 2.0 cm^2, 24-well culture plates; 4×10^5 cells per 35 mm Petri dish; and 4×10^4 cells per well in 0.32 cm^2, 96-well culture plates.

Although the phase-dark process-bearing cells settle onto the surface of the culture flask, adhere and begin to send out processes almost immediately, it is best not to disturb flasks for 24 h. At this time add new culture medium. These cultures are approximately 98% pure. Further purification can be accomplished by adding new culture medium and slapping the bottom of the flask sharply with the palm of the hand. Place the flasks in an upright position for drainage and subsequently filter the cell suspension through

Nitex 33. Re-plate the resulting filtrate in 75 cm^2 culture flask or other culture plates after proper density is determined. We estimate the purity of these cultures to be 99% oligodendrocytes by immunochemical, biochemical and ultrastructural criteria (11,27).

After oligodendrocyte removal, the remaining astrocyte cultures are estimated 95% pure by phase contrast microscopy.

(i) Further purify the cultures by washing twice with fresh SSM, adding fresh culture medium and shaking at 100 r.p.m. for 24–48 h.
(ii) At this time cultures usually have fewer than 10 phase-dark cells per microscopic field ($\times$100). If not, wash the cells with fresh SSM (Section 2.2.2) and re-shake until phase-dark cells are less than 10 per microscopic field.
(iii) Wash the purified cultures with fresh culture medium and trypsinize with trypsin solution II (Section 3.3.2).
(iv) Place the cultures at 37°C until the cell layer begins to flow down the side of the flask when tilted on edge. At this point add 5 ml of fresh SSM.
(v) Triturate the cell suspension several times with a 10 ml pipette, transfer it to a sterile tube with cap and centrifuge at 40 *g* for 3 min at room temperature. Pour off the supernatant and add fresh SSM.
(vi) Resuspend the cell pellet in culture medium and count the cells.

After plating for experimentation, astrocyte cultures have been estimated to be 98–99% pure by biochemical, ultrastructural and immunocytochemical criteria (11,28). Optimal plating density has been found to be 1×10^6 cells per 75 cm^2 culture flask; 2×10^4 cells per 2.0 cm^2 well, 24-well culture plate; and 4×10^5 per 35 mm Petri dish.

2.2.5 *Chemically defined media*

The unknown complement of serum components has limited the elucidation of molecular requirements for cell attachment, survival, proliferation and differentiation. Serum is manufactured by pooling sera from many animals, so batches have inherent variation depending upon age, sex and nutritional status of the animal used. Serum also contains unknown growth factors and enzymes which may metabolize hormones and hence, change their effect. Finally, serum is pathological, a by-product of blood coagulation and not present in the body except at sites of vascular and epidermal wound healing. To create an unambiguous environment to study adhesion, survival, proliferation and differentiation, we and others have developed serum-less, chemically defined media for neurones, astrocytes and oligodendrocytes.

(i) *Oligodendrocytes.*

(1) After final filtration of the oligodendrocyte cell suspension, seed the cells in SSM at a density of 1×10^5 cells per well of a 24-well plastic culture plate.
(2) Allow the cells to attach for a 18–24 h period and subsequently determine the number of cells in a few wells to establish plating efficiency.
(3) Wash the remaining cultures three times with SFM and add fresh oligodendrocyte-defined medium (ODM). ODM is prepared by adding insulin (5 μg/ml), transferrin (500 ng/ml) and fibroblast growth factor (75 ng/ml) to SFM. If SFM is stored, fresh pyruvate (110 μg/ml) is added to the basal medium. The order of supplement addition is not important. Together these compounds synergistic-

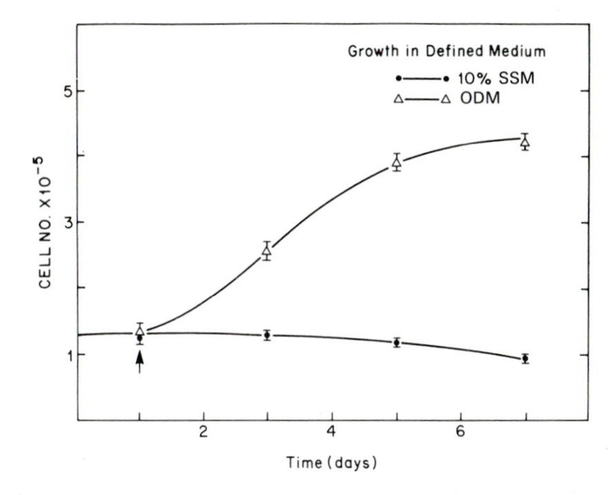

Figure 2. Growth curve of isolated oligodendrocyte progenitor cells. Cells were grown in either SSM (●) or ODM (△). Arrow indicates the time at which medium is changed to either SSM or ODM. Data represent cell numbers per well, values are the means of six wells ± SEM (27).

Table 1. Immunocytochemical characterization of isolated oligodendrocytes.

Specific antiserum	*% positive cells*	
	ODM	*SSM*
GPDH	95–98	95–98
Myelin basic protein	90	10
Galactocerebroside	60–70	20–30
A2B5	70	30
RBA1 (vimentin)	0	30
RBA2 (desmin)	0	30
GFAP	2–5	20–30
Fibronectin	0	0

Percentage of cells within the culture that were stained by indirect immunofluorescence was determined by counting cells within four separate fields, with a minimum of 100 cells counted. Isolated cells were grown in culture for 5 days at the time of immunocytochemical staining. Values represented are the means of four separate experiments (27).

ally act to produce a 3-fold increase in cell number, while sister cultures grown in SSM do not significantly increase in cell number over the same growth period (*Figure 2*).

(4) Change the medium (ODM) every 2 days, adding fresh supplements at each medium change.

Recently we have found that pre-plating the surface of the culture dish with fetuin eliminates the need for the 18–24 h serum pre-incubation. When oligodendrocyte progenitor cells were seeded in SFM, the number of cells attached after 20 h was only 38% of that found in SSM. However, in the presence of fetuin (100 μg/ml) and ODM, the number of cells attached after 20 h is equivalent to the number found in culture medium. The influence of fetuin is dose-dependent and optimal at 25 μg/ml (unpublished data). If seeding directly in ODM, the cell pellet produced after the centrifugation step

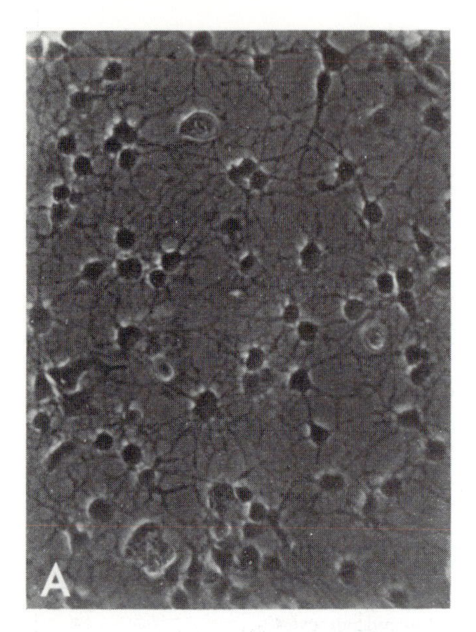

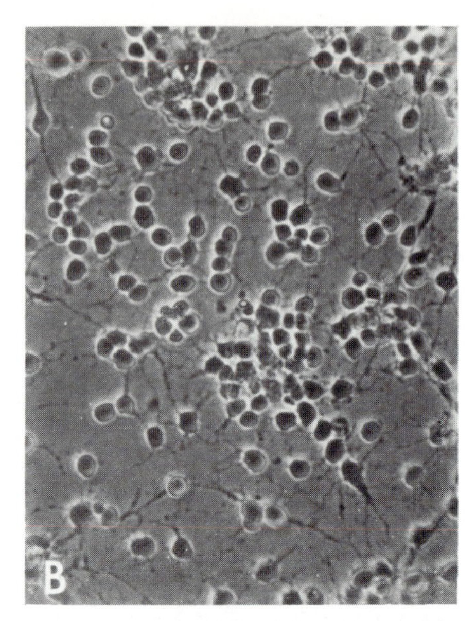

Figure 3. Morphology of isolated oligodendrocytes. Phase-contrast micrographs of isolated oligodendrocytes grown for 5 days in SSM (**A**) or in ODM (**B**) (×190).

is washed three times with SFM to remove serum. After the third wash, cells are resuspended in ODM and plated.

Oligodendrocyte progenitor cells cultured in ODM are morphologically less differentiated with respect to certain markers and more differentiated with respect to others (*Table 1*). The cells appear to have less elaboration of processes when cultured in ODM (*Figure 3*). However, by immunological, biochemical and ultrastructural criteria, cells grown in ODM are oligodendrocytes. Cells are serologically negative for the astroglial-specific antigen, GFAP and fibroblast and meningeal marker, fibronectin (29). These cells, after a 5-day growth period, are immunocytochemically stained for the oligodendrocyte-specific markers, GPDH (95−98%) and myelin basic protein (90%: *Figure 4*). They are inducible by the proper treatments for the enzymes glycerol phosphate dehydrogenase (EC 1.1.1.18), lactate dehydrogenase (EC 1.1.1.37) and 2′,3′-cyclic nucleotide phosphodiesterase (EC 3.1.4.37). The induction of these enzymes is specific to oligodendrocytes in the CNS (30−32). In addition, transmission electron microscopy reveals ultrastructure identical to immature oligodendrocytes *in vivo*.

(ii) *Astrocytes.*

(1) After removal of oligodendrocytes, trypsinize the remaining astrocytes with trypsin solution II, dilute with culture medium and centrifuge at 40 *g* for 5 min.
(2) After resuspending the cell pellet in SSM, count the cells, dilute to the appropriate cell density and plate at 2×10^4 cells per well of a 24-well culture plate.
(3) After an incubation period of 18−24 h, wash the cells three times with SFM and subsequently add fresh astrocyte-defined medium (ADM) to each well.

ADM is prepared by adding to SFM, insulin (50 μg/ml), putrescine (100 mM), prostaglandin $F_{2\alpha}$ (500 ng/ml), hydrocortisone (50 nM) and pituitary fibroblast growth factor

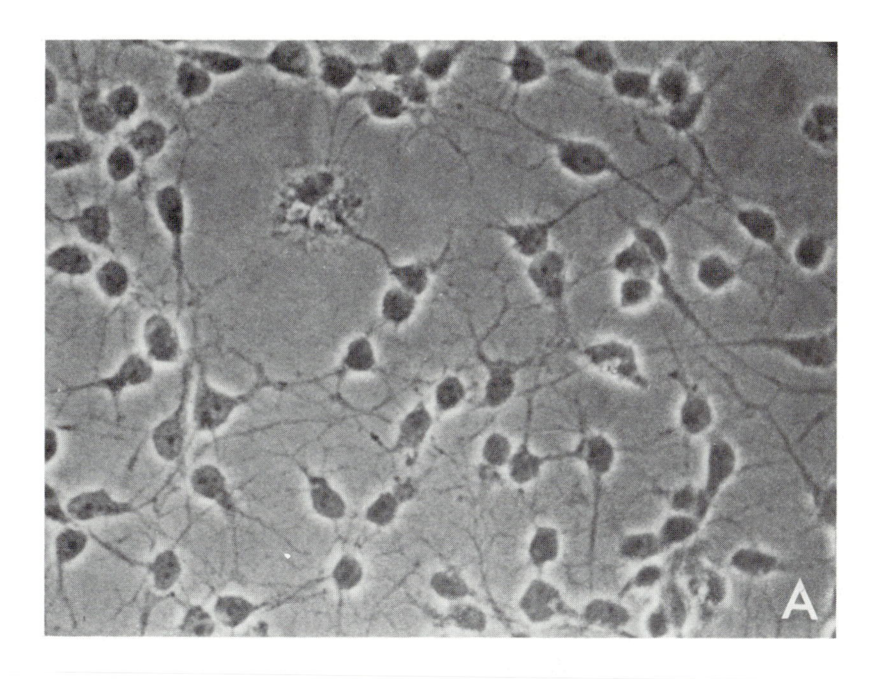

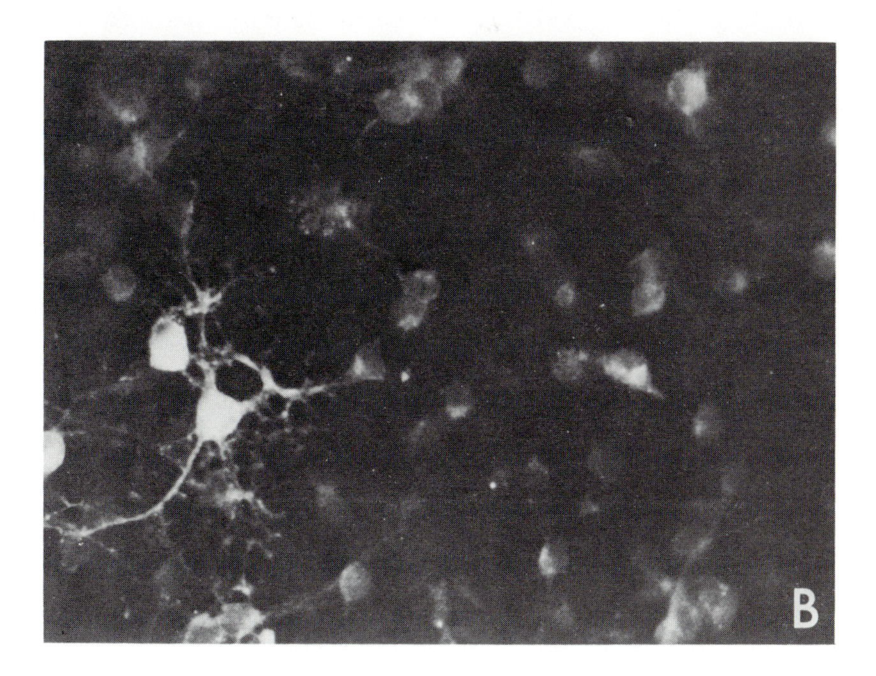

Figure 4.

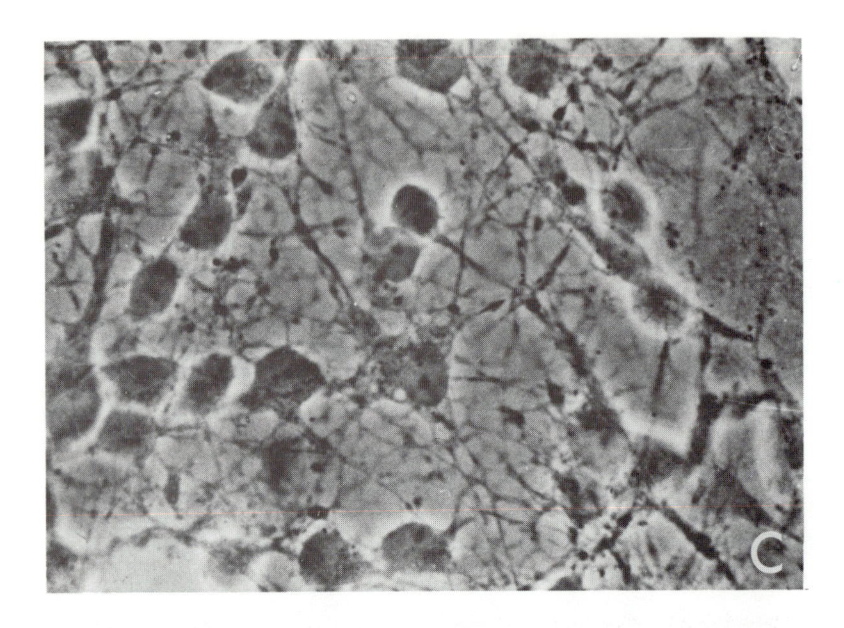

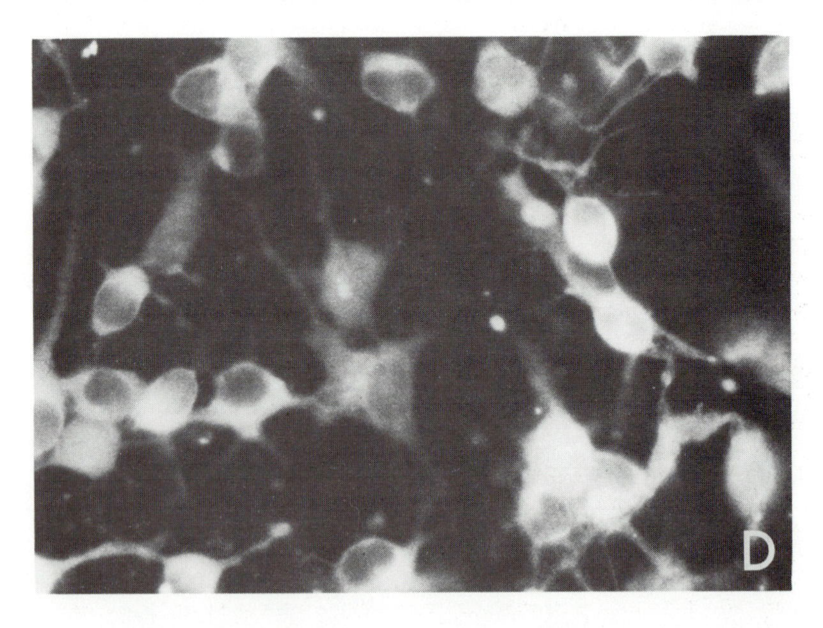

Figure 4. Myelin basic protein and GPDH immunofluorescence of oligodendrocyte cultures. Cultures were grown in ODM for 5 days. More than 90% of the cells seen in phase contrast (**A**) stained positive for myelin basic protein (**B**). Control sera (not shown) did not stain. (×365). Cultures were grown in culture medium and stained for GPDH. Isolated oligodendrocytes are visualized by phase contrast microscopy (**C**) and indirect immunofluorescence for GDPH (**D**) (×680).

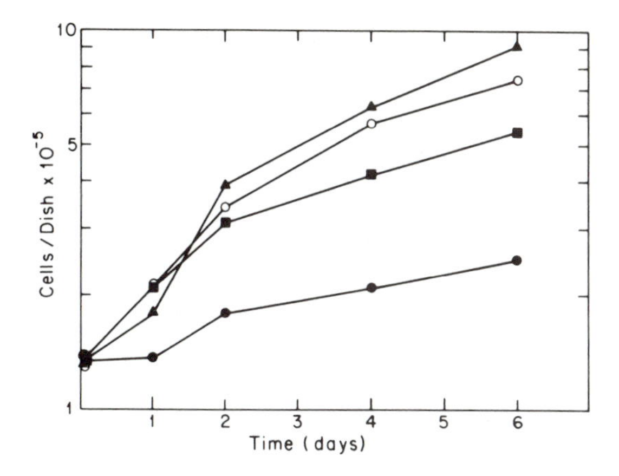

Figure 5. Growth curve of isolated astrocytes. Cells were grown in one of the following media: serum-free (●), serum-supplemented (△), chemically defined (○), or chemically defined without hydrocortisone (□). Data represent cell numbers per 5 mm Petri dish; day 0 is the end of the serum pre-incubation. SEM was <10% for all points (28).

(100 ng/ml). Each component by itself has little effect on astrocyte proliferation, however, in combination these compounds act synergistically to produce a 3-fold increase in cell number over SFM (27, *Figure 5*). Subsequently we found epidermal growth factor to be a potent mitogen for astrocytes (33). Astrocytes can be directly seeded in ADM without the pre-incubation in SSM.

The need for serum is eliminated by pre-coating the culture vessel with fibronectin (10 μg/ml), an attachment protein found in serum. When plating directly in chemically defined media, wash the cell pellet produced after trypsinization three times with SFM and then resuspend in ADM.

The components of chemically defined medium induce a dramatic change in astrocyte morphology (*Figure 6*). Astrocytes cultured in SSM are flat and polygonal in appearance, possess few processes and contain the astrocyte marker, GFAP (28; *Figure 7*). In the presence of chemically defined medium, astrocytes express a smaller somal diameter, extend long branching processes and express increased amounts of GFAP (28,34).

2.3 Establishment of neuronal primary cultures

The less complex organization and cell heterogeneity of the peripheral nervous system allowed the culture of pure neurones long before their counterparts in the CNS (for review, see 19). It is only recently that the technology has sufficiently advanced to allow successful growth of pure CNS neurones. The advent of chemically-defined media together with use of highly adhesive sub-strata have enabled highly pure neuronal cultures from both embryonic and post-natal tissues to be cultured. The use of serum promotes the growth of contaminating glial cells and therefore precludes its use for long-term culture of pure neurones. Although other laboratories have developed defined media for neuronal cultures (35) which support purified neuronal populations from different CNS areas (36), we will discuss two model systems developed in our laboratory.

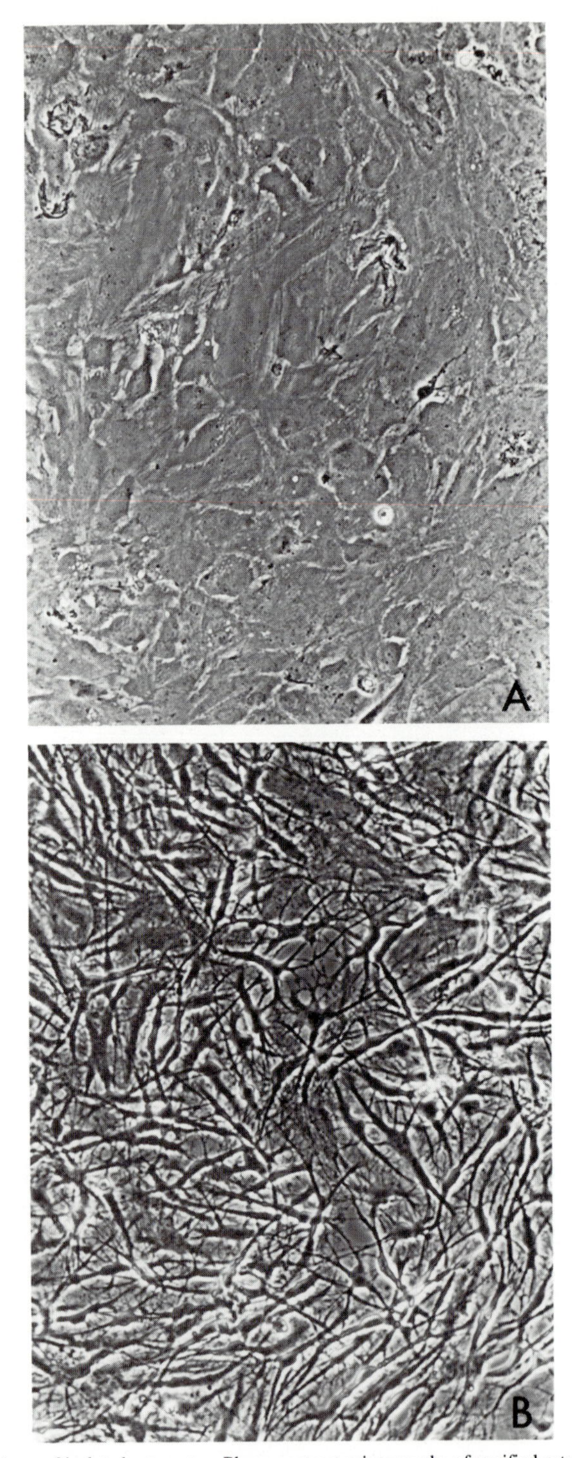

Figure 6. Morphology of isolated astrocytes. Phase-contrast micrographs of purified astrocyte cultures grown 5 days in SSM (**A**) and chemically defined medium (**B**).

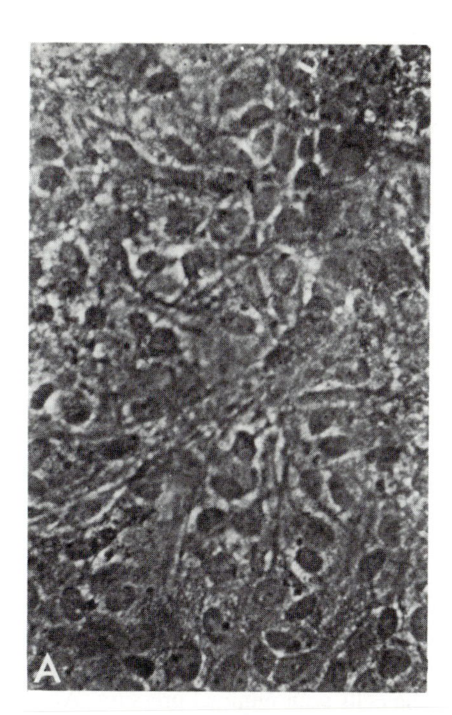

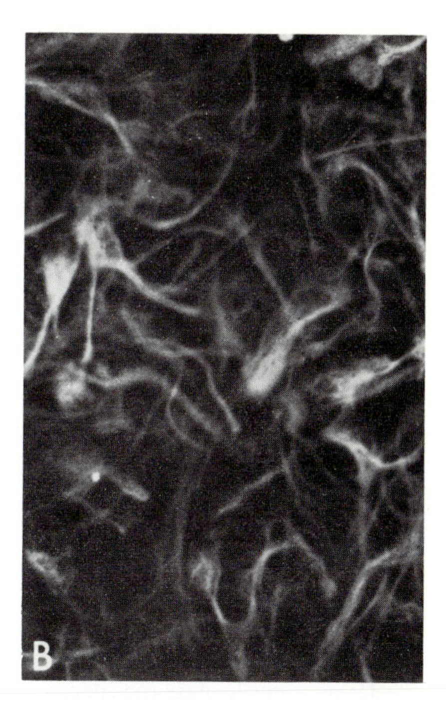

Figure 7. GFAP: immunohistochemical staining of purified astrocyte cultures. Cultures were grown in SSM for 5 days. The same field from a representative culture as visualized by phase contrast microscopy (**A**) and indirect immunofluorescence for GFAP (**B**).

2.3.1 *Embryonic cortical neuronal cultures*

(i) Sacrifice pregnant rats of 16 day gestation (the day where the vaginal plug is found is designated as day zero) by chloroform asphyxiation. This can be accomplished by soaking several gauze sponges (4 × 4″ −8 ply) with chloroform. Place the soaked gauzes in a plastic bag and subsequently place the rat inside the bag and close the bag immediately. Open the bag when the rat becomes quiescent and lay the rat on its back exposing the abdomen. Place the chloroform-soaked gauze over the muzzle of the rat to prevent consciousness.

(ii) Wipe the abdominal area with 70% alcohol, and using a pair of forceps to lift the skin at the median line just below the diaphragm, use a pair of scissors to make a transverse cut across the abdomen. Make a sagittal cut up the midline until the untouched ventral surface of the abdominal wall is exposed. Fold the skin back and sterilize the abdominal wall with 70% ethanol. From this point on, it is advisable to use sterile instruments.

(iii) Make a cut longitudinally along the median line with scissors, revealing the viscera. Expose the uteri filled with embryos posteriorly and dissect out using a pair of forceps pulling the 'string' of embryos out of the abdominal cavity. Cut the point of attachment, the uterine horns, with scissors releasing the uteri and place the embryos in a large 100 mm Petri dish. All of the above procedures should be done outside the sterile hood to prevent contamination.

(iv) Rinse the intact uteri with fresh saline solution I and transfer to another sterile 100 mm Petri dish. Use a sterile pair of forceps and scissors to free the embryo from the uterus. Caution needs to be exercised to prevent distortion of the uterus and bring too much pressure to bear on the embryo.

(v) Hold one end of the uterus by the pair of forceps while the incision is made into the uterus, being careful to avoid the embryo. As the embyro is freed from the uterus, detach the membranes and placenta by pulling the amniotic membrane away with forceps and pinching off the placenta with the edge of forceps against the bottom of the Petri dish. Transfer the embryo to a fresh Petri dish for dissection.

(vi) Use two pairs of forceps to dissect the brain out of the calvarium. Use one pair of forceps to lift the skin and soft cranium, the other to tear away tissue leaving the brain exposed. Once the entire brain is exposed, excise the brain with a scalpel.

(vii) In the same manner described for cortical removal in Section 2.2.1, remove the cortices and place them in sterile saline solution I. After all of the cortices are collected, carefully remove the meninges with another pair of forceps. Better results are obtained when meningeal membranes are removed from each cortex in a separate 35 mm Petri dish containing saline solution I.

(viii) Once the meninges have been removed, transfer the tissue to a fresh Petri dish containing saline solution I. After all of the cortices have been processed, use a 10 ml pipette to transfer the cortices into a nylon mesh bag (mesh size: 210 μm) in a 100 mm sterile Petri dish containing 15–20 ml of SFM.

(ix) Using a sterile glass rod, gently disrupt the cells with minimal pressure. After the tissue is completely broken up, wash the nylon bag with a stream of SFM as it is held above the Petri dish.

(x) Collect the cell suspension in a 10 ml pipette and pass it through a Nitex 130 filter (mesh size: 130 μm).

(xi) Centrifuge this filtrate at 40 *g* and re-disperse the cell pellet with fresh SFM.

(xii) Pass this suspension through Nitex 33 (mesh size: 33 μm) and count the cells in the suspension.

(xiii) Add insulin (5 μg/ml) and transferrin (100 μg/ml) to the suspension to form the neurone-defined medium (NDM). Then seed the cells at a density of 1×10^5 cells per well of a 24-well culture plate which has been previously pre-coated with polylysine [2.5 μg/ml, Section 3.2.2(i)]. Both SFM and supplements are used fresh and should not be stored. If SFM is stored, fresh pyruvate (110 μg/ml) should be added.

These cultures are judged as containing more than 95% neurones by the immunological criteria of expressing the marker, neurofilament protein (*Figure 8*) while not expressing the biochemical and immunological markers for astrocytes and oligodendrocytes.

2.3.2 *Post-natal cortical neuronal cultures*

The initial suspension of cells used for seeding neuronal cultures is prepared as described in Sections 2.2.1 and 2.2.2. However, the age of the pup used for cortical neuronal cultures is more critical. Newborn to 1 day post-natal give best results, older ages producing a larger population of glia and fewer surviving neurones. After the final centrifugation step (Section 2.2.1), count the cells and seed them at a density of 1×10^6 cells in 35 mm Petri dishes with SSM supplemented with cytosine arabinoside (5 μM).

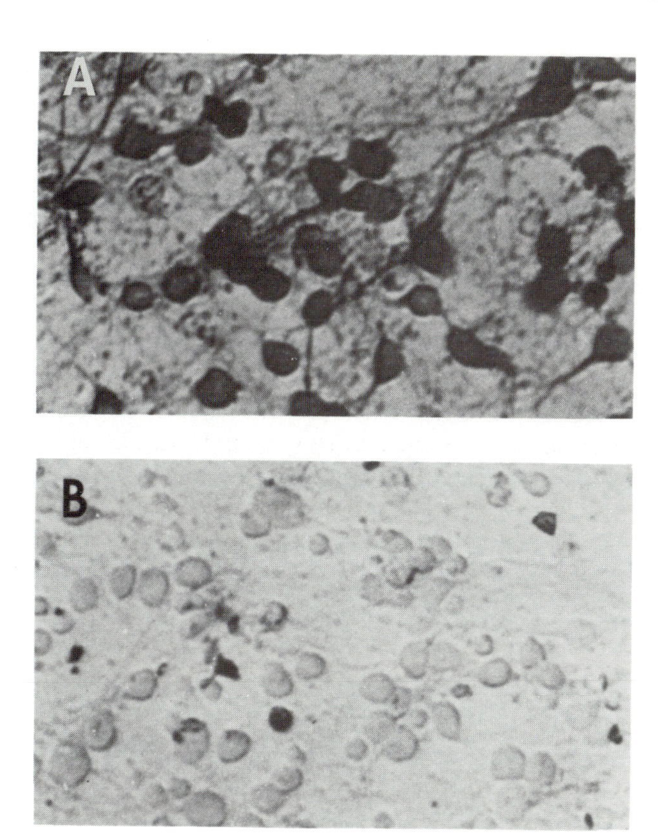

Figure 8. Neurofilament protein immunohistochemical staining of embyronic neurones. Cultures of embryonic cortical neurones are grown in chemically defined medium (**B**) and immunohistochemically stained using the avidin–biotin complex technique for neurofilament protein (**A**). Control sera (not shown) did not stain.

Eighteen hours after initial plating, rinse the cultures twice with SFM and supplements of ADM [Section 2.2.5(ii)] containing 5 μM cytosine arabinoside to kill dividing non-neural cells. The milieu of the chemically defined medium is able to support post-natal derived neurones (37, *Figure 9*) although the roles of the components, with the exception of the fibroblast growth factor, remain to be investigated. The change in developmental age requirements may reflect the varying requirements for embryonic and post-natal cortical neurones. These cultures (>95%) express the neuronal markers, neurofilament protein and neurone-specific enolase.

2.3.3 *Conclusions*

The varying nutritional requirements for the different CNS cell populations suggest the unique extrinsic control of adhesion, survival, proliferation and differentiation. The use of defined media has allowed investigators to begin elucidating the factors specific for each cell population (*Table 2*). As more experimentation and refinement of cell detection methodology become available, identification and isolation of subpopulations within the major brain cell populations should be possible. Furthermore, the change

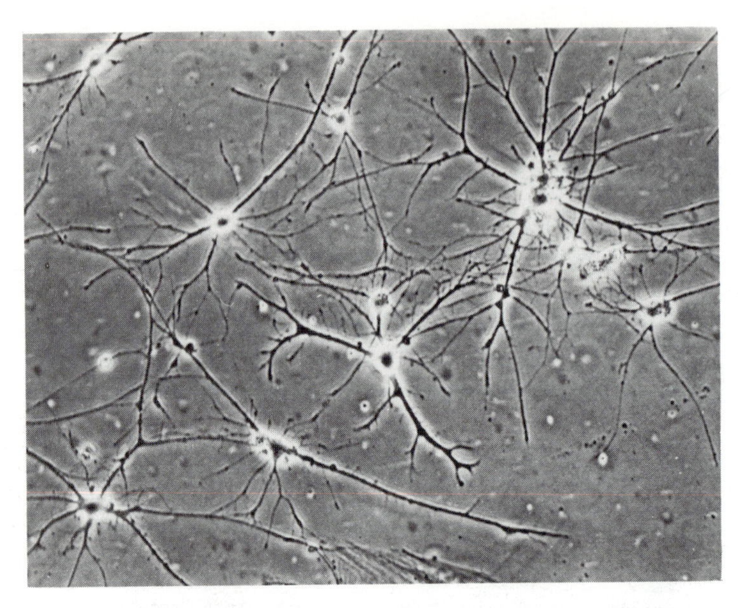

Figure 9. Phase-contrast micrographs of primary neuronal cultures. Cultures of post-natal cortical neurones are grown in chemically defined medium for 10 days (37).

Table 2. Defined medium supplements for primary neural cells.

Compound	*Optimal concentration of supplements*			
	Neurones		*Oligodendrocytes*	*Astrocytes*
	Embryonic	*Post-natal*		
Hydrocortisone		50 nM		50 nM
Putrescine		100 μM		100 μM
Prostaglandin $F_{2\alpha}$		500 ng/ml		500 μg/ml
Insulin	5 μg/ml	50 μg/ml	5 μg/ml	50 μg/ml
Fibroblast growth factor		500 ng/ml	75 ng/ml	100 ng/ml
Transferrin	100 μg/ml		500 ng/ml	

in environmental requirements as cells differentiate will begin to be elucidated. As these data appear in the literature, a better understanding of environmental control of developmental gene activity will increase our knowledge of cellular interactions *in vivo*.

2.4 Evaluation of cell types

As previously discussed (Section 1.3.2) the use of isolated primary culture is dependent on cell identification. The characterization of cell types is more difficult *in vitro* than are the cell counterparts *in vivo*. Morphological features which are useful *in vivo* as diagnostic criteria are of limited value *in vitro*. Topographical relationships are disrupted in the artificial environment where cells develop in the flat interphase between solid and liquid. Hence every effort to identify cell types based upon multiple criteria needs to be accomplished for meaningful measurement and standardization of data between laboratories.

2.4.1 *Ultrastructure*

Electron microscopic (EM) evaluation of cultures is time-consuming. However, it allows the unequivocal identification of cell types. In fact, EM is still the best criterion to distinguish glial cell types. Astrocytes have a light, irregular oval nucleus, and contain glycogen granules and bundles of 9 – 10 nm filaments located in the cytoplasm around the nucleus and in the processes (37, *Figure 10*). In contrast, the oligodendrocyte is characterized by its abundant cytoplasm, usually darker staining than the nucleus, a small round nucleus with clumps of heterochromatin, stacked cisternae of rough endoplasmic reticulum and a predominant peri-nuclear Golgi apparatus (39, *Figure 10*). Neurones can be identified by polysomes containing regions (Nissl substance) and arrays of microtubules interspersed with neurofilaments (40).

Cultures are best prepared for EM using 35 mm Petri dishes. This allows the manipulations of solutions and removal of plastic from the embedded cells. Twenty four well culture plates can be used but are more difficult to handle during fixation, dehydration and processing. We use a procedure developed by Brinkley and co-workers (41).

2.4.2 *Biochemical properties*

Specific cell types within the CNS have identifiable biochemical characteristics or parameters. The number of cells required for biochemical assay precludes these criteria for culture purity. However, together with ultrastructure and immunocytochemistry this parameter is useful for purposes of cell identification.

The evaluation of biochemical parameters is also useful in identification of the contributions of neuronal and glial cells to the extracellular compartment. This overlooked compartment is important in understanding the spatial and temporal organization of the brain. Little is known concerning cellular interactions during development. The elucidation of biochemical specificity and therefore the extracellular contribution will be a valuable tool in understanding CNS development and function.

(i) *Neurones*. Neurotransmitter metabolizing enzymes that are not part of common metabolic pathways can be used as neuronal markers. Useful markers include: glutamate decarboxylase (GAD: EC 4.1.1.15) for GABAergic neurones (42), choline-acetyltransferase (ChAT: EC 2.3.1.6) for cholinergic neurones (43), and tyrosine hydroxylase (TH: EC 1.14.16.2) for adrenergic neurones (44).

(ii) *Oligodendrocytes*. There are three enzyme markers that are useful for oligodendrocyte identification. Two of these marker enzymes are likely to be involved in myelination. The enzyme 2′,3′-cyclic nucleotide 3′-phosphodiesterase (CNPase: EC 3.1.4.37) is oligodendrocyte specific and inducible by N^6,O^2-dibutyryl cAMP (dbcAMP) in ODM, while GPDH (EC 1.1.1.8) is induced by hydrocortisone exclusively in oligodendrocyte progenitor and mature cells (11,18,27,30). Although the presence of lactate dehydrogenase (LDH: EC 1.1.1.27) is found in both glial cell types, its inducibility by dbcAMP is a specific oligodendrocyte marker (27,31).

The specific activity of these enzymes varies for cells grown in culture medium and ODM. However, the presence of inducibility does not change between culture media with the exception of CNPase. CNPase is only inducible in early culture times of oligodendrocytes when cells are cultured in SSM while remaining inducible when cultured in ODM (27).

(iii) *Astrocytes*. The absence of specific oligodendroglial enzyme activities and their induction is used as evidence of astrocyte culture purity. The presence of these activities may reflect oligodendrocyte progenitor cell contamination, however, this should only be used as an indication of culture purity. Other cell types, such as meningeal and endo-

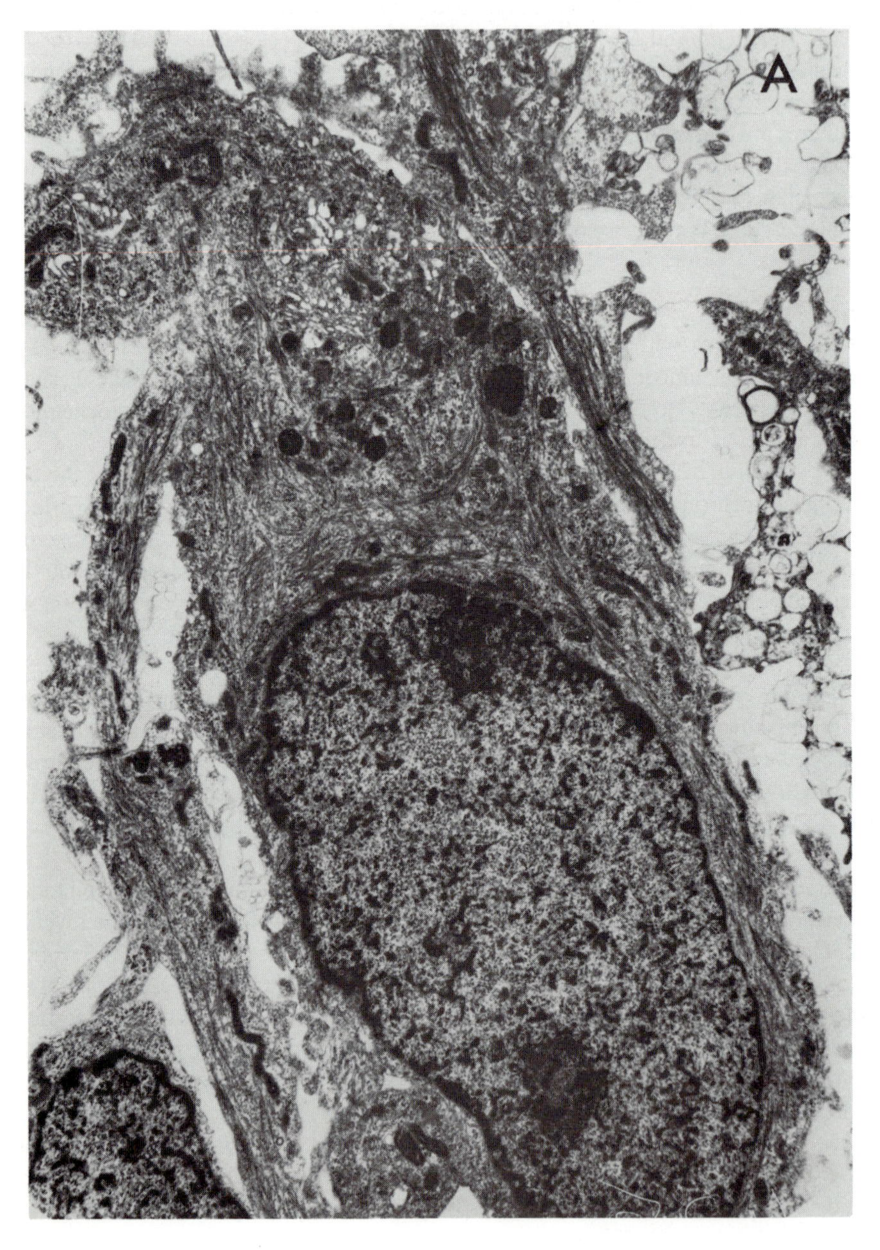

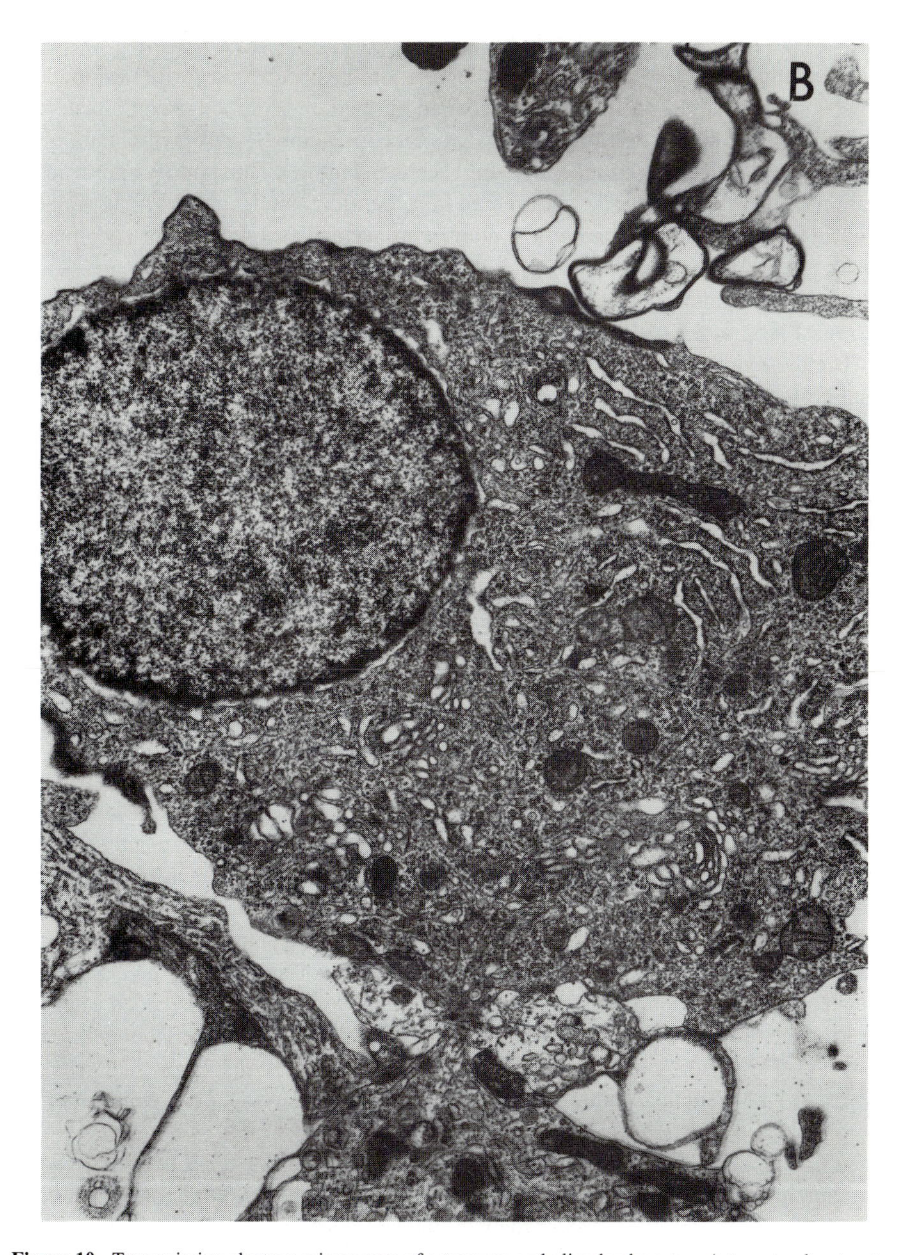

Figure 10. Transmission electron microscopy of astrocytes and oligodendrocytes. Astrocyte ultrastructure as revealed by this electron micrograph (**A**: ×9000) and typical oligodendrocyte ultrastructure represented (**B**: ×16 200).

thelial cells may also be contaminating the culture and not be detected by such criteria. A positive marker for astrocyte cultures is the presence and induction of glutamine synthetase (GS: EC 2.6.1.53) by hydrocortisone. Although GS is present in both serum-supplemented and chemically defined media, the hydrocortisone-induced activity is present only in chemically defined media (45).

2.4.3 *Immunocytochemical properties*

The use of antibodies to specific epitopes on antigens has been a powerful tool to identify cell types. The strong interaction between antigen and antibody, combined with methodology that allows detection of low antigen copy number, makes immunological techniques sensitive enough to identify single cells in culture. Thus, purity and/or identification of cell types within a culture can be accurately determined. The usefulness of this technique is only limited by the number of antibodies available for specific antigens.

Although the availability of polyclonal sera makes their use convenient and accessible to most laboratories, there are shortcomings that limit their widespread contribution. The biological diversity of B-cell differentiation and genetic differences between animal hosts, precludes universal antisera that contain monospecific determinants and homogeneous titres. These parameter differences make standardization difficult. In addition, the need for absolute purity of the antigen used to develop an antiserum makes the range of polyclonal antibodies small. The development of monoclonal antibody technology has allowed the generation of large amounts of a single antibody targeted to a specific antigenic determinant, without having to purify the antigen. This facilitates standardization of antisera between experiments and laboratories.

The techniques of immunocytochemical detection are well standardized (46; see Chapter 3). We routinely used both indirect immunofluorescence (27) and avidin–biotin (47) procedures. The avidin–biotin procedure is a more sensitive procedure. Both the presence and absence of markers for all the cell types need to be determined to calculate purity of cultures.

(i) *Neurones*. The purity of neuronal cultures is routinely ascertained in our laboratory by the presence of two neuronal-specific markers, neurofilament protein and neurone-specific enolase (EC 4.2.1.11). Both of these markers are general neuronal markers. It should be noted that neurofilament protein is expressed throughout development while neurone-specific enolase is expressed late in development. Thus, depending on the developmental stage of the culture, neurone-specific enolase may not be expressed. Tetanus toxin and the monoclonal antibody A2B5 are no longer considered good neuronal markers since they are expressed on glial progenitor cells and certain glial populations. Negative data are also a good parameter for purity. The absence of specific glial markers (see above) and meningeal marker protein, fibronectin (29), allow the culture purity to be partially assessed.

The identification of neuronal subpopulations is much more difficult *in vitro* because of the morphological changes that occur in culture. However, the immunocytochemical detection of neurotransmitters and/or their receptor proteins allow some sub-populations of neurones to be identified. The detection of neurotransmitter-synthesizing enzymes can also be used to detect sub-populations of neurones.

(ii) *Oligodendrocytes*. The majority of oligodendrocyte-specific markers are involved in myelination. The expression of these myelination-associated proteins and lipids indicates a cell type that is differentiated. These markers are myelin basic protein, proteolipid protein, galactocerebroside and myelin-associated glycoprotein. In cultures grown in serum, these markers are found to be sequentially expressed. Galactocerebroside is found to be present before the other markers. The expression of galactocerebroside is considered to be the point at which oligodendrocyte progenitor cells become differentiated or mature oligodendrocytes.

The markers that detect immature or progenitor cells are A2B5, GPDH and CNPase. In the method described in Section 2.2.4, the phase-dark process-bearing cells can express a different spectrum of oligodendrocyte markers depending on the culture environment (27). However, in both SSM and ODM the cultures do not totally express galactocerebroside and therefore we classify these cells as progenitor cells (*Table 1*). It should be mentioned, in the environment of serum, that some of the cells within a culture express both GPDH and the astrocyte GFAP. We feel that the environment of serum allows the expression of the astrocyte marker due to the cell not being fully committed to the oligodendrocyte lineage. In this case, the GFAP gene is 'leaky' and expressed. However, we cannot rule out the possibility that this bi-potential cell type is an astrocyte type II as described by Raff and co-workers (48). Furthermore, the environment of serum is pathological and hence we are not sure whether this phenomenon is merely a culture artifact. (For reviews of this subject, see 27,48.)

(iii) *Astrocytes*. Both *in vivo* amd *in vitro*, the predominant feature of astrocytes is the presence of 9–10 nm glial filaments containing GFAP. Recently Kimelberg and co-workers (49) have co-localized carbonic anhydrase II and GFAP in the same cell type. *In vivo*, as astrocytes become more differentiated, they stop expressing vimentin and begin expressing GFAP (50). However, in culture they continue expressing vimentin together with GFAP. Thus in culture, vimentin can be considered as a marker for astrocytes. We have developed monoclonal antibodies to vimentin and desmin from purified astrocyte cultures (50). Astrocytes do not express fibronectin, thus the presence of fibronectin would be a negative indication of culture purity. Contamination of the cultures with fibroblasts and/or meningeal cells would make cultures positive for fibronectin.

2.4.4 *In situ hybridization*

The presence of protein, as detected by immunocytochemistry, is not incontrovertible evidence of protein-specific synthesis within that cell. One cannot rule out the possibility of protein uptake. Although pulse-labelling experiments could be undertaken to examine this possibility, low protein copy number would mask differential protein synthesis. A new powerful technique to address this dilemma is *in situ* hybridization. This method involves the use of cDNA or riboprobes (anti-sense RNA) for the specific mRNA coding for the protein in question. This technique can detect low mRNA abundance at the single cell level and therefore the site of protein synthesis can be determined unequivocally.

This technique is analogous to immunocytochemistry except that the detection probe is a nucleic acid probe specific for mRNA instead of an antibody specific for protein. For detailed procedures see ref. 51.

2.5 **Cell lines**

As discussed in Section 1.2, clonal tumour cell lines represent a useful model system to study CNS function. Cell lines are genetically homogeneous (this should be re-evaluated periodically) and can be cultured in large amounts which allows biochemical and molecular studies to be carried out. However, their tumour origin makes rigorous characterization necessary before *in vivo* correlations are meaningful.

2.5.1 *Neuroblastoma*

To date, there are no homogeneous populations of pure cultured CNS neurones available. Pure cultures of neurones are possible (Sections 2.3.1 and 2.3.2) but not homogeneous populations of neurones. Until homogeneous populations are developed, cell lines represent the best model system to study biochemical and molecular processes. The various types of CNS cell lines and their studied properties have been extensively reviewed (52,53).

2.5.2 *Glioma*

There are various glioma cell lines that have been derived from many different species. The majority of these cell lines express both astrocyte and oligodendrocyte properties. Nevertheless, their use has profoundly influenced the understanding of glial physiology (53). For example, one of these cell lines, C_6 glioma, has been helpful in elucidating hydrocortisone induction of both GPDH and GS. The large amounts of cells capable of being grown have proved helpful in elucidating both the biochemical and molecular mechanisms involved in induction (18,54).

2.5.3 *Neuroblastoma–glioma hybrids*

The vast majority of investigations utilizing neuroblastoma × glioma hybrids have dealt with opiate receptor expression. The best characterized cell line is NG108-15 developed by Klee and Nirenberg (55) which has been shown to express the enkephalin or delta opiate receptor. However, although binding studies reveal the presence of opiate receptors on a variety of hybrid cell lines, their biological significance remains obscure. In fact, binding studies have shown these specific receptors are often not apparently functional. However, regulation of receptor expression may be studied in such systems.

3. TISSUE CULTURE REAGENTS

3.1 **Media**

3.1.1 *Water*

An important component of basal medium, used for both defined media or SSM, is high grade water. We recommend the use of the highest purity rated pyrogen-free water. We use a water purification system obtained from Millipore Corporation (Bedford, MA); Milli-Q Reverse Osmosis System with added pyrogen and organic removal systems. Although not absolutely necessary, the addition of a final 0.22 μm filtration unit allows the purified water to be used for h.p.l.c. solutions. In addition, the filtration helps prevent the sterilization filter from clogging when sterilizing the media.

3.1.2 *Synthetic basal medium*

For economical and time efficiency reasons we prefer to use powder basal medium instead of pre-mixed medium. The medium we use in Dulbecco's modified Eagles' and Ham's F-12 medium in a 1:1 (w/w) mix with added pyruvate (Irvine Scientific,

Santa Ana, CA). Powdered medium mix is dissolved in high quality water at room temperature. We routinely add 1.2 g of $NaHCO_3$/l and 15 mM Hepes buffer. After mixing, the pH is adjusted to 7.3 and sterilized. We use a pressure filtration unit with dispensing bell (Millipore Corp.). Twenty litre amounts of basal medium are made and stored in 400 ml bottles at −20°C. When bottles are needed, they are thawed at room temperature for several hours and refrigerated. Medium is checked for sterility before freezing by culturing a small volume (1 ml) of medium at 37°C for 24 h. Care should be taken in utilizing the media storage bottles only for media use.

3.2 **Culture conditions**

3.2.1 *Culture vessels*

We have found that culture vessel efficiency varies from manufacturer to manufacturer. Adhesion, cell growth and survival have been found to vary even within the culture vessel line from an individual manufacturer. We find that optimal mixed glial cell growth occurs using Corning 75 cm^2 flasks (Houston, TX), while isolated glial cell populations adhere and grow well in Falcon 24-well culture plates (Oxnard, CA). Neuronal cell cultures grew best using Falcon 24-well culture plates. However, all our neuronal cultures are seeded on polylysine-coated plates and it is likely that the type of plate is not important. When cultures are grown in 35 mm Petri dishes we routinely use Lux dishes (Miles Laboratories, Naperville, IL).

3.2.2 *Substrata*

Anchorage to a solid substratum in dispersed culture has long been recognized as a requirement for survival. Only recently have the molecular properties of specific substrata been found to affect survival, shape, transmitter activities and general growth properties. Until the use of defined medium made it necessary to study such parameters, adhesion molecules or proteins found in serum were necessary and limiting for cells in culture.

(i) *Modification with poly-D-lysine*. Tissue culture plastics are negatively charged and may not be suited for the attachment of certain cell types. The use of the positively-charged poly-D-lysine has been found to facilitate neuronal attachment. Coat culture plates (24-well) with 2.5 μg/ml polylysine (mol. wt. 30 000−70 000; Sigma Chemical Corp., St. Louis, MO) for 10 min or longer at room temperature. Subsequently, wash the culture plates either with sterile water or SFM. Pre-coated plates can be stored for short periods of time, we have used plates stored at 4°C for 48 h without loss of adhesion activity.

(ii) *Modification with fibronectin*. Fibronectin is purified from outdated human plasma by affinity chromatography (56). The affinity ligand is gelatin coupled to activated Sepharose. This affinity gel is also commercially available (Bio-Rad, Richmond, CA). The isolation procedure is performed at room temperature. Pass the plasma through the prepared column and wash thoroughly with 4 volumes of phosphate-buffered saline (PBS). Elute the fibronectin into polypropylene tubes with PBS containing 4 M urea. Fibronectin is absorbed onto glass surfaces, therefore glass containers should be avoided.

Combine the peak protein fractions, dialyse overnight in PBS containing 1 M urea, filter sterilize and store at 4°C in polypropylene tubes. Fibronectin is stable in these conditions for 3 months. Purity is approximately 98% as judged by polyacrylamide gel electrophoresis. When added to culture dishes or plates, 15 min before addition of cells is sufficient for binding. We generally add fibronectin to plates or dishes containing SFM. Care should be taken to tip the dish in several directions to ensure even coating of the plating surface.

(iii) *Modification with fetuin*. We have found that pre-treatment of culture plates and dishes with polylysine (Section 3.2.2) facilitates fetuin attachment of oligodendrocytes. Fetuin (CalBiochem, San Diego, CA) is made as 1 mg/ml stock solutions in SFM and stored frozen at −20°C. Coat the culture plates with polylysine and wash twice with SFM. Add fetuin at 25 μg/ml and allow to incubate overnight at 37°C. Before adding cells, wash the plates twice with SFM.

3.3 **Enzymes used for tissue dissociation**

3.3.1 *Primary culture*

(i) Make up the trypsin solution for primary culture in saline solution I.
(ii) Add hog pancreas trypsin (2.5 g; ICN Pharmaceuticals, Cleveland, OH; 1:250) to 100 ml of saline solution I containing two drops of liquid phenol red (pH indicator).
(iii) Stir this solution at 4°C for 15 min, centrifuge at 5000 r.p.m. to remove undissolved trypsin and freeze the solution in 1 ml aliquots at −20°C. When needed, thaw the trypsin solution, adjust the pH to approximately 7.0 with 0.5 M NaOH, filter sterilize (0.22 μm filter) and add to the cell suspension.

This trypsin solution is stable for many months at −20°C, we have used it after 18 months of storage with equivalent results to those obtained with freshly made solution.

3.3.2 *Sub-culture techniques*

(i) Detach the cells to be passaged with trypsin solution II (1% Enzar-T; 40× concentrate and 0.1 mM EDTA in Hanks' balanced salt solution that is Ca^{2+}- and Mg^{2+}-free).
(ii) Add the trypsin solution II to culture vessels and incubate at 37°C until cells begin to detach as a cell sheet.
(iii) Add culture medium and triturate the cell clumps several times using a 10 ml pipette.
(iv) Centrifuge the cell suspension at 1000 r.p.m. for 2 min and discard the supernatant.
(v) Resuspend the cells in culture medium and count for plating.

If SFM is used, use soybean trypsin inhibitor (0.5 mg/ml) to inactivate trypsin after the cells become detached. The cells also need to be washed and centrifuged twice more to remove any loosely attached serum proteins or components.

3.4 Defined medium supplements

There have been many chemically defined media developed for cell lines and these have been useful for primary cultures of CNS cell types (for review; 57). Our rationale has been to custom design the simplest but most efficacious milieu needed for growth of a particular cell type, reasoning that cell lines are largely autocrine and differ fundamentally from normal cells with respect to growth control. Defined medium is always made fresh prior to utilization with basal medium and concentrated supplements. If fresh basal medium is not used, stored medium should be replenished with pyruvate (100 mg/l). Defined medium is never stored, since we have found loss of activity upon storage. All supplements are sterilized using a 0.22 μm syringe filter. The order in which the supplements are added makes no difference, nor is there a need to re-check the pH after the components have been added.

3.4.1 *Stock solutions* (see Appendix)

Each of the supplements added to the defined media (*Table 2*) is commercially available. Hydrocortisone is obtained as the hemisuccinate sodium salt and kept as a 100 μM stock solution. The Merck index (Merck and Co. Inc., New Jersey) recommends that in solution, hydrocortisone be kept no more than 5 days at 4°C. Putrescine is made up in water as a 100 mM stock solution and stored at 4°C for up to 4 months without loss of activity. Prostaglandin $F_{2\alpha}$ can be conveniently purchased in 1 mg vials. One hundred percent ethanol is added per vial to yield a stock of 1 mg/ml. This stock solution is sterile and stored at −20°C for up to 4 months. We routinely use bovine pancreas insulin made up as a stock solution of 10 mg/ml dissolved in 0.05 M HCl. The addition of small amounts of this insulin solution does not alter the pH of the medium. Insulin solutions are stored at 4°C and fresh solutions should be made up every 2 weeks. Transferrin (human) is prepared as a 10 mg/ml stock solution in SFM and stored at −20°C. Fresh stock solutions should be made up every week. The fibroblast growth factor used in all of our defined media is obtained from Collaborative Research (Lexington, MA) in sterile 10 μg aliquots. SFM is added to make a 10 μg/ml stock solution and stored at −20°C. This sample can be thawed and re-frozen up to four times without loss of activity.

3.5 Cryopreservation

Long-term storage of viable cells in liquid nitrogen allows cell maintenance without frequent intervals of subculture. Some clonal cell lines have been found to change in biochemical properties after prolonged passage. Therefore, if clonal lines are used for experimentation, passage number should be noted. The freezing down of clonal cell lines facilitates future experiments with cells of the same passage number. Cells survive long-term storage best at temperatures below −90°C. Cells to be stored are suspended in SSM containing 10% dimethyl sulphoxide (DMSO) or a solution of 90% fetal calf serum and 10% sterile glycerol. We routinely use the latter condition because of the possible effects of DMSO on cell physiology. One ml aliquots of cell suspension are placed in freezing vials and the temperature is slowly lowered. Our step-down procedure is 4°C for 30 min, −20°C for 1.5 h, −80°C for 1 h and then storage in the vapour phase of a liquid nitrogen storage tank. When needed, stored cells should be thawed quickly, by gently shaking the vial in a 37°C water bath and subsequently plated. If DMSO is used it should be removed after the cells have adhered to the culture vessel, approximately 18 h after seeding.

4. BIOASSAY OF CULTURE SYSTEMS

Qualitative observations and quantitative assays of isolated primary cultures provide a powerful means to dissect cellular and molecular events. The homotypic composition of an isolated culture lends itself to the understanding of cell-specific physiology. However, the reduction of complexity and intercellular interactions necessitates the integration and confirmation with *in vivo* physiology. In this section, we give some bioassay protocols we use to ascertain cell-specific function.

4.1 Growth factors

Factors regulating astrocyte and oligodendrocyte proliferation can be detected and characterized by two methods:

(i) increases in actual cell number over a period of time and
(ii) incorporation of radiolabelled thymidine into acid-precipitable material (27,47,58).

We feel that both procedures should be done to confirm that a factor is inducing proliferation. There are potential pitfalls to the exclusive use of radiolabelled thymidine as the sole criterion *in vitro* for cell proliferation (59). We routinely use both culture medium and chemically defined medium to ascertain potential growth factors. Together these criteria rule out the possibility that the factor is trivial. These methods can be used to detect mitogenic activity of known growth factors and novel growth factors found in tissues (*Figures 11* and *12*).

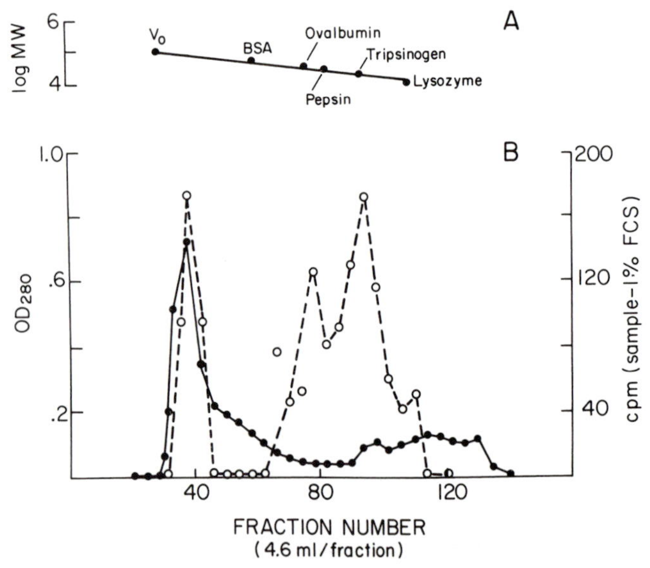

Figure 11. Brain-derived mitogens for astrocytes. Elution profile of mitogenic activity in adult brain extract fractionated on Sephadex G-100 columns. The figure is representative of duplicate experiments.

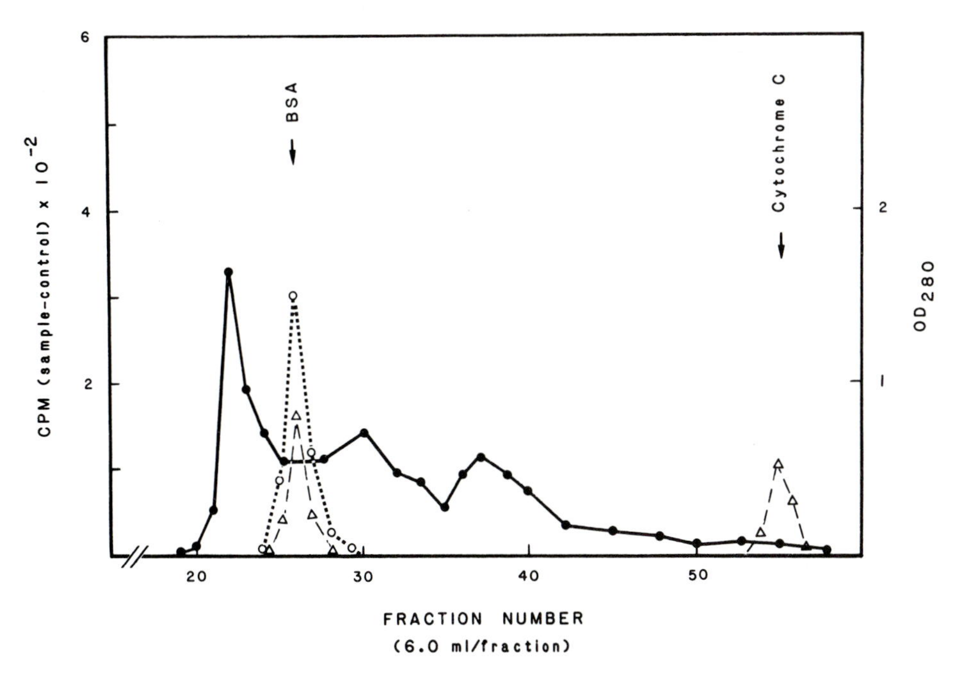

Figure 12. Brain-derived mitogens for oligodendrocytes. Elution profile of mitogenic activity in adult brain extract fractionated on Sephadex G-100 columns. The figure is representative of triplicate experiments. Mitogenic activity was determined in the presence of ODM (○) and SSM (▲).

4.1.1 [^{3}H]*Thymidine incorporation*

Mitogen stimulation of astrocyte growth is detected utilizing secondary cultures (sub-cultured purified primary cultures) grown in 24-well culture plates.

(i) Seed the cells at 2×10^4 cells/well and grow to confluency in culture medium.
(ii) Wash the cells three times with SFM and subsequently culture for 48–72 h in SFM. Serum starvation has been found to enhance the signal-to-noise ratio in growth factor detection.
(iii) Add mitogens to the cells in SFM containing 1% fetal calf serum.
(iv) After 22 h add radioactive thymidine (1 μCi/ml) in a 2 h pulse and subsequently remove the medium. Long exposure times may not be a valid indication of mitogen activation of mitogenesis but may reflect repair, radiolabel lability and methyl group exchange. We find the 2 h pulse convenient and reproducible in mitogen detection.
(v) Wash the cells three times with ice-cold PBS and add 1.0 ml of a 10% trichloroacetic acid (TCA) solution for 30 min (or longer) at 4°C.
(vi) Wash the cells three times with PBS, allow to dry for at least 15 min and add 1.0 ml of a 0.5 M NaOH solution.
(vii) After an overnight (6 h minimum) incubation at room temperature, neutralize an aliquot with 0.5 M HCl (this can be done in the scintillation vial) and determine the radioactivity. If chemically defined medium is used, radioactive label can be added any time after the medium switch from serum to defined medium.
(viii) Process the cells in the same manner as described.

Mitogen detection utilizing radioactive thymidine in oligodendrocyte cultures is determined as with astrocyte cultures.

(i) Seed the cells at 10^5 cells per well in 24-well plates in SSM.
(ii) After 18–24 h to allow the cells to attach, switch the medium to defined medium. Serum pre-incubation eliminates the possibility that the factor in question has a differential survival action on a sub-population of cells.
(iii) Wash the cells three times with SFM and add defined medium. The mitogen can be added at this time or later times in culture.

The protocol is similar to that described for astrocytes. If SSM is used as the bioassay medium, SFM containing 1% fetal calf serum should be used (58).

4.1.2 *Cell proliferation*

Actual increases in cell number are performed by adding mitogens to SSM (1% fetal calf serum) or chemically defined medium. The limited serum concentration is useful in that it allows survival of cells but limits the high concentration of serum proteins that might interact with the mitogen. Cells are plated and counted after 18–24 h and this cell density is used as the initial plating density or cell count. Initially, cell viability should be checked, trypan blue exclusion indicates whether a cell is alive. We routinely count cells every 2 days as an indication of cell proliferation.

4.2 **Receptors**

4.2.1 *Steroid receptors*

Purified cultures of astrocytes, oligodendrocyte progenitor cells and C_6 glioma cells are ideal to study glucocorticoid receptors. The induction of GPDH and GS by hydrocortisone in nervous tissue has been shown to involve all the intracellular steps found in steroid induction of specific proteins *in vivo* (for review, see 18). However, the molecular mechanisms of gene activation remain unknown. The use of cell culture paradigms has allowed meaningful insights into this process. For example we have shown that specific lectins can down-regulate glucocorticoid receptors (60) and that steroid-receptor activation of GPDH is controlled at the transcriptional level (54).

4.2.2 *Membrane receptors*

The task of receptor determination, ligand–receptor interaction and subsequent cascade of biochemical effects would be limited without isolated cell culture. Although single cell analysis is possible by the combination of autoradiography with immunocytochemistry, this technique is basically limited to receptor identification. Cell culture provides the number of cells necessary for biochemical analysis of receptor–ligand interaction, receptor kinetics, receptor number and studies of the regulation of receptor expression. Credibility of data generated from such studies is directly related to the purity and identification of cell homogeneity of the culture. We have utilized isolated primary cultures to estimate epidermal growth factor receptors on the various CNS cell types (*Figure 13*) and the spectrum of α and β agonists and antagonists on glial cell populations (*Table 3*).

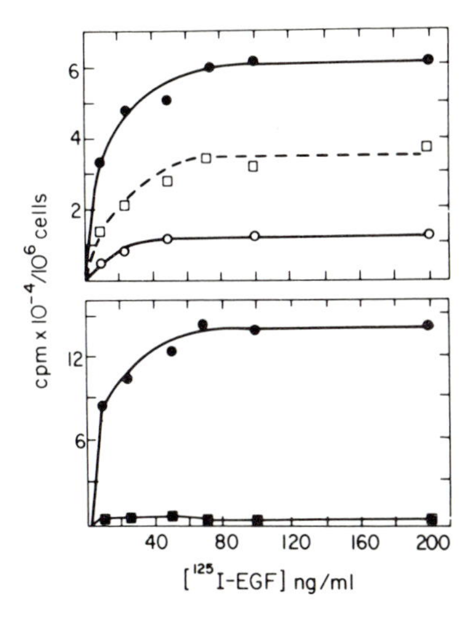

Figure 13. Binding of [^{125}I]epidermal growth factor to cultured astrocytes, oligodendrocytes and neurones. The astrocytes and oligodendrocytes used for these assays had been cultured for 14 days while neurones were cultured for 9 days. Binding assays were performed at 4°C (33). Values are averages of triplicate determinations at each concentration: astrocytes (●), oligodendrocytes (○), neurones (■) and 3T3 cells (□).

Table 3. Regulation of cAMP.

Treatment	*cAMP (pmol/mg protein)*	
	Astrocytes	*Oligodendrocytes*
Control	8.7 ± 0.2	8.9 ± 0.4
Norepinephrine	28.7 ± 2.9	50.1 ± 9.0
Norepinephrine + phentolamine	180.6 ± 10.1	62.0 ± 4.4
Adenosine	44.3 ± 1.3	14.3 ± 0.7
Prostaglandin E_1	67.6 ± 8.8	187.4 ± 31.2

The concentration of all drugs, except adenosine, was 3 μM. Adenosine was used at 100 μM. All incubations were carried out for 5 min. Each value represents the mean of three cultures ± SEM (11).

4.3 Released proteins and cellular interactions

The close juxtaposition of glia with neuronal elements, their sequential embryological development and their predominance in the neuronal micro-environments suggest a glial influence on CNS development. One method of studying such interactions is the use of conditioned media from purified cultures and ascertaining the biological influences such media have on other cell types.

(i) To obtain astrocyte-conditioned medium, grow primary astrocytes to confluency in culture medium.

(ii) Wash the cultures extensively with SFM and incubate in fresh SFM for 24 h at 37°C.

(iii) Remove conditioned medium, centrifuge, filter (0.22 μm filter) and freeze at −70°C.

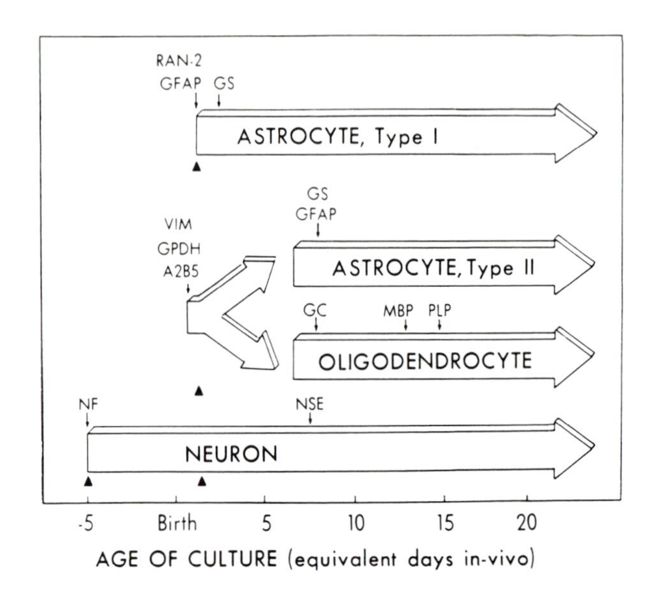

Figure 14. Temporal expression of CNS markers found in primary culture. This figure represents the approximate *in vitro* time when all specific markers are expressed. We have found that these markers are regulated by extrinsic factors (27). Hence, the time of marker expression will vary according to culture conditions. Therefore, comparison of marker expression in culture will vary with brain area cultured, age of animal used for culture and culture media. This figure presents marker expressions for rat cerebral cultures. Astrocyte type II and oligodendrocytes have been proposed to arise from a common precursor cell (48). Depending on culture conditions, the phase-dark process-bearing cells can mature into oligodendrocytes (27) or a mixture of astrocyte type II and oligodendrocyte cells (27,48). The cultures are initiated (▲) at embryonic day 16 for neuronal culture or post-natal day 1–3 for mixed glial culture or post-natal neuronal culture. Abbrevations: GPDH; glycerol phosphate dehydrogenase; VIM, vimentin; GFAP, glial fibrillary acidic protein; GS, glutamine synthetase; GC, galactocerebroside; MBP, myelin basic protein; PLP, proteolipid protein; NF, neurofilament; and NSE, neuronal specific enolase.

Oligodendrocyte-conditioned medium cannot be obtained in SFM (27). However, the unambiguous environment of chemically defined medium lends itself well as a source of conditioned medium. Conditioned medium is obtained as described above. Neuronal-conditioned medium is obtained in the environment of chemically defined medium and processed similarly to astrocyte-conditioned medium. Care should be taken to store conditioned medium in plastic containers to avoid protein adherence to glass. Conditioned medium can also be lyophilized and stored.

The use of conditioned media has allowed the detection of biological communication between the different CNS cell types. We have found that conditioned media from astrocytes and the C_6 glioma cell line possess mitogenic activity for oligodendrocyte progenitor cells (57,58). This medium also contains a number of neuronal active factors (61). Oligodendrocyte progenitor cell conditioned medium possesses mitogenic activity for chick telencephalon neuroblasts (unpublished data). Making the circle of cellular interaction complete, neuronal-conditioned medium contains an activity that increases expression of myelin basic protein in both mixed glial cultures and isolated oligodendrocytes (62).

Cellular interactions are also possible by creating model systems to study events found *in vivo*. Oligodendrocytes have been shown to produce 'myelin-like' membranes reminiscent of compact myelin found *in vivo* (63). Wood *et al.* (64) have used this observation, and by combining oligodendrocytes with purified dorsal root ganglion cells, found myelin formation. Thus an *in vivo* process can begin to be dissected and manipulated *in vitro* to elucidate cellular and molecular events controlling myelination.

4.4 Cell-specific RNA

The advent of molecular biological techniques has allowed insight into the genetic mechanisms of biological function. The large number of purified cell populations in primary cultures allows the generation of both cytoplasmic and nuclear RNA. These RNAs can be used to detect unique RNA species specific for a cell type or a differentiated state. Probably the most effective use of molecular biological techniques is the elucidation of gene regulatory events controlling CNS function and structure. Detection of small transitory changes in synthetic rates allows precise and unambiguous identification of gene activity. The controlled and homogeneous environment of tissue culture allows such a detailed study. Utilizing both primary culture and molecular techniques we have undertaken the elucidation of events involved in steroid-mediated induction of GPDH activity. We have been able to show that hydrocortisone acts at the transcriptional level in increasing new RNA species and thus the subsequent proteins (54). These techniques adapted to CNS function are a means of understanding the diversity and functions of cells within the CNS at the genomic level.

4.5 Generation of cell-specific markers

The paucity of specific markers for developmental stages (*Figure 14*) remains a major constraint in defining specific events of CNS cell commitment and differentiation. Markers to define these events would be useful in analysing gene expression during development, both qualitatively and quantitatively. Markers are also needed to ascertain and identify subpopulations of CNS cell types. Primary culture and clonal cell lines will be useful in analysing and producing cell-specific markers. We have generated monospecific monoclonal antibodies RBA_1 and RBA_2 (50), with specificity for astrocytes and 217c specific for C_6 glioma cells and transformed glia (65).

5. CONCLUSIONS

The use of primary cultures of purified CNS cells has greatly contributed to our understanding of cellular and molecular mechanisms involved in development. However, our understanding of their functions and unique properties have only begun to be elucidated. Together with the development of new techniques, primary culture of CNS cell types will have a more important role in defining the unique characteristics of the brain. We hope that the reader has received some feeling for the usefulness of primary culture in studying CNS function.

6. ACKNOWLEDGEMENTS

We would like to thank Michele Bloom for her help in typing the manuscript. Special acknowledgement is given to Ruth Cole who perfected many of the primary culture techniques. This work was supported by the National Institution of Health Grant HD06576 and Department of Energy Contract DE-AC03-76-0012.

7. REFERENCES

1. Cajal,S.R. (1933) *Histology*. 10th edition, Wood, Baltimore.
2. Fewster,M.E. and Mead,J.F. (1968) *J. Neurochem.*, **15**, 1303.
3. Poduslo,S.E. and Norton,W.T. (1972) *J. Neurochem.*, **19**, 727.
4. Sensenbrenner,M. (1977) In *Cell, Tissue and Organ Culture*. Federoff,S. and Hertz,L. (eds), Academic Press, New York, p. 191.
5. Harrison,R.G. (1907) *Proc. Soc. Exp. Biol. Med*, **4**, 140.
6. Ingebrigtsen,R. (1913) *J. Exp. Med.*, **17**, 182.
7. Lumsden,C.E. (1968) In *The Structure and Function of Nervous Tissue*. Bourne,G.H. (ed.), Academic Press, New York, p. 67.
8. Hosli,L. and Hosli,E. (1978) *Rev. Physiol. Biochem. Pharmacol.*, **81**, 135.
9. Shein,H.M. (1966) *Exp. Cell Res.*, **40**, 554.
10. Hertz,L., Juurlink,B.H.J. and Szuchet,S. (1985) In *Handbook of Neurochemistry*. Lajtha,A. (ed.), Plenum Publishing Corporation, New York, Vol. **8**, p. 603.
11. McCarthy,K.D. and de Vellis,J. (1980) *J. Cell. Biol.*, **85**, 890.
12. Meier,D. and Schachner,M. (1982) *J. Neurosci. Res.*, **7**, 135.
13. Szuchet,S., Stefansson,K., Wollman,R.L., Dawson,G. and Arnason,B.G.W. (1980) *Brain Res.*, **200**, 151.
14. Barkley,D.S., Rakic,L.L., Chaffee,J.K. and Wong,D.L. (1973) *J. Cell Physiol.*, **81**, 271.
15. Morris,J.E. and Moscona,A.A. (1971) *Dev. Biol.*, **25**, 420.
16. Moscona,A.A. (1961) *Exp. Cell Res.*, **22**, 455.
17. Arenander,A.T. and de Vellis,J. (1983) In *The Clinical Neurosciences*. Rosenberg,R. (ed.), Churchill Livingstone, New York, p. 533.
18. Saneto,R.P. and de Vellis,J. (1985) In *Cell Culture in the Neurosciences*. Bottenstein,J. and Sato,G. (eds), Plenum Press, New York, p. 125.
19. Varon,S. and Adler,R. (1981) *Adv. Cell Neurobiol.*, **2**, 115.
20. Levi-Montalcini,R. and Hamburger,V. (1951) *J. Exp. Zool.*, **116**, 321.
21. Muller,R., Curran,T., Muller,D. and Guilbert,L. (1985) *Nature*, **314**, 546.
22. Ramony Cajal,S. (1913) *Trab. Lab. Invest. Biol. Univ. Madrid*, **11**, 255.
23. del Rio Hortega,P. (1928) *Mein. R., Soc. Esp. Hist. Nat.*, **14**, 5.
24. Jacobson,M. (1978) *Development Neurobiology*. Plenum Press, New York and London.
25. Freshney,R.I. (1983) *Culture of Animal Cells: A Journal of Basic Techniques*. Alan R. Liss, Inc., New York.
26. Kruse,P.F. and Patterson,M.K. (1973) *Tissue Culture: Methods and Applications*. Academic Press, New York.
27. Saneto,R.P. and de Vellis,J. (1985) *Proc. Natl. Acad. Sci. USA*, **82**, 3509.
28. Morrison,R.S. and de Vellis,J. (1981) *Proc. Natl. Acad. Sci. USA*, **78**, 7205.
29. Schachner,M., Schoonmaker,G. and Hynes,R.O. (1978) *Brain Res.*, **158**, 149.
30. Leveille,P.J., McGinnis,J.E., Maxwell,D.S. and de Vellis,J. (1980) *Brain Res.*, **196**, 287.
31. Kumar,S. and de Vellis,J. (1981) *Dev. Brain Res.*, **1**, 303.
32. McMorris,F.A. (1983) *J. Neurochem.*, **41**, 506.
33. Simpson,D.L., Morrison,R., de Vellis,J. and Herschman,H.R. (1982) *J. Neurosci. Res.*, **8**, 453.
34. Morrison,R.S., de Vellis,J., Lee,Y.L., Bradshaw,R.A. and Eng,L.F. (1985) *J. Neurosci. Res.*, **14**, 167.
35. Bottenstein,J. (1983) *Adv. Cell Neurobiol.*, **4**, 333.
36. Adler,R., Magistretti,P.J., Hyndman,A.G. and Shoemaker,W.J. (1982) *Dev. Neurosci.*, **5**, 27.
37. Morrison,R.S, Sharma,A., de Vellis,J. and Bradshaw,R.A. (1986) *Proc. Natl. Acad. Sci. USA*, **83**, 7537.
38. Kruger,L. and Maxwell,D.S. (1966) *Am. J. Anat.*, **118**, 411.
39. Peters,A., Palay,S.L. and Webster,H.de F. (1976) *The Fine Structure of the Nervous System: The Neurones and Supporting Cells*. W.B.Saunders, Philadelphia.
40. Metuzals,J. and Mushynski,W.E. (1974) *J. Cell Biol.*, **61**, 701.
41. Brinkley,B.R., Murphy,P. and Richardson,L.C. (1967) *J. Cell Biol.*, **35**, 279.
42. Chude,O. and Wu,N.J. (1976) *J. Neurochem.*, **27**, 83.
43. Kobayashi,Y. and de Vellis,J. (1985) *J. Neurosci. Res.*, **13**, 509.
44. Kilpatrick,D.L., Ledbetter,F.H., Larson,K.A., Kirshner,A.G., Slepetes,R. and Kirshner,N. (1980) *J. Neurochem.*, **35**, 679.
45. Morrison,R.S. and de Vellis,J. (1983) *Dev. Brain Res.*, **9**, 337.
46. Brockes,J. (1982) *Neuroimmunology*. Plenum Press, New York and London.
47. Saneto,R.P., Altman,A., Knobler,R.L., Johnson,H.M. and de Vellis,J. (1986) *Proc. Natl. Acad. Sci. USA*, **83**, 9221.
48. Raff,M.C., Miller,R.H. and Noble,M. (1983) *Nature*, **303**, 390.

49. Kimelberg,H.K., Stieg,P.E. and Mazurkiewicz,J.E. (1982) *J. Neurochem.*, **39**, 734.
50. Pixley,S.K.R., Kobayashi,Y. and de Vellis,J. (1984) *Dev. Brain Res.*, **15**, 185.
51. Berger,C.N. (1986) *EMBO J.*, **5**, 85.
52. Harvey,A.L. (1984) *The Pharmacology of Nerve and Muscle Tissue Culture*. Alan R.Liss, Inc., New York.
53. Schubert,D. (1984) *Developmental Biology of Cultured Nerve, Muscle and Glia*. John Wiley and Sons, New York.
54. Kumar,S., Sachar,K., Huber,J., Weingarten,D.P. and de Vellis,J. (1985) *J. Biol. Chem.*, **260**, 14743.
55. Klee,W.A. and Nirenberg,M. (1974) *Proc. Natl. Acad. Sci. USA*, **71**, 3474.
56. Engrall,E. and Ruoslahti,E. (1977) *Int. J. Cancer*, **20**, 1.
57. Bottenstein,J.E. (1985) In *Cell Culture in the Neurosciences*. Bottenstein,J.E. and Sato,G. (eds), Plenum Press, New York, and London, p.3.
58. Saneto,R.P. and de Vellis,J. (1985) *Dev. Neurosci.*, **7**, 340.
59. Mauer,H.R. (1981) *Cell Tissue Kinet.*, **14**, 111.
60. McGinnis,J.F. and de Vellis,J. (1981) *Proc. Natl. Acad. Sci. USA*, **78**, 1288.
61. Arenander,A.T. and de Vellis,J. (1982) In *Proteins in the Nervous System: Structure and Function*. Haber,B., Perez-Polo,J.R. and Coulter,J.D. (eds), Alan R.Liss, New York, p.243.
62. Bologa,L., Aizenman,Y., Chaippelli,F. and de Vellis,J. (1986) *J. Neurosci.Res.*, **15**, 521.
63. Rome,L.H., Bullock,P.N., Chiappelli,F., Cardwell,M., Adinolfi,A.M. and Swanson,D. (1986) *J. Neurosci. Res.*, **15**, 49.
64. Wood,P. and Williams,A.K. (1984) *Dev. Brain Res.*, **12**, 225.
64. Peng,W., Bressler,J., Tiffany-Castigliani,E. and de Vellis,J. (1982) *Science*, **215**, 1102.

8. APPENDIX

Solutions used in cell culture.

Abbreviation	*Name*	*Defined in Section*
	Saline solution I	2.2.1
SFM	Synthetic basal (serum-free) medium	3.1.2
SSM	Serum-supplemented medium	2.2.2
ODM	Oligodendrocyte-defined medium	2.2.5(i)
ADM	Astrocyte-defined medium	2.2.5(ii)
NDM	Neurone-defined medium	2.3.1
	Trypsin solution I	3.3.1
	Trypsin solution II	3.2.2

CHAPTER 3

Immunocytochemical techniques for the localization of neurochemically characterized nerve pathways

JOHN V.PRIESTLEY

1. INTRODUCTION

In the last 10 – 15 years there has been a great increase in our knowledge of the distribution of putative transmitters and neuronal markers in the peripheral and central nervous systems. In this time the subject of chemical neuroanatomy has developed as a distinct topic and this development is reflected by the recent publication of several comprehensive review volumes devoted to this field (1,2). Part of this growth has been due to a parallel growth in the sophistication of histochemical procedures and one of the most important growth areas has been in the realm of immunocytochemistry. For mapping neurochemically characterized neuronal pathways, immunocytochemistry may be used either alone or in combination with various tract-tracing procedures. The aim of this chapter is to provide a basic practical guide to the use of immunocytochemistry in chemical neuroanatomy. For specific examples of the contribution that immunocytochemistry has already made to the field of neuroscience, the reader is referred to other recent reviews (3,4).

The simplest approach to analysing the distribution of an antigen in the nervous system is to localize it using a light microscopic (LM) immunocytochemical procedure. The term 'immunocytochemistry' refers to any technique for localizing a substance in a histological preparation using a specific antibody which recognizes the substance. The antibody concerned is referred to as the primary (I°) antibody. This definition embraces a wide variety of different methods and these may include different types of primary antibody, different types of histological preparation and different methods of visualizing the I° antibody. In the main part of this chapter some of these variations are considered. Section 2 covers general background theory and then protocols for two of the main visualization techniques (immunofluorescence and peroxidase – anti-peroxidase procedure) are presented in Section 3. Questions of specificity in immunocytochemistry are covered briefly in Section 4.

More sophisticated immunocytochemical procedures are normally required in order to go beyond a simple distribution analysis and some of these techniques are described in the final portion of the chapter. Double immunocytochemical staining techniques may be used to compare directly the distribution of two different substances (Section 5) and this approach has been especially useful in studies of co-existence. Information on the association of antigens with specific neuronal pathways may be obtained from routine LM immunocytochemistry if a transmitter or neuronal antigen is distributed throughout the cell body, axon and dendrites of a neuron. However, this situation is rare and normally such information can only be obtained by combining immunocytochemistry with tract-tracing procedures. Some of these approaches are presented in Section 6. Finally, examination of synaptic contacts between cells requires work at the ultrastructural level. Electron microscopic (EM) immunocytochemistry is described in Section 7.

2. IMMUNOCYTOCHEMICAL METHODS, GENERAL BACKGROUND THEORY

2.1 **Primary antibodies**

Macromolecules, especially proteins and polysaccharides, are naturally immunogenic. That is, they induce the production of antibodies following their introduction into a suitable host. The sites on a molecule recognized by the antibodies are called 'antigenic determinants'. Many smaller molecules possess suitable antigenic determinants but they do not induce antibody formation unless the molecule is conjugated to a large carrier protein. Such molecules are referred to as haptens. One of the first examples of immunocytochemistry applied to neurochemistry was the production and immunofluorescent localization of antibodies against the adrenergic vesicle components dopamine-β-hydroxylase (DBH) and chromogranin A by Geffen and colleagues in 1969. In the following 10 years numerous studies were published using antibodies raised against moderately immunogenic molecules such as enzymes, peptides and hormones (for review see 3). However in the last few years two important advances have occurred in techniques of antibody production and these have greatly increased the range of I° antibodies available to workers in the field of neuroscience. These new developments will be outlined below because an understanding of them is necessary in order to appreciate the present scope of immunocytochemistry and the correct choice and use of its primary reagents. However, protocols for antibody production are not included. This is a complex subject which is outside the scope of the present chapter and the reader is referred to other reviews which cover this theme (5–7).

2.1.1 *Antibodies against transmitters*

One important development has been in methods for conjugating haptens to carriers. Until recently it was not possible to produce specific antibodies against 'classical' transmitters such as acetylcholine (ACh), 5-hydroxytryptamine (5-HT), noradrenaline (NA), adrenaline (AD), dopamine (DA), γ-amino butyric acid (GABA), glutamate (Glu), aspartate (Asp) and glycine (Gly). This is because the transmitters are too small to be naturally very immunogenic and antibodies raised against hapten carrier conjugates are often not specific for the original hapten. However, it is now recognized that this change in specificity can in fact be advantageous in the production of antibodies which are to be used for immunocytochemistry. Several groups have recently raised antibodies against transmitters by immunizing with transmitter–carrier conjugates which mimic the complex formed between transmitter and tissue following fixation (8, 9). Generally these antibodies do not bind with high affinity to the unmodified transmitter and they may even cross-react strongly with other related transmitter molecules when tested using a liquid phase assay such as radioimmunoassay (RIA). However, when applied to fixed tissue in immunocytochemistry the transmitter antibodies show high affinity and specificity. This phenomenon is probably best illustrated by using an example. Many currently available 5-HT antibodies cross-react with DA in liquid phase assays although they do not stain DA-containing neurons in paraformaldehyde-fixed tissue sections. This is because they are raised against a conjugate consisting of 5-HT coupled to a suitable carrier [e.g. bovine serum albumin (BSA), keyhole limpet haemocyanin] using formaldehyde and have very high affinity for a modified derivative of 5-HT (possibly a β-carboline) which is formed in the tissue during paraformaldehyde fixation. The antibodies have much lower affinity for either free DA or fixed DA than for the 5-HT derivative (9). Two important points are derived from this example which are also true for antibodies against other small molecules.

(i) The fixation conditions in immunocytochemistry with these antibodies are much more stringent than for most immunostaining. Conditions must be used which mimic those used for production of the original immunogen (hapten–carrier complex).
(ii) RIA data do not give an accurate indication of the specificity of the antibodies when used for immunocytochemistry. Therefore specificity has to be assessed using other criteria such as model systems which mimic the tissue situation (see Section 4.2).

These two points place certain limitations on the use of the new transmitter-directed antibodies, but within these limitations they are very valuable reagents which have already been used to great effect and will undoubtedly continue to be very important.

2.1.2 *Monoclonal antibodies*

A second important development has been the introduction of monoclonal antibodies. Monoclonal antibodies produce immunostaining with much less non-specific background than is obtained with polyclonal antisera and this aspect will be considered further in Section 4. The other main advantage of monoclonal antibody techniques is that they allow the production of types of primary antibody which would be very difficult or impossible to produce using traditional immunization procedures. Generally, the ease with which an antibody can be produced depends on the purity and the antigenicity of the immunogen. If the antigen of interest represents only a minor fraction of the preparation used for immunization, the antisera produced are likely to be dominated by antibodies against the contaminants. This problem is neatly illustrated by the recent history of attempts to produce antibodies against choline acetyltransferase (ChAT). Production of specific antibodies required a 10^6-fold purification of the enzyme to produce approximately 80% purity (10). Use of less pure preparations for immunization produced antisera of undefined and disputed specificity (11). With monoclonal antibody techniques this problem is avoided. The various antibodies in the serum of a typical antigenic response are produced by separate lines of lymphocytes and their derived plasma cells. In the monoclonal antibody approach splenocytes are fused with myeloma cells and the resulting hybrids cultured. Each hybrid produces a single type of antibody. The mixed population of hybrid cells can be subdivided, tested and cultured to produce homogeneous populations derived from a single parent (a clone) and secreting only the antibody of interest. Thus, even if an antibody represented only a minor compo-

Table 1. Immunocytochemical transmitter-related markers.

Tables 1 and *2* contain lists of nervous system components to which antibodies have been raised and which may be used in immunocytochemistry. A full bibliography can be found in references 3, 4 and 16. Numbers in parentheses refer to additional papers not quoted in the above reviews.

Catecholamines	
Biosynthetic enzymes	Tyrosine hydroxylase, Dopamine-β-hydroxylase, Phenylethanolamine N-methytransferase
Transmitters	Dopamine, adrenaline, noradrenaline
Receptor-related proteins	α_1 and β Adrenergic receptor, DARPP-32 (17)
Indoleamines	
Biosynthetic enzymes	Tryptophan hydroxylase
Transmitter	5-HT (9)
Acetylcholine	
Biosynthetic enzyme	Choline acetyltransferase
Transmitter	ACh
Receptors and related proteins	Nicotinic and muscarinic receptor subunits, cholinesterase
Synaptic constituents	Nicotinic proteoglycan vesicle antigen
Amino acids	
Biosynthetic enzymes	Glutamic acid decarboxylase, Glutaminase, Aspartate aminotransferase (18)
Degradative enzymes	GABA-transaminase (19)
Transmitters	GABA, glutamate, aspartate (8)
Peptides	
Processing and degradative enzymes	Various endo- and exo-peptidases, endopeptidase-24.11 (20)
Transmitters	Numerous peptides and precursors
Other transmitter markers	
Biosynthetic enzymes	Histidine decarboxylase, S-Adenosylhomocysteine hydrolase, Adenosine deaminase (21)
Transmitters	Histamine
Second messengers	Cyclic nucleotides (22)

nent of a mixed serum it can be obtained in pure form. Using this approach it has been possible to obtain monoclonal antibodies against ChAT (12) and against other enzymes and receptors using only partially purified preparations as immunogens. Another interesting and very important approach has been to immunize animals with crude membrane preparations and then use immunocytochemistry to select interesting monoclonal antibodies such as those specific for particular neuronal sub-populations (13,14).

Another advantage of the monoclonal techniques is that once a monoclonal antibody has been developed it can be produced in very large quantities and maintained virtually indefinitely. This is in marked contrast to antisera which are normally available in only limited quantities and whose characteristics vary over the period of immunization. In addition to the obvious advantage of having a reliable supply of pure antibody, this has allowed the establishment of banks of monoclonal antibodies which are available to workers of different interests. Unexpectedly, various monoclonal antibodies against non-neuronal antigens have turned out to label specific neuronal sub-populations and

Table 2. Immunocytochemical nervous system markers.

1. *Markers of major cell classes*	
Neurones	Tetanus toxin receptor, neurone-specific enolase (NSE), N-CAM, A4, 38/D7, NILE glycoprotein, BSP-2, GQ ganglioside, Thy-1
Oligodendrocytes	Galactocerebroside, 01 to 04 (23)
Schwann cells	Ran-1
Astrocytes	Glial fibrillary acidic protein (GFAP), S-100, α_2-glycoprotein, non-neuronal enolase, M1 and C1, Ran-2, vimentin, stage-specific embryonic antigen (SSEA-1), glutamine synthetase
Epithelial cells, fibroblasts and ependymal cells	Vimentin, fibronectin
2. *Markers of specific neuronal subpopulations*	
Subsets of I° afferents	Lacto- and globo-series oligosaccharides (24)
Purkinje cells	Human T lymphocyte antigen (15)
Subsets of vertebrate neurones	CAT 301, CE 5 (25) and other monoclonals (13,14,26)
3. *Markers of specific membrane complexes and organelles*	
Voltage-dependent Na^+ channel, Na^+,K^+-ATPase, Ca pumps	Monoclonal and polyclonal antibodies to various subunits and to different forms
Cytoskeletal elements	Actin, tubulin, intermediate filament-associated proteins, microtubule-associated proteins (MAP1, MAP2), spectrins
Organelles	Neurone-specific mitochondrial protein, synaptic vesicle proteins (synapsin 1), pre- and post-synaptic membrane proteins
Extracellular matrix	Collagen, basal lamina fractions
4. *Developmental markers*	
Avian neural crest and mesenchymal markers	NC-1 and E/C8 monoclonals
cell adhesion molecules	N-CAM, Ng-CAM and L-CAM
Stage-specific embryonic antigens	SSEAs (27)

these antibodies are another important tool in neurochemical research (15).

Many workers will not want to, or need to, raise their own I° antibodies. *Tables 1* and *2* list some of the presently available antibodies which are likely to be of interest to the neuroscientist. Some of these antibodies are only available from the particular scientists involved in their production and in this case the original papers should be consulted for information on the antibody origin. Other antibodies are available commercially and Appendix 1 lists some sources of both I° antibodies and developing reagents.

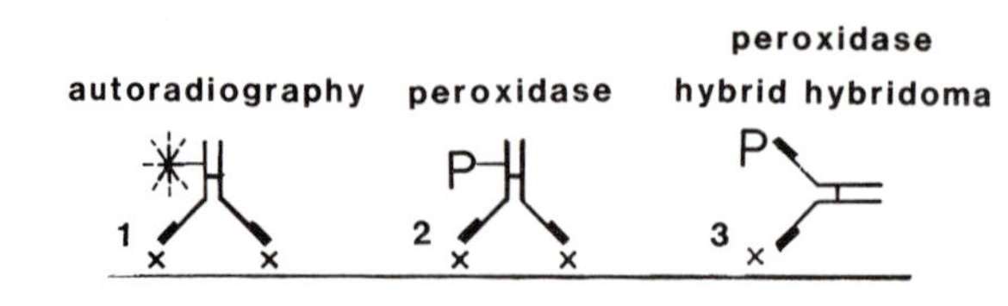

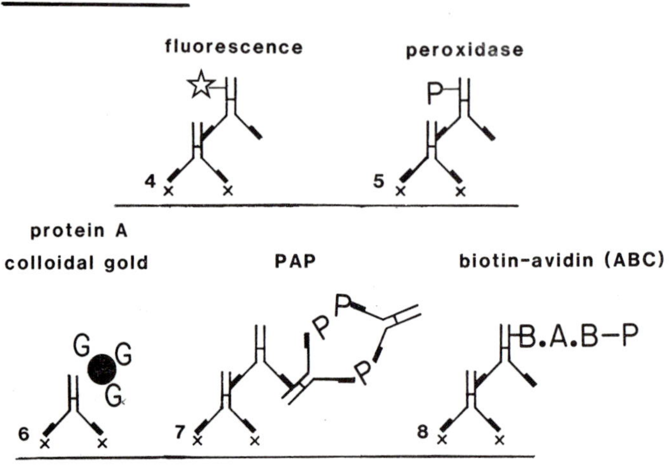

Figure 1. All immunohistochemical procedures involve the binding of an antibody (the I° antibody) to a tissue antigen (x). **1–8** show a number of different ways of visualizing the I° antibody. **1–3**. In direct methods the I° antibody is directly labelled. A label such as peroxidase (P) may be covalently coupled to the antibody (**2**). Alternatively, using monoclonal techniques it is possible to produce antibodies internally labelled with radioactive [^{3}H]amino acids. (**1**) or to produce hybrid antibodies with binding sites for both a tissue antigen and a label such as peroxidase (**3**). **4–8**. In indirect methods the I° antibody is not itself labelled but is visualized using a second labelled reagent. Anti-Ig antibodies which have been covalently labelled with a fluorochrome (**4**) or an enzyme (**5**) may be used to bind to the I° antibody (indirect labelled procedures). Alternatively, labels may be bound by methods other than direct covalent linkage to an antibody (**6–8**). Protein A will bind to certain classes of antibody and can be labelled with a suspension of colloidal gold (**6**). In the PAP procedure peroxidase enzyme is bound by anti-peroxidase antibodies (**7**). Avidin can be used to bind biotinylated peroxidase to a biotinylated anti-Ig antibody (**8** the avidin–biotin peroxidase complex method, ABC). The immunofluorescence (**4**) and PAP (**7**) procedures are shown in more detail in *Figure 2*.

2.2 Developing antibodies

A large number of different methods are available for localizing the I° antibody and the various reagents involved are generally termed 'developing antibodies'. Some of the methods are illustrated in *Figure 1*. Methods in which a label is bound directly to the I° antibody are referred to as 'direct' methods. 'Indirect' methods use a general second reagent to bind to the I° antibody and the second reagent is then labelled in some way. Direct methods involve the special preparation of labelled antibodies for each I° antibody being used and, since such preparation can be complicated, direct methods are normally only used in rather specialized circumstances. In contrast the second reagents in the indirect methods can be bought commercially and each reagent recognizes all I° antibodies belonging to a certain immunological class (e.g. rabbit IgG).

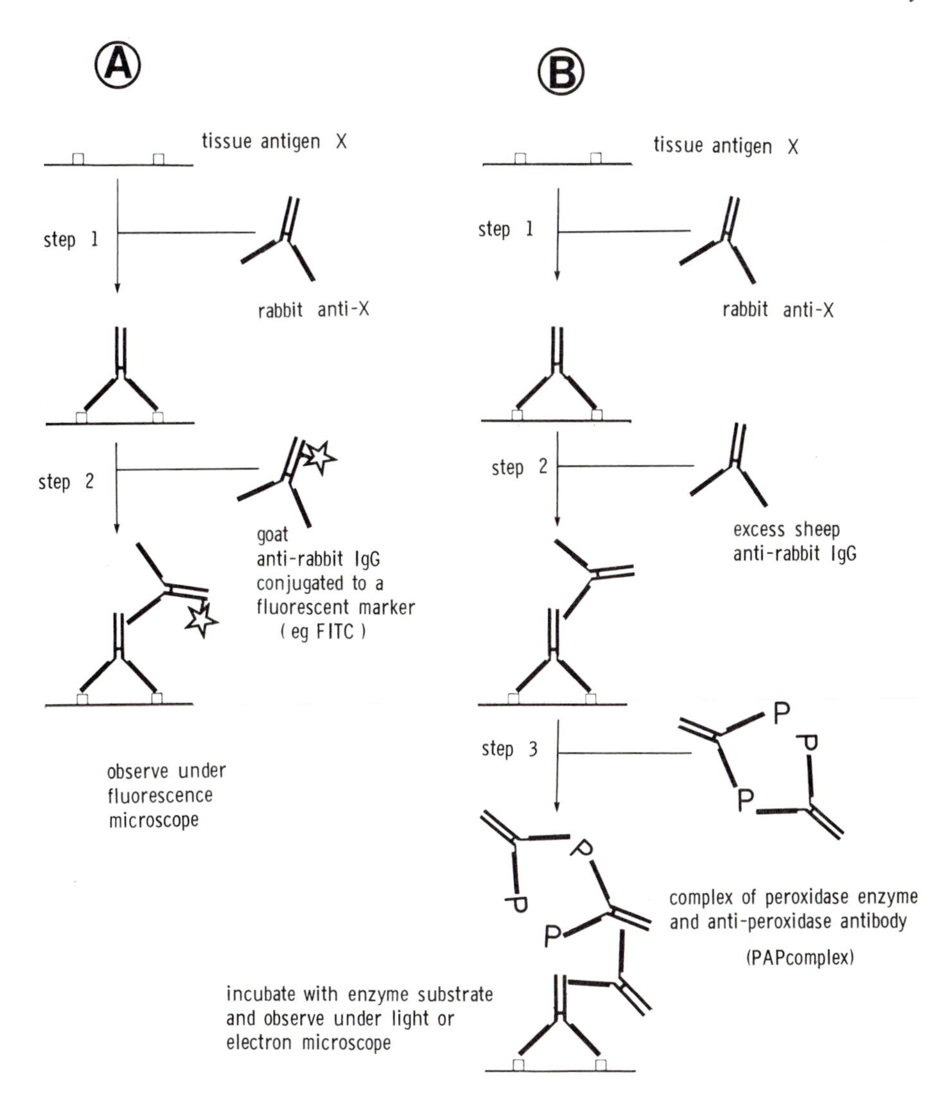

Figure 2. Immunofluorescence (**A**) and PAP (**B**) procedures. (**A**) In the immunofluorescence procedure antibodies are added in two steps. The second antibody is conjugated to a fluorescent marker and binds to the first (I°) antibody. (**B**) In the PAP procedure antibodies are added in three steps. The second antibody binds to the first (I°) antibody but is added in excess so as to leave one of its binding sites unoccupied. This site subsequently binds the PAP reagent added in the third step. The reagent consists of a soluble complex of peroxidase enzyme (P) and anti-peroxidase antibody. The anti-peroxidase antibody should be raised in the same species as the I° antibody. Rabbit-raised I° antibodies are shown in the figure. Other species may be used (e.g. rat) but then the appropriate second and third antibodies must be used (e.g. rabbit anti-rat IgG and rat PAP). In some protocols a stage prior to step 1 is used consisting of incubation in normal serum of the same species as the second antibody. This reduces non-specific binding of the second antibody by the tissue.

In practice this means that only a few different types of second reagent are required to localize all the different I° antibodies likely to be used in a typical laboratory.

The first indirect method to be introduced was the immunofluorescence technique of Coons and colleagues and this is still one of the most widely used procedures for light microscopic studies. A detailed protocol is presented in Section 3.2. The technique uses a fluorescent labelled second antibody to bind to the I° antibody (*Figure 2*) and is an example of a class of techniques known as indirect labelled procedures. Commonly used indirect labels include different coloured fluorochromes [fluorescein isothiocyanate (FITC) and tetramethylrhodamine isothiocyanate (TRITC)] and the enzymes horseradish peroxidase and alkaline phosphatase. With these reagents most types of single and double immunocytochemistry at either light or electron microscopic level can be carried out. However the labelled procedures have the disadvantage that chemical conjugation of a label onto an antibody can damage both the antibody and the label and lead to loss of both specificity and sensitivity. In order to avoid this problem Mason and colleagues and Sternberger and Cuculis introduced an unlabelled antibody method and Sternberger and colleagues subsequently developed this into the so-called peroxidase–anti-peroxidase (PAP) procedure (28). The principle of the PAP procedure is that the final label (peroxidase enzyme) is bound immunologically by anti-peroxidase antibodies. The peroxidase and anti-peroxidase antibodies are added together as a soluble complex and this complex is bound to the I° antibody by an intermediate link antibody (*Figure 2*). Both the I° antibody and the anti-peroxidase antibodies of the PAP complex must have been raised in the same animal species. The PAP procedure is claimed to be more specific and more sensitive than the indirect labelled procedures and is probably the most widely used immunocytochemical staining technique. A detailed protocol is given in Section 3.3. A large number of variations on the PAP protocol have been described in the literature in an effort to increase further the specificity and the more important of these are outlined at the end of Section 3.3.2.

In the last few years various other immunocytochemical staining methods have been introduced and some of these are shown in *Figure 1*. Only the immunofluorescence and PAP procedures will be described in detail in this chapter but a few comments will be made here indicating the areas of application of other important techniques. Cuello and colleagues (29) have introduced various ways of producing labelled monoclonal antibodies and these are useful agents for double immunocytochemical staining, especially at the EM level. Another important method for EM immunocytochemistry is to label with small (3–60 nm) gold particles (30). The gold will adsorb to the surface of macromolecules such as antibodies or protein A (an antibody-binding protein from the cell walls of *Staphylococcus aureus*) and can be used in direct or indirect methods. Finally, methods based on biotin–avidin interactions are coming into wide use. Biotin is a vitamin which is bound with very high affinity to the glycoprotein avidin (K_A 10^{15}/M). Antibodies, enzymes and fluorochromes may be conjugated to biotin or avidin and then used as labels in a variety of different ways (30). The most commonly used method is referred to as the ABC (avidin–biotin peroxidase complex) procedure and is illustrated in *Figure 1*.

2.3 **Choosing an immunocytochemical method**

The previous section has given some indication of the wide range of immunocytochemical procedures that are available. However, the choice of method for a particular neuroanatomical project will not be quite so wide and will be limited by both the nature of the project and the equipment available.

The immunofluorescence technique is normally not sensitive enough for use on sections cut from embedded materials (paraffin or plastic) and is not suitable for material which will require long or repeated examination using high magnification objectives (50 − 100×). The fluorescence fades with storage and with exposure to u.v. light during observation. A fluorescence microscope is required and the procedure cannot be used for EM localization. In contrast, localization procedures which involve an enzyme (probably peroxidase) give permanent preparations which can be examined on a standard transmitted light microscope using objectives even up to 100× magnification. The indirect procedures (peroxidase-conjugated second antibody, PAP, ABC) are sensitive enough for localizing antigens in embedded material. However, the immunofluorescence procedure is easier than the enzyme-based techniques and arguably is as sensitive and specific if applied to the right type of preparation. Immunofluorescence can be combined easily with other fluorescent dyes such as retrograde tracers or intracellularly injected dyes. It is therefore the method of choice for intermediate magnification (10−40×) analysis of unembedded materials (e.g. cryostat sections, tissue culture) and for various specialized double staining techniques (see Section 5.1 and 6.2). Of the enzyme-based procedures, the PAP technique is probably the best method to choose when initially setting up immunocytochemistry because it is the most widely used and most flexible of the techniques. However it is also the most complex procedure and the most difficult to correct if there are problems. Peroxidase-conjugated antibodies are simpler to use and in some situations are quite adequate. The biotin−avidin procedures are generally comparable with PAP, and are becoming increasingly popular.

With each localization procedure (fluorescence, PAP, etc.) an appropriate tissue preparation and sectioning method has still to be chosen and these variables are discussed in Section 3. Thus it is not possible to give a single general immunocytochemical staining protocol. *Table 3* lists a number of possible neuroscience projects and indicates the type of protocol most suitable for each project.

Table 3. Choice of immunocytochemical protocol.

Project	*Suggested protocol*
1. Mapping of CNS antigen distribution	Immunofluorescence on 15 μm cryostat sections or PAP on 20 μm cryostat or 40 μm sliding microtome frozen sections
2. Detailed analysis of cells and processes	PAP on 40 μm vibratome or sliding microtome sections
3. Antigen distribution in peripheral tissues	Immunofluorescence on 15 μm cryostat sections or PAP on 20 μm cryostat or 40 μm sliding microtome frozen sections. Where appropriate, PAP or immunofluorescence on whole mounts
4. Relative distribution of two different antigens	Dual colour immunoperoxidase
5. Co-existence	Staining of serial 5 μm paraffin or cryostat sections. Alternatively, dual colour immunofluorescence

3. IMMUNOCYTOCHEMICAL PROTOCOLS

Sections 3.2 and 3.3 contain detailed protocols for immunofluorescence and the PAP procedure. In Section 3.1 procedures and recipes common to both techniques are given. References 3, 28, 31–33 also contain useful general protocols.

3.1 **General recipes and procedures**

3.1.1 *Fixation*

The process of fixation has two main purposes, namely to preserve tissue structure and to retain antigen *in situ* without significant loss of antigenicity. The initial fixation is one of the most important stages in immunocytochemistry and one of the most frequent causes of bad immunostaining. Inadequate or uneven fixation produces loss of antigen and tissue damage. Over-fixation leads to loss of antigenicity. The end result of both over- and under-fixation is therefore decreased sensitivity in the immunostaining. For each particular antigen there will be a 'best fixative' which provides the most suitable compromise between preserving tissue structure and antigen location while retaining antigenicity. *Table 4* lists some commonly used fixatives and in any study it

Table 4. Immunocytochemical fixatives.

1. 4% paraformaldehyde, 0.0–0.5% glutaraldehyde in 0.1 M phosphate buffer, pH 7.4
2. 2% paraformaldehyde, 0.01 M periodate, 0.075 M lysine in 0.04 M phosphate buffer, pH 6.2
3. 2% paraformaldehyde, 0.15% picric acid in 0.1 M phosphate buffer, pH 7.3
4. Aqueous saturated picric acid, 37% formaldehyde, glacial acetic acid in a ratio of 150:50:10 (Bouin's solution)
5. 4% paraformaldehyde, 0.2% picric acid, 0.05% glutaraldehyde in 0.1 M phosphate buffer, pH 7.3 (for EM)
6. 4% paraformaldehyde in 0.1 M sodium acetate, pH 6.5, followed by 4% paraformaldehyde in 0.1 M sodium borate, pH 11
7. 4% paraformaldehyde, 0.5% zinc salicylate in 0.45% NaCl, pH 4.0 or pH 6.5
8. 5% acrolein in 0.1 M phosphate buffer, pH 7.2
9. 1% 1-ethyl-3(3-dimethylaminopropyl) carbodiimide (EDC), 0.3% glutaraldehyde in 0.1 M phosphate buffer, pH 7.0 (for EM)
10. 0.4% parabenzoquinone in 0.1 M sodium cacodylate, pH 7.4

Table 5. 4% Paraformaldehyde in 0.1 M phosphate buffer, pH 7.4.

0.2 M phosphate buffer (pH 7.4)

Prepare 0.2 M phosphate buffer stock solution as follows:

A. 0.2 M monobasic sodium phosphate
(31.2 g of $NaH_2PO_4.2H_2O$ in 1 litre of water)

B. 0.2 M dibasic sodium phosphate
(71.7 g of $Na_2HPO_4.12H_2O$ in 1 litre of water)

Add x ml of A to B until pH is reached (~19 ml of A to 81 ml of B).
(Dilute the stock 1:1 with water to obtain 0.1 M buffer).

Paraformaldehyde

Prepare 8% paraformaldehyde stock solution as follows:

Add 40 g of paraformaldehyde to 500 ml of distilled water, heat to 58–60°C.
Add 1 M NaOH dropwise until clear. Leave to cool, then filter.
For 4% fixative dilute 1:1 with 0.2 M buffer stock.

may be necessary to test a variety of different fixatives. In practice, however, one or both of the bi-functional cross-linking reagents paraformaldehyde and glutaraldehyde are suitable for most purposes. Paraformaldehyde is good for many peptides, receptors, enzymes and nucleic acids. My colleagues and I have used 4% paraformaldehyde (*Table 5*) for successful immunocytochemical localization of various neuropeptides (substance P, enkephalin, neurotensin and somatostatin) and for localization of glucagon, 5-HT and tyrosine hydroxylase. In some cases better immunostaining is obtained by including a small amount of glutaraldehyde (e.g. 0.2%) but normally glutaraldehyde decreases the immunoreactivity of peptides and enzymes. However, for retention of amino acids such as glutamate or GABA, high concentrations of glutaraldehyde are needed (e.g. 1%). For EM studies small amounts of glutaraldehyde (0.02–0.5%) are needed to preserve ultrastructure.

The method of fixation is as important as the fixative used. In some situations it may only be possible to immerse the tissue in fixative but whenever possible perfusion fixation (*Figure 3*) should be carried out.

(i) *Perfusion fixation.*

(1) Anaesthetize the animal (for a rat: 0.1 ml/100 g of a 60 mg/ml solution of sodium pentobarbital) and place it on its back with its four limbs tied down. Expose and cannulate the trachea and respire the animal on 95% 0_2, 5% CO_2. This can be accomplished using either a small animal respirator or a piece of tubing with a T-junction near one end. The end near the junction is attached to the animal and the other end to a gas cylinder and the open limb of the T-junction alternatively opened and closed by an assistant using his/her finger (*Figure 3A*).

(2) Make a mid-line incision through the abdominal wall and then laterally along the caudal border of the rib cage, exposing the diaphragm.

(3) With a pair of artery forceps firmly hold the xiphoid process, cut along the top (ventral) edge of the diaphragm and then forward through the sides of the rib-cage. Lift and fold forward the slab of tissue so formed. The forceps should hold it in place, leaving the heart and lungs exposed. The stages from cutting into the thoracic cavity to commencing the perfusion should be carried out as quickly as possible.

(4) With fine scissors remove any fat which may be obscuring the exit of the ascending aorta.

(5) Inject into the left ventricle a small amount of an anti-coagulant (for a rat, 0.05 ml of a 5000 units/ml heparin solution).

(6) Holding the apex of the heart with a pair of forceps, cut the right atrium. Then make a cut in the side of the left ventricle and introduce the perfusion cannula through the ventricle and up into the aorta. The cannula tip should just be visible in the aorta (*Figure 3C*). Do not advance it too far or it is likely to block the exit of the carotid arteries and this will lead to uneven perfusion. If the lungs immediately go pale this is a sure indication that the cannula has entered the left atrium and/or the pulmonary vein and this will also give poor perfusion. If only the brain is required, the descending aorta may be clamped.

(7) For a rat, perfuse with 20–50 ml of vascular rinse (*Table 6*) followed by 200–500 ml of fixative. Run the vascular rinse slowly from the cannula during

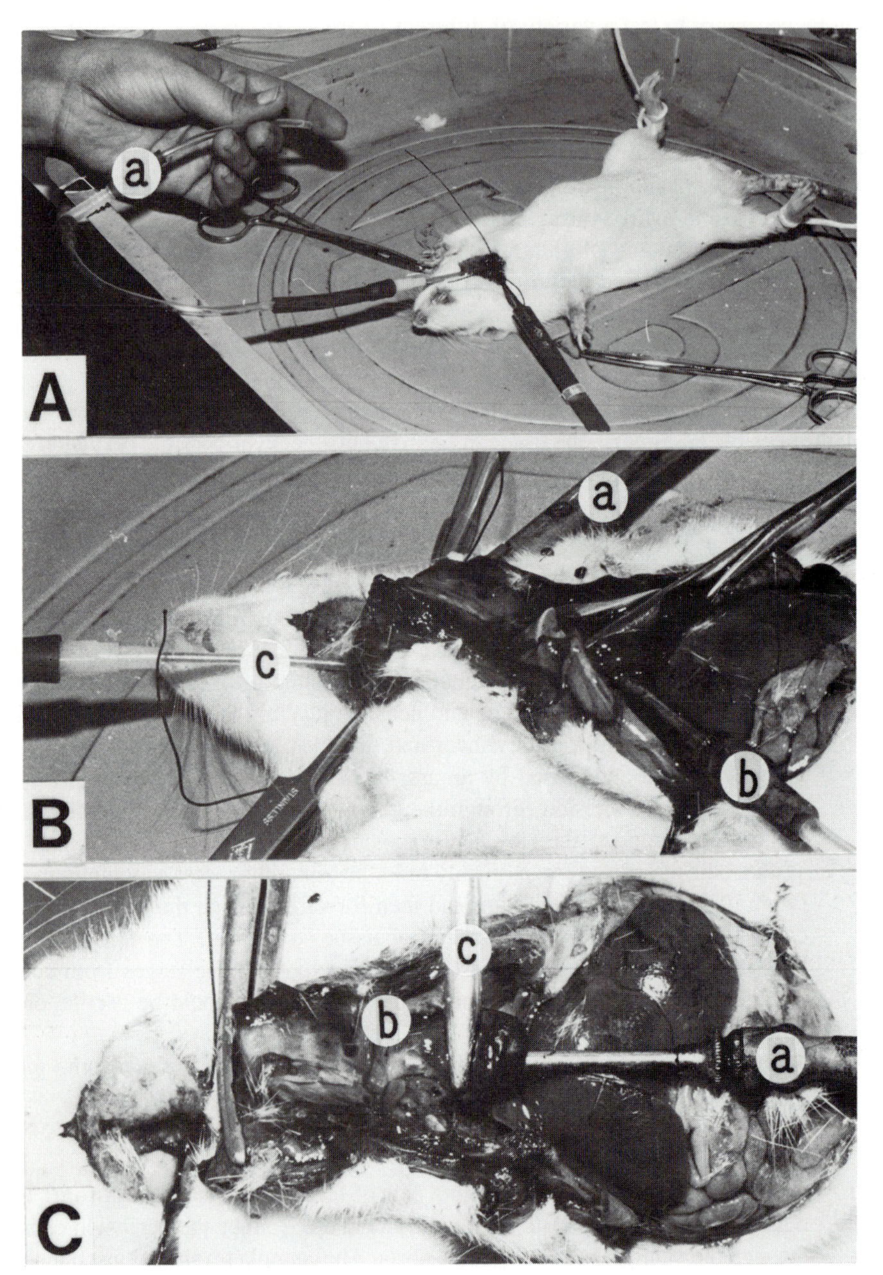

Figure 3. Surgical procedure for vascular perfusion of a rat. (**A**) Before opening the thoracic cavity the trachea is cannulated and the animal respired on 95% 0_2, 5% CO_2. Respiration can be maintained manually by simply opening and closing the third limb (a) of a T-junction inserted in the line connecting the animal and the gas cylinder. (**B**) While maintaining manual respiration the thoracic cavity is opened. The cut slab of rib cage may be held out of the way with a pair of large artery forceps (a) and the perfusion cannula (b) introduced into the left ventricle. The tracheal canulla is indicated (c). (**C**) The perfusion cannula (a) should be pushed through the left ventricle and up until the tip is just visible in the aorta (b). The cannula may then be clamped in place with another pair of artery forceps (c). Once perfusion has started the tracheal cannula may be removed.

Table 6. Vascular rinse solution (34).

For 500 ml mix:	475 ml of water
	25 ml of 0.2 M phosphate buffer, pH 7.4
	4 g of NaCl
	0.125 g of KCl
	0.25 g of $NaHCO_3$
Prior to perfusion, bubble the solution with 95% O_2 5% CO_2 until saturated	

its entry into the left ventricle and aorta. This helps prevent the entry of air bubbles into the vascular system.

(8) Carry out perfusion by syringe either manually or on a syringe pump, by peristaltic pump or air pressure, or by gravity feed from drip bottles. The last of the methods is the easiest and probably also one of the most reliable.

(9) For a rat, carry out perfusion at 80–120 cm H_2O (rat blood pressure = 120 cm H_2O, 90 mm Hg) and over a period of about 30 min (for 500 ml of fixative).

For immunocytochemistry, tissue should be left in fixative for an additional 3 h after perfusion. Some authors leave the tissue in fixative until needed (i.e. several days) but the effect of prolonged fixation should be tested for each type of antigen under study. For short periods of fixation (3–12 h) the tissue (presumably brain) can be left *in situ* and the body stored at 4°C. This is said to give better preservation of morphology, especially for EM studies. However, I prefer to dissect out the brain after perfusion and transfer to a beaker containing fixative. This allows one to see immediately how successful was the perfusion. If perfusion was poor, post-perfusion fixation can be increased by cutting the area of interest into a block and by increasing the time in fixative. Following fixation, tissue is transferred to 0.1 M phosphate-buffered containing 5–30% sucrose depending on the degree of cryoprotection required (see Section 3.1.2). In situations where the length of time in fixative must be kept short (see Section 6.3) it may be advantageous to leave the animal attached to the perfusion apparatus and perfuse with phosphate-buffered sucrose after a period of fixation.

Some workers use cold fixative and cover the animal with ice in order to reduce anoxic and autolytic effects. However, at lower temperatures the rate of penetration of fixative is reduced. In situations where perfusion via the heart is not possible other routes may be used successfully. For example, following a chronic physiological experiment, perfusion may be carried out via a carotid cannula which may have been used to measure blood pressure.

3.1.2 *Sectioning*

Sectioning methods may be divided into two broad classes, namely methods for sectioning embedded material and those for unembedded material. For LM the most commonly used embedding material is paraffin wax. Relatively routine paraffin embedding and sectioning procedures may be used and these will not be described here. The advantage of using paraffin sections is that material once embedded can be kept indefinitely and sectioned or stained whenever required. Staining usually has very low background and thin sections can be obtained with better morphological preservation than that obtained with unembedded (e.g. cryostat) sections. The disadvantage of paraffin sections

Table 7. Embedding compounds used in immunocytochemistry.

1. Paraffin wax
2. Methacrylates [e.g. 2-hydroxyethyl (gycol) methycrylate]
3. Polyethylene glycol
4. Epoxy resins subsequently etched with sodium alkoxide or oxidized with hydrogen peroxide, potassium permanganate or periodic acid
5. Aliphatic and aromatic cross-linked arcylics (e.g. Lowicryl, LR White, LR Gold)

is that antigenicity is frequently lost during dehydration and paraffin wax embedding. A large number of other embedding materials are available and some of the most important are listed in *Table 7*. Some of these are water-soluble cold-cure resins which are claimed to preserve antigenicity. However their use is complex and they are generally more important for EM than for LM studies (35, see also Section 7).

Methods for sectioning unembedded material include vibrating blade instruments such as the Vibratome® and frozen section procedures. These methods are described in detail below. Good immunocytochemistry can be carried out on unembedded material and in addition various histochemical procedures (e.g. Section 6.3) can be carried out which are not applicable to embedded material.

(i) *The Vibratome®* . The Vibratome® *(Figure 4)* is a vibrating blade instrument originally developed for cutting unfixed unembedded tissue for histochemistry. It is also very suitable for cutting fixed material and has the advantage of being able to cut 20–200 μm sections without the necessity of freezing or embedding the material. The tissue, cut into a shaped block, is attached to a horizontally oriented chuck. The chuck is contained in a small tank which can be filled with liquid (such as a buffer) and the chuck is lowered or raised by means of a knob calibrated in microns. A vibrating razor blade advances horizontally through the block (*Figure 4B*). To cut a section the block is raised the required amount (e.g. 40 μm) and the blade advanced slowly through the tissue. The section floats off into buffer and can be collected from the bath. Before cutting another section the blade must be withdrawn and the specimen raised the required amount. The amplitude of blade vibration, the cutting angle and rate of blade advance can all be adjusted. Instruments which work in the same basic way as the Vibratome® but which are made by other manufacturers are also available (see Appendix 2).

To cut successful sections on the Vibratome® the following basic points need to be followed.

(1) Securely attach the tissue to the chuck, normally with a rapidly drying cyanoacrylate adhesive. First blot dry the tissue surface to be attached with filter paper, but do not allow the whole tissue block to dry.
(2) For brain, set the blade at an angle of 15–20° at near maximum vibrating amplitude (#8 or 9) and the minimum advance speed (#1).
(3) Clear all tough connective tissue from around the tissue block. The block must not be so tall that it bends when touched by the advancing blade. For blocks of small cross-sectional area (e.g. rat spinal cord or brain stem) the block cannot be more than 3.5 mm tall.

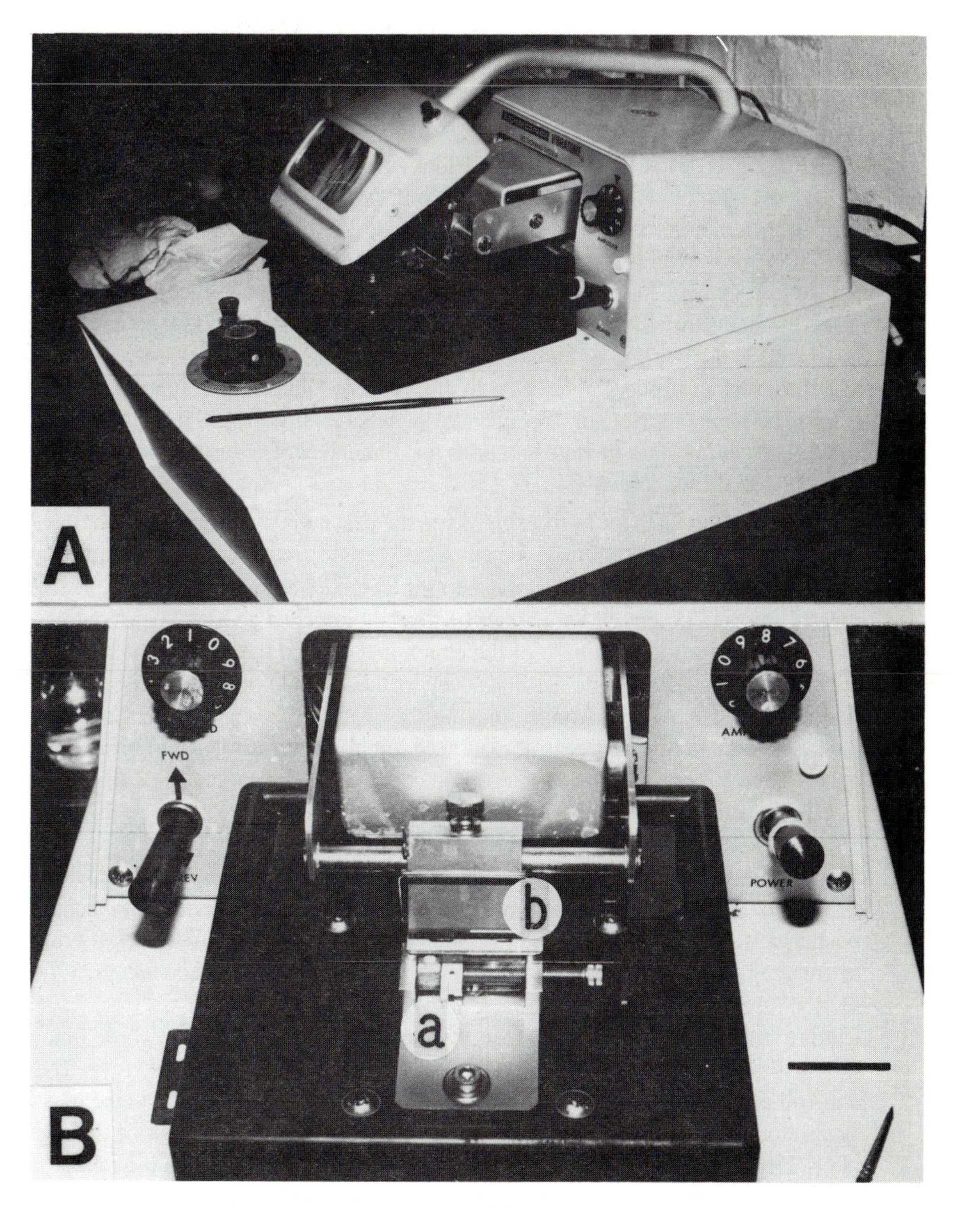

Figure 4. (**A**) The Oxford Instruments Vibratome® . (**B**) Close up of the cutting tank. The tissue, glued to a small block, is held in a clamp (a). The tissue is cut by a vibrating razor blade (b) which moves forward across the tank. The tissue is raised a set number of μm each time a section is to be cut. Scale bar = 25 mm.

Vibratome sections have better morphological preservation than frozen sections and so are the sections of choice where high resolution is required as in high magnification LM and in EM. Sections are normally stained free-floating and are not mounted until after all the immunocytochemical stages are completed. However, the Vibratome® cannot handle large slabs of tissue and is extremely slow to use. It can take several hours to cut a few sections from three or four blocks. Where large amounts or large size material have to be processed, a frozen sectioning procedure based on the cryostat or sliding microtome must be used.

(ii) *The cryostat.* A cryostat is a microtome which operates within a cold cabinet, normally at −12 to −22°C. The tissue block and knife are thus both cold, in contrast to the sliding microtome method (see below) where only the tissue is cold. The cryostat is probably the most widely used sectioning method in immunocytochemistry and histochemistry. It can cut 5−50 μm thick frozen sections which can then be collected in liquid or mounted directly onto slides. A modern motorized cryostat is easy and quick to use and is the method of choice in situations where sections have to be slide-mounted before further processing (e.g. with most types of autoradiography, or where sections must be kept in strict serial sequence) or where thin (5 μm) frozen sections are required. For cutting slide-mounted sections for immunocytochemistry, the following basic protocol should be followed.

(1) Freezing the tissue. Prior to freezing, keep the tissue in 0.1 M phosphate buffer, pH 7.4, containing 7% sucrose. To cut good sections freeze the tissue as quickly as possible and preferably directly onto the chuck. The best way is to use the type of microtome chuck which can be directly attached to a CO_2 gas source. Gas is forced through the base of the chuck and around the tissue block. More sophisticated rapid freezing methods such as those based on liquid nitrogen are not necessary. However, simply placing the block in the cryostat or spraying the block with an instant freeze aerosol (e.g. dichloro-difluoroethane) is not adequate! The 7% sucrose partially cryoprotects the tissue and reduces damage due to crystal formation. Higher concentrations of sucrose may be used (up to 20%) but may make the sections difficult to handle in the cryostat. Following freezing keep the tissue in the cryostat for a while until it equilibrates with the chamber temperature (12−15°C). Some types of tissue need to be surrounded with a compound such as O.C.T. embedding medium (Tissue-Tek, Miles Laboratories) prior to freezing. The O.C.T. compound helps maintain the shape and integrity of sections during cutting.

(2) Cutting and mounting sections. Most cryostats have an anti-roll plate whose function is to prevent the sections from rolling up as they are cut by the knife. The anti-roll plate is normally the most difficult thing to adjust but it is also the most important. Most difficulties in sectioning are due to poor adjustment of the roll plate. Details of adjustment differ from cryostat to cryostat and so the maker's instruction manual must be referred to. When correctly adjusted cut sections should lie flat and uncurled on the knife. Thaw mount cut sections onto subbed slides (*Table 8*). Keep the slide just warmer than the section such that a wave

Table 8. Preparation of subbed slides.

Mix 1.0 g of gelatin
0.1 g of chrome alum (chromic potassium sulphate)
200 ml of water

Bring the solution slowly to 60°C and leave it to cool to room temperature.
Dip each slide in the solution and leave to drain and dry in a slide box at 37°C.

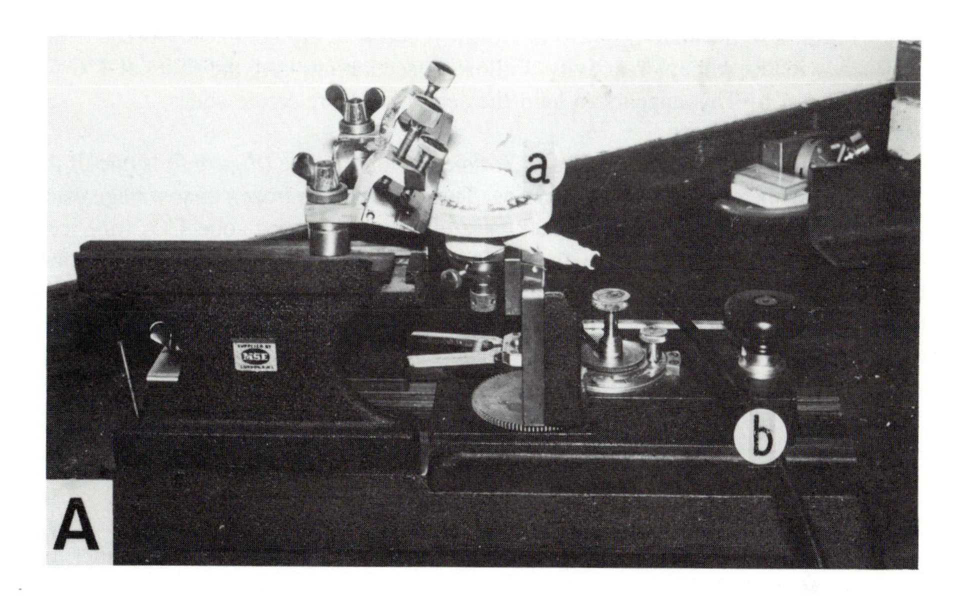

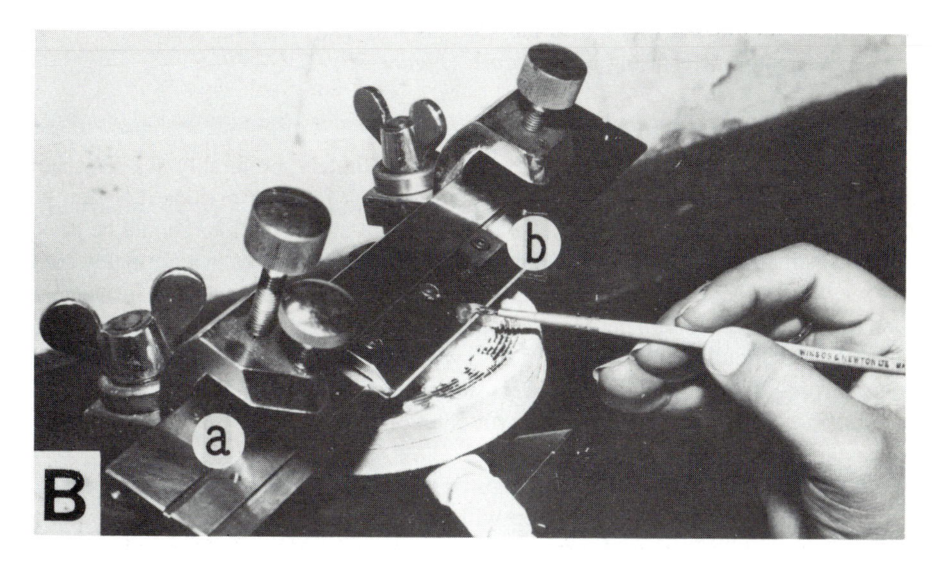

Figure 5. (**A**) MSA sliding microtome. The microtome is fitted with a freezing stage (a) which is attached to a CO_2 cylinder via a flexible tube (b). Following initial freezing the CO_2 is turned off and the tissue kept frozen with dry ice piled onto the freezing stage. (**B**) Close up of a section being cut. As the section is cut it thaws and may be lifted from the knife with a fine paint brush. In this illustration the microtome is fitted with a knife back (a) which takes disposable Feather blades (b).

of melting moves across the section without trapping air bubbles or creating folds in the section. Do not leave either tissue block or cut sections in the cryostat for longer than a few hours. We normally keep the slides on ice in a box outside the cryostat cabinet. Introduce the slide to mount a section and then immediately return it to the slide box. Do not allow sections to dry completely because this may reduce immunoreactivity. Following sectioning, keep the slides at 4°C for about 1 h. This appears to help the sections adhere to the slides.

(iii) *The sliding microtome.* The sliding (or sledge) microtome (*Figure 5*) is one of the simplest (and cheapest) sectioning devices. Tissue blocks are frozen on to a stage using CO_2 gas and then kept frozen by packing around with crushed solid CO_2 (dry ice). The microtome itself (including the knife) is not cooled and is operated at room temperature. With such a device good 20–50 μm sections can be cut easily and quickly. Sections thaw and fold immediately they touch the knife and so the method is suitable only for relatively homogeneous tissue (such as brain) and for methods where sections will be stained free floating. Sectioning is simple as long as the following points are followed.

(1) Freeze tissue blocks as quickly as possible, preferably by CO_2 gas. Cryoprotect tissue to be frozen in 0.1 M phosphate buffer–20% sucrose (pH 7.4).
(2) Once frozen keep the tissue block frozen with crushed dry ice. Good sections may be cut over a wide range of temperatures, but not if the block is either too warm or too cold. If the block is too cold (e.g. the temperature of the dry ice) it will fracture on cutting. If the block is actually allowed to thaw the block will move on the chuck and also the sections will tear. Transfer sections from the knife into buffer using a fine brush (*Figure 5B*).

3.1.3 *Antibodies*

(i) *Storage.* Lyophilized antibodies may be kept indefinitely, preferably in a desiccator at 4°C. Once hydrated, liquid antibodies may be kept for long periods if frozen but repeated freezing and thawing must be avoided. Thus stock solutions should be divided into small aliquots before storage in a −70°C serum freezer or in a normal −20°C freezer. Some domestic freezers have automatic defrost cycles and such instruments should not be used. Certain peroxidase-conjugated second antibodies and some preparations of PAP should not be frozen. Thus, with commercially obtained reagents it is advisable first to check the manufacturers storage instructions. To avoid freezing, some workers store antisera in a 50% glycerol solution at −20°C.

(ii) *Purification.* If monoclonal antibodies or well characterized high affinity sera are being used, an initial purification step will probably not be necessary. However, in other situations purification of sera may be necessary and this can take two forms. Affinity purified antibodies may be produced by elution of antibodies from immune complexes formed with the antigen to be localized. Thus specific antibodies are selected and isolated. Alternatively, unwanted antibodies may be removed by pre-adsorption with possible contaminating antigens. The second of these methods is widely used and will be discussed in more detail below. Many workers do not affinity purify primary antibodies because the sera they are working with contain high affinity antibodies which can be selected for by using the appropriate dilution. However, where low affinity antibodies are being used affinity purification should be employed (*Table 9*).

It should be noted that many commercially available I° and developing antibodies

Table 9. Affinity purification (modified from 36).

1. Couple antigen to Sepharose-48 – CNBr (Pharmacia) according to manufacturers' instructions. Usually 5 – 10 mg of antigen is coupled per gram of dry gel to give a total gel volume of 3.5 ml.
2. Add this volume to 15 ml of antiserum and incubate for 2 – 3 h at room temperature with gentle shaking.
3. Pour serum/gel into a column and wash with 10 mM Tris-buffered saline (TBS) at pH 7.6 until the absorbance of the effluent is zero.
4. Wash the gel with TBS and 1 M NaCl to elute a small peak of non-specific absorbed material.
5. Remove the salt from the gel by washing with TBS.
6. To elute the specifically bound antibody wash with 0.1 M glycine-HCl buffer, neutralize the eluant with 10% Tris-HCl, pH 8.5.
7. After elution, remove the gel from the column and wash with 10^{-2} M HCl (pH 2.0) on a sintered glass filter.
8. Re-equilibrate with TBS with 2×10^{-2} M NaN_3 and store the gel at 4°C.

Table 10. Pre-adsorption using antigen-linked beads.

1. Couple antigen to Sepharose-4B – CNBr (Pharmacia) according to the manufacturers' instructions.
2. Incubate 1 ml of diluted antisera with 10 – 50 µl of gel for 1 h, at 37°C and then at 4°C overnight.
3. Centrifuge and keep the supernatant. A range of antibody – antigen concentrations must be tested in order to determine the concentration range for effective adsorption.

are available either as whole serum, IgG fraction (that is, serum purified by ammonium sulphate precipitation), or affinity-purified serum. Whenever possible, affinity-purified reagents should be obtained.

The second method of antibody purification is the method of pre-adsorption with possible contaminating antigens. This, of course, pre-supposes that one knows what the contaminants are most likely to be. In practice there are two main sources. If the carrier is a hapten, likely contaminants are antibodies to the carrier protein to which the antigen was initially conjugated for the purpose of immunization (see Section 2.1). These are removed by pre-adsorption with the appropriate carrier (e.g. BSA, keyhole limpet haemocyanin, thyroglobulin). The second important source of contamination is low affinity antibodies against unidentified but widespread antigens in the experimental tissue. These are removed by adsorption with tissue powders (see below). When adsorbing I° antibodies care has to be taken to ensure that the tissue powder does not contain the specific antigen under study.

Pre-adsorption with specific cross-reacting antigens should be carried out in a way that will remove the antigen – antibody complex from the serum. Otherwise antibody remaining as part of soluble immunocomplexes in the adsorbed serum may still interfere with immunostaining. The best method is to use antigen attached to a solid support for pre-adsorption and the simplest method is to use CNBr-activated Sepharose beads (*Table 10*). Antigen linked to such beads should also be used when preparing control serum pre-adsorbed with the antigen under study (see Section 4.1)

For soluble adsorption, incubate with excess antigens at 37°C for 1 h and then at 4°C overnight. Centrifuge (2000 *g* for 15 min at 4°C) and use the supernatant. For pre-adsorption with tissue powders, 10 – 20 mg of acetone-extracted tissue powder should be added per 500 µl of diluted serum. Incubate at 37°C for 1 h, centrifuge at 2000 *g* and Millipore filter the supernatant.

(iii). *Buffers.* The most commonly used buffer in immunocytochemistry is phosphate-buffered saline (PBS), pH 7.4 (*Table 11*).

PBS is suitable for most purposes and, especially when first attempting immunocytochemistry, it should not be necessary to try other buffers. However, alternative buffers are described in the literature and some are listed in *Table 12*. Increasing the salt concentration decreases non-specific ionic interactions and also decreases specific binding. Increased salt is also claimed to enhance antibody penetration.

The buffers can be used both for diluting antibodies and for washing sections between incubations. Antisera which are to be kept for some time should have 0.1% azide (NaN_3) included in the buffer. However, azide should not be added to peroxidase-labelled antisera (peroxidase-conjugated second antibody, PAP complex, ABC complex) because it inhibits the enzyme. Thimerosal (0.01%) may be used as an alternative to azide. I° antisera may be collected after an incubation and re-used over a period of several months.

Developing antibodies may bind non-specifically to tissue sections for various reasons (see Section 4.1). To prevent this it is normal to add to the incubation buffers substances which are designed to saturate the non-specific binding sites. Substances used include 1% normal serum of the same species as the developing antibody and 0.5% lambda carrageenan (34). It is also common to add detergent (0.2% Triton X-100) to the incubation buffers. The main effect of the detergent is to break down membranes and thus enhance enzyme penetration. However detergent has the additional property of helping to reduce non-specific hydrophobic interactions.

A typical buffer solution for antibody incubations might be PBS, 0.1% azide. 0.2% Triton X-100, 0.1% normal serum.

Table 11. Phosphate-buffered saline (PBS, 0.01 M phosphate).

For 2 litres mix:
- 100 ml of 0.2 M buffer stock (see *Table 5*)
- 17.52 g of NaCl
- 400 mg of KCl
- 1900 ml of water

Table 12. Alternative buffers for immunocytochemistry.

(a) *0.04 M Tris phosphate (34)*
- 5.0 g of Tris
- 1.2 g of Na_2HPO_4
- 0.25 g of $NaH_2PO_4.2H_2O$
- 7.0 g of NaCl

Make up to 1 litre and adjust the pH to 7.8 with HCl

(b) *0.075 M PBS (32)*

Make 0.5 M phosphate buffer (pH 7.3) by mixing dibasic and monobasic 0.5 M phosphate. Take 300 ml of the buffer, 100 ml of 9% NaCl and dilute to 2000 ml.

(c) *0.5 M Tris saline (33)*

Add 9.0 g of NaCl to 1000 ml of pH 7.6 Trizma® base (Sigma)

3.1.4 *Enzyme labels*

The most commonly used enzymic label in immunocytochemistry is horseradish peroxidase (HRP). The other main label is alkaline phosphatase. However, there are many more types of peroxidase utilizing developing antibodies available than with alkaline phosphatase and there is a wider choice of chromogens to act as a catalyst in the peroxidase reaction. HRP is also widely used in tract-tracing studies (Section 6). The most commonly used chromogen is 3′,3-diaminobenzidine.4 HCl (DAB). DAB acts as an electron donor for peroxidase and, on oxidation, DAB forms a polymer which precipitates at the reaction site. The oxidized DAB (ODAB) product is insoluble in alcohols, is clearly visible and, in addition, chelates with OsO_4 to form a highly electron-dense product. Addition of different heavy metals gives products of different colours. DAB is therefore a very suitable histochemical stain. The most commonly used procedures for localization of peroxidase using DAB as a chromogen are given in *Table 13*.

The pH of the buffer in the standard DAB protocol is not optimal for the enzyme. The reaction is more sensitive when performed at pH 3.3–5.0 but at acid pH various staining artifacts are observed in the tissue sections. Imidazole may be used to increase the sensitivity of the reaction when performed at neutral pH (0.04% DAB, 0.015% H_2O_2 and 0.01 M imidazole in 0.05 M Tris-HCl buffer pH 7.4).

DAB is thought to be carcinogenic and so must be handled with extreme care. Hanker and Yates have developed a non-carcinogenic DAB substitute based on the oxidation of aromatic amines and phenols to form osmiophilic melanin-like polymers (38). The protocol is as follows.

(i) Mix 7.5–15 mg of *p*-phenylenediamine.4HCl/pyrocatechol reagent (Polysciences), 10 ml of 0.1 M Tris buffer (pH 7.6) and 0.1 ml of 1% H_2O_2.

Table 13. Localization of peroxidase, using DAB as chromogen.

(a) *Standard DAB protocol*

1. Just before use make up 0.06% DAB in 0.05 M Tris (pH 7.6).
2. Incubate sections in DAB for 5–10 min and then add hydrogen peroxide solution (H_2O_2) to give a final concentration of 0.01% H_2O_2.
3. Stain sections for 3–15 min depending on the intensity of staining required and stop the reaction by washing with buffer (either 0.1 M phosphate buffer, PBS or Tris).

(b) *Cobalt and nickel-intensified DAB (37)*

1. Make up 1% solutions of cobalt chloride ($CoCl_2.6H_2O$) and nickel ammonium sulphate [$(NH_4)_2SO_4.NiSO_4.6H_2O$].
2. Make up a 0.05% solution of DAB in 6 ml of 0.1 M phosphate buffer (pH 7.3). This solution must be kept agitated whilst adding the cobalt and nickel solutions.
3. Add dropwise 150 μl of cobalt chloride and then 120 μl of nickel ammonium sulphate.
4. Incubate the tissue in this solution for about 15 min, then add 20 μl of 3% hydrogen peroxide.
5. Incubate tissue for a further 15–20 min and wash with phosphate buffer.

(c) *Osmium tetroxide-intensified DAB*

1. Stain the sections lightly as for the regular DAB method.
2. Mount onto slides from Tris buffer.
3. Dehydrate the sections and de-fat them in xylene.
4. Re-hydrate the sections and stain them for a few minutes with 0.2% osmium tetroxide.
5. Wash thoroughly and mount as normal.

Table 14. 4-Cl-naphthol protocol.

1. Dissolve 100 mg of 4-Cl-napthol in 10 ml of ethanol.
2. Add 190 ml of 0.05 M Tris, pH 7.6, and then H_2O_2 to give 0.005% H_2O_2.
3. Stain the sections for 5–15 min and then wash them in Tris buffer.
4. The reaction precipitate is soluble in alcohol and so sections must be mounted in a water-soluble medium such as glycerine jelly.

(ii) Incubate sections in reagent mixture for 5–15 min and then stop the reaction by washing in Tris buffer. The Hanker Yates procedure has not yet been widely used for immunocytochemistry but is used for localizing retrogradely transported HRP (see also below).

Uncontrasted ODAB is brown. For double antigen immunocytochemistry (see Section 5.2) the standard DAB protocol is sometimes combined with DAB contrasted with heavy metals (black) or with the alternative substrates 4-Cl-1-naphthol (grey–blue, *Table 14*) or α-naphthol/pyronin (red–purple).

For retrograde HRP labelling tetramethylbenzidine (TMB), benzidine dihydrochloride and *o*-dianisidine hydrochloride are in common use as substrates. Of these compounds TMB is widely regarded as being the most sensitive but the oxidized TMB reaction product is soluble in alcohols and so great care must be taken during dehydration and coverslipping. None of these substrates are normally used for immunocytochemistry.

3.2 **Immunofluorescence**

3.2.1 *Basic equipment*

Immunofluorescence is normally carried out on 10–20 μm cryostat sections. Thicker sections (cryostat or microtome) have too much background tissue fluorescence. Thus, both a cryostat and a fluorescence microscope are required in order to set up fluorescence immunocytochemistry. If possible, a top opening motor-driven cryostat should be obtained since these are easier to use than the more traditional angled front-opening models. In a fluorescence microscope the following features are important.

(i) The brightness and specificity of the final immunofluorescence image depends on factors such as the emission spectrum of the light source, the degree of separation in the final image between excitation and emission spectra and the amount of emitted fluorescence captured by the microscope and transmitted to the eye pieces. Generally these requirements are met by using a 50 W mercury high pressure lamp, correct combination of primary and secondary filters and high numerical aperture objectives and condenser. Most modern fluorescence microscopes use incident light excitation (epi-fluorescence) since this gives a brighter image than transmitted light excitation, except at lower magnification. With epi-fluorescence the image brightness depends on the fourth power of the objective aperture and so fluorite or planapochromatic objectives should always be used and immersion objectives should be used if possible.

(ii) Most modern epi-fluorescent illuminators use filter blocks which contain combinations of primary and secondary filters. Filter blocks appropriate for specific fluorochromes can be bought and are easily interchangeable. The most commonly used fluorochrome in immunocytochemistry is FITC which fluoresces green.

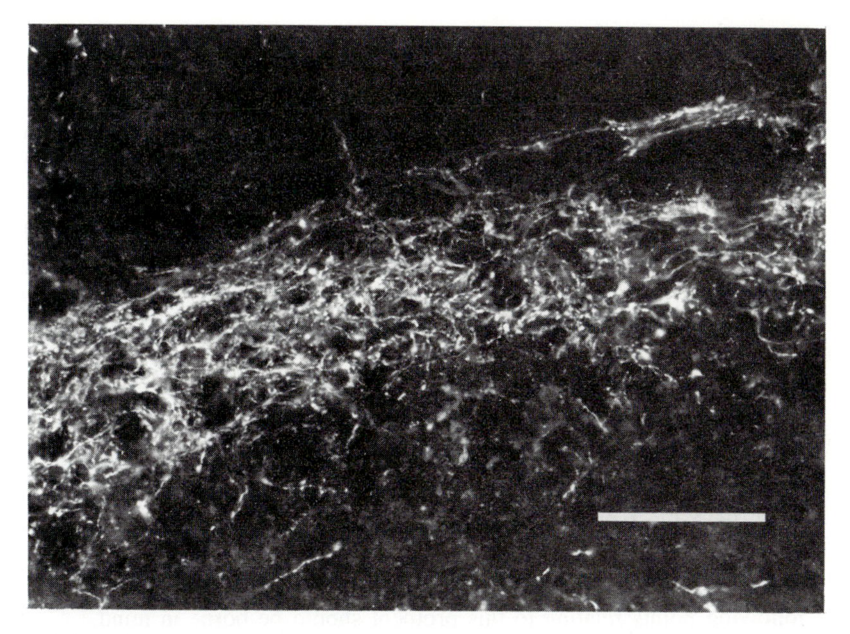

Figure 6. Example of an immunofluorescence-stained cryostat section. Substance P in the spinal trigeminal nucleus of the rat. Monoclonal I° antibody. The staining is seen in a band (the substantia gelatinosa) and consists of fine varicose fibres. Scale bar = 100 μm.

Red fluorescing TRITC is frequently used in combination with FITC in double immunostaining (see Section 5.1) but generally gives weaker fluorescence than FITC and is not recommended for use alone.

(iii) Accurate photography of fluorescence preparations requires a camera with a high sensitivity exposure meter. High speed films such as Kodak Tri X or Ilford FP4 and HP5 should be used for black and white photography, and the films may need to be pushed one or two stops during development. Exposure times of 30 sec to 2 min are often necessary.

With all types of immunostaining it is imperative that sections are not allowed to dry out once staining has been commenced. Staining of slide-mounted sections such as cryostat sections is normally carried out by applying drops of antibody to horizontally held slides incubated in a humid environment. This can be accomplished in Petri dishes or in containers such as immunoelectrophoresis tanks.

3.2.2 *Immunofluorescence protocol*

The reader will have gathered from Sections 2 and 3.1 that a very large number of variations are possible in immunocytochemistry. The protocol given here is therefore an example of only one possibility and involves the use of a monoclonal antibody to localize the neuropeptide substance P in the rat spinal trigeminal nucleus (*Figure 6*). Important comments concerning possible variations in the protocol are placed together at the end.

(i) Perfuse a rat with 4% paraformaldehyde, dissect out the brain and leave in fixative for a further 2.5 h.

(ii) Transfer to 0.1 M phosphate/7% sucrose buffer (pH 7.4) and leave overnight.
(iii) Dissect out the brain stem and freeze onto a cryostat chuck with gaseous CO_2.
(iv) Cut 10 μm sections onto subbed slides.
(v) Store the slides at 4°C for 1 h (see Section 3.1.2), then wash the sections in PBS 0.2% Triton X-100 solution (PBS TX) by incubating for 30 min in a Coplin jar (or equivalent).
(vi) Dry the back of each slide and the front of each slide around the area with the tissue sections.
(vii) Apply a few drops (100 μl/slide) of 1:50 monoclonal rat anti-substance P in PBS TX and incubate in humid chambers overnight at 4°C.
(viii) Dilute FITC rabbit anti-rat IgG in PBS TX (1:20) and incubate for 1 h at 37°C with rat brain powder.
(ix) Centrifuge and Millipore filter the supernatant.
(x) Wash the slides in PBS for 1 h at room temperature.
(xi) Apply drops of FITC anti-rat IgG to the slides and incubate in humid chambers in the dark for 1 h at 37°C.
(xii) Wash the sections in PBS and mount in PBS/glycerol (2:1).
(xiii) Store the slides in the dark at 4°C.

The following points relating to this protocol should be borne in mind.

(1) The dilution of I° antibody in this protocol is low (i.e. very concentrated). For immunofluorescence I° antibodies must be used at lower dilutions than for PAP and (for both fluorescence and PAP) monoclonal antibodies (provided as tissue culture supernatant) must be used at much lower dilution than polyclonal antisera. Typical dilutions must be:

Monoclonal using immunofluorescence	1× – 50×
Monoclonal using PAP	10× – 500×
Serum using immunofluorescence	50× – 500×
Serum using PAP	500× – 10 000×

(2) Incubations in I° antibody may be varied from 1 h at 37°C to 10 days at 4°C. With longer incubations the antibody may be used at higher dilution and this is normally associated with greater specificity. However, slide-mounted sections incubated for much longer than a day run a severe risk of drying out.
(3) Exposure of sections to u.v. light required for specimen observation and photography leads to loss of fluorescence. This loss can be reduced by adding 25 mg/ml 1,4 diazobicyclo(3-2-2)octane to the mounting medium. However, this should be done with care because excessive exposure of sections to such additives can itself lead to loss of fluorescence.

3.3 Peroxidase–anti-peroxidase (PAP) procedure

3.3.1 *Basic equipment*

The PAP procedure is applicable to almost any type of material and so any of the sectioning procedures outlined in Section 3.1.2 may be used. Unlike immunofluorescence (Section 3.2.1), PAP staining may be set up without major outlay on equipment. The simplest approach is to stain frozen tissue free-floating and cut on a basic sliding microtome or simple freezing microtome. If frozen tissue must be directly mounted on slides before processing, a cryostat will be required (see Section 3.2.1). A Vibratome® may be needed for cutting unembedded tissue without freezing. For plastic or paraffin-embedded material, routine sectioning methods may be used which will not be covered in this chapter.

The final experimental result will not be better than the microscope used to examine the sections so, as with immunofluorescence, a microscope must be bought which is adequate for the task. This does not necessarily imply great expense but the following basic points should be noted.

(i) The most important parts of any microscope are the optics. High numerical aperture objectives (fluorite or planapochromatic) should be obtained along with equivalent quality condenser and eyepieces.
(ii) It is often advantageous to be able to view the PAP-stained sections using darkfield illumination (see Section 3.3.2 below). This can most conveniently be achieved by using a universal condenser since this allows easy change-over between brightfield and darkfield illumination. For darkfield illumination a high power light source (50 or 100 W tungsten halogen) is recommended.
(iii) The simplest photographic system is to have a SLR camera back attached to a trinocular tube. For darkfield photomicrography, a more sensitive metering system will be required.

3.3.2 *PAP protocol*

No one 'PAP protocol' exists because of the large variety of different types of preparation to which it applies (whole mount, cryostat, free floating frozen or vibratome sections, paraffin sections) and also because of the large number of variables in immunocytochemistry (Section 3.1). So the protocol given here (staining in free-floating Vibratome® sections using a polyclonal 5-HT serum, *Figure 7*) is just one specific example and has been chosen to contrast as much as possible with the immunofluorescence protocol given in Section 3.2.2. Other appropriate PAP protocols (e.g. staining of cryostat sections with a monoclonal antibody) may be constructed by comparison of the two protocols (Sections 3.2.2 and 3.3.2) and reference to Sections 2.3 and 3.1.

(i) Perfuse a rat with 4% paraformaldehyde, dissect out the brain and leave in fixative for a further 2.5 h.
(ii) Transfer to 0.1 M phosphate/5% sucrose buffer (pH 7.4)and leave overnight.
(iii) Cut out a small block of tissue containing the spinal trigeminal nucleus and glue on to a Vibratome® chuck using superglue.
(iv) Cut 40 μm sections into 0.1 M phosphate buffer (see Section 3.1.2).
(v) Transfer the sections to PBS in disposable plastic test tubes or scintillation vials kept on ice. The incubations may be carried out in these vessels and solutions added or removed with Pasteur pipettes.
(vi) Incubate the sections for 1 h in 10% normal goat serum in PBS 0.2% Triton X-100, 0.1% azide solution (PBS TXa).
(vii) Wash in PBS for 1 h, incubate in 1:5000 rabbit anti-5-HT antibody for 4 days at 4°C. The antibody is diluted in PBS TXa, 0.5% lambda carrageenan.
(viii) After the incubation wash the sections for several hours in PBS.
(ix) Incubate for 2 h in 1:50 goat anti-rabbit IgG in PBS TXa and wash for 1 h.

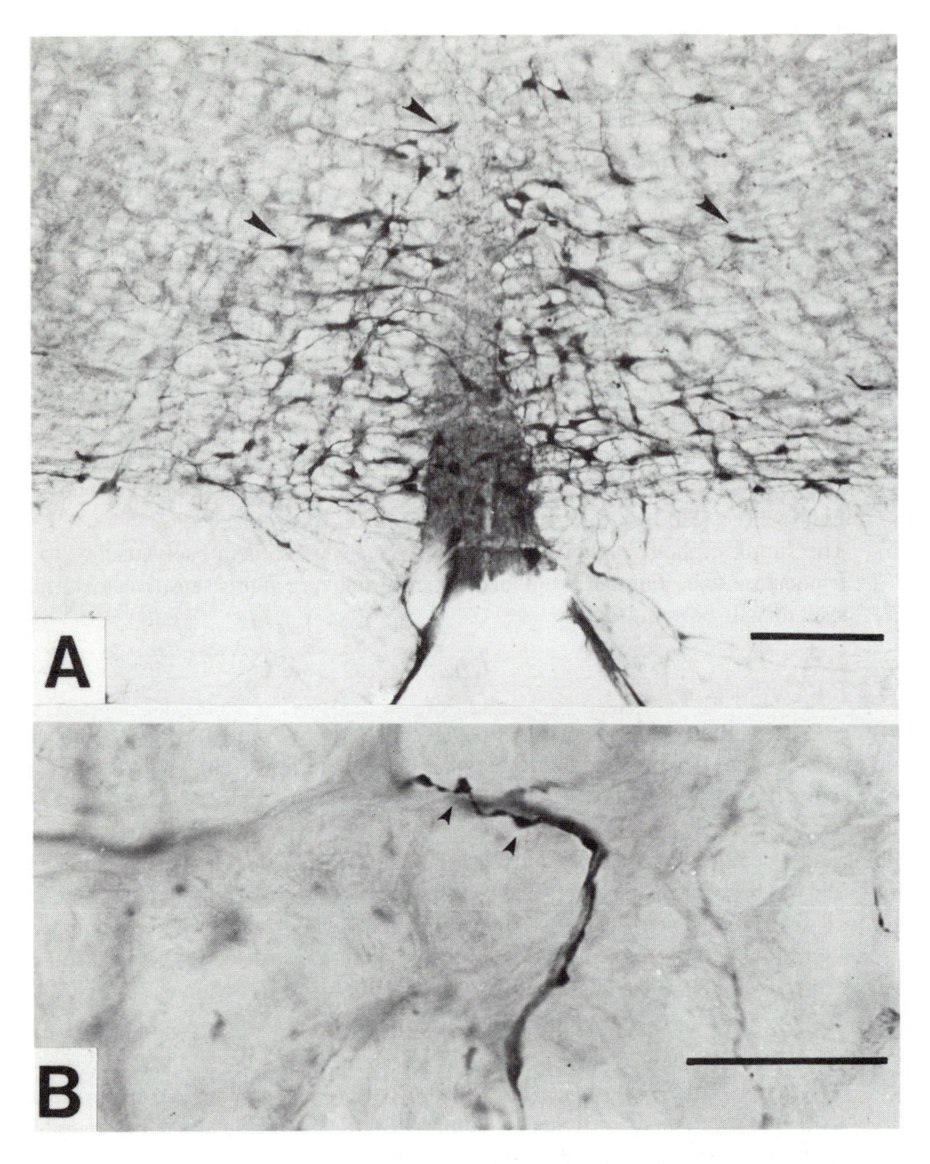

Figure 7. Example of a PAP-stained vibratome section. 5-HT in the nucleus raphe magnus of the rat. Polyclonal I° antiserum. (**A**) The staining consists of cell bodies (arrow heads) and also fine fibres which are too small to see at this magnification. Scale bar = 100 μm. (**B**) Higher magnification micrograph of **A** taken with 100 × oil immersion objective. An immunoreactive fibre with numerous varicosities (arrow heads) is seen contacting a lightly stained dendrite. Scale bar = 10 μm.

(x) Incubate for 2 h in 1:200 rabbit PAP in PBS TX then wash for 1 h.

(xi) Incubate the sections for 10 min in 0.06% DAB in 0.05 M Tris buffer, pH 7.6, and then add H_2O_2 to give 0.01% H_2O_2 (see Section 3.1.4).

(xii) Stain for 5–20 min, wash the sections in PBS overnight and then mount on gelatin chrome alum-subbed slides. Intensify some sections with OsO_4 (Section 3.1.4).

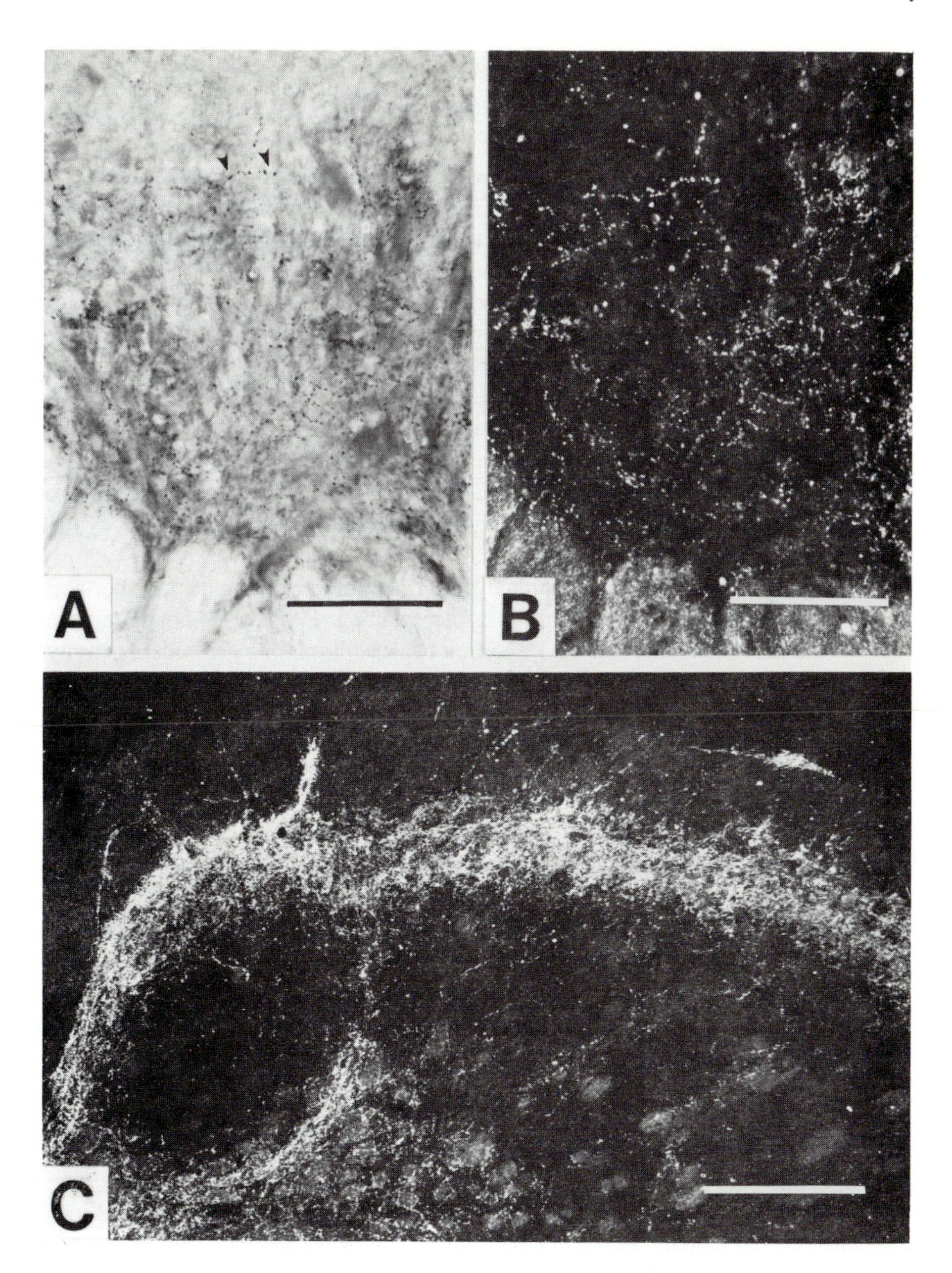

Figure 8. Use of darkfield illumination to view PAP-immunostained material. **A, B**. An identical field of 5-HT immunoreactive fibres in the ventral horn of the rat spinal cord viewed using brightfield (**A**) or darkfield (**B**) illumination. Fibres (arrow heads) generally show up more clearly using darkfield. This is partly because of the enhanced contrast of the bright fibres against a dark background and partly because darkfield illumination has greater depth of field than brightfield illumination. Scale bars = 100 μm. (**C**) Example of a low magnification darkfield micrograph. Enkephalin-immunoreactive fibres in the spinal trigeminal nucleus of the rat. Staining is seen mainly in a band in the region of the substantia gelatinosa. Compare with *Figure 6*. Under darkfield illumination, myelinated fibre tracts appear as light patches against a dark background. Scale bar = 250 μm.

(xiii) Dehydrate and coverslip. Observe using brightfield and darkfield illumination (*Figure 8*).

The following points concerning this protocol should be noted.

(1) The antibody used here is directed against a small transmitter antigen and is of the type discussed in Section 2.1.1. Tissue fixation conditions are very important and addition of glutaraldehyde to the fixative is likely to alter the staining specificity of the antibody.

(2) Since a good quality antiserum is being used in conjunction with the PAP method a high dilution of the I° antibody may be used.

(3) Long incubations in the I° antibody increase the sensitivity, however sections must be washed thoroughly after long incubations. If possible, sections should be agitated during incubation. The lamba carrageenan in the I° antibody serves both to decrease non-specific staining and to make the solution more viscous and this has the effect of keeping the sections floating. The 10% normal goat serum decreases non-specific binding by the goat anti-rabbit link antibody (see Section 4.1).

(4) The 10% normal serum, I° antibody and link antibody are collected after each incubation and may be re-used for further staining. PAP should not be re-used. Azide is omitted from the PAP solution because it inhibits peroxidase activity.

(5) The staining may be intensified by modifying the DAB protocol in various ways (Section 3.1.4, *Table 13*). Alternatively the PAP procedure may be made more sensitive by repeating the incubations in the link antibody and PAP reagent before carrying out the DAB staining.

4. SPECIFICITY

The antibody–antigen interaction is highly specific in the sense that each antibody recognizes a small number of closely related antigenic determinants. Since this specific molecular interaction is the basis of immunocytochemistry, immunocytochemistry itself has the potential of being a highly specific and sensitive procedure. However, an immunocytochemical staining sequence is far removed from the theoretical situation of the behaviour of a single antibody and the establishment and interpretation of specificity is a crucial problem. Two different types of specificity are usually defined (40, 41). To establish serum specificity one must show that the staining is due to the I° antibodies reacting only with the compound to which they are meant to be specific and not due to interactions with other related or unrelated compounds. Method specificity refers to the identification of all the other types of staining artifact which arise caused by mechanisms other than the interaction between I° antibodies and tissue antigens. These two different areas of specificity will be dealt with in turn.

4.1 **Method specificity**

Method specificity will be dealt with first because it is far easier to establish than serum specificity. Omission of the I° antibody (or antiserum) or use of pre-immune serum should give no staining, thus indicating that all staining is due to the I° antibody. An alternative approach is to use a I° antibody dilution series and show that staining disappears with increasing dilution. In practice such controls never completely abolish staining but they do allow one to establish which components of the staining are due to non-specific interactions by the developing antibodies. Some causes of non-specific staining and methods used to reduce such staining are listed below.

(i) Developing antisera are normally mixed populations of antibodies of differing affinity and specificity. Antibodies may be present which bind to tissue components or to other antigenic sites in addition to those on the I° antibodies. Such cross-reactions can be greatly reduced by using monoclonal or affinity-purified developing antibodies (*Table 9*, *Figure 9*). In addition, developing antibodies may be pre-adsorbed with acetone-extracted tissue powders (Section 3.1.3).

(ii) Tissue sections may possess binding sites for I° and developing antibodies. Free aldehyde groups remaining after fixation can be reduced by treating sections with 0.05% sodium borohydride in PBS and sections should always be washed well before commencing immunostaining in order to remove unbound excess fixative. Other binding sites may be saturated by incubation with substances such as BSA, gelatin (e.g. λ carageenan) or normal serum of the same species as the developing antibody. These may be applied in a pre-incubation step (e.g. 10% normal serum) or may be added to the primary and developing antibodies (e.g. 1% normal serum, 0.5% λ carageenan). Non-specific ionic and hydrophobic interactions may be reduced by use of appropriate immunocytochemical buffers (see Section 3.1.3).

(iii) In peroxidase-based methods endogenous peroxidase and pseudoperoxidase activity can be abolished by pre-incubation with 0.1% phenylhydrazine (pH 7.0, 1 h at 37°C) or with 1% hydrogen peroxide in methanol (10 min).

4.2 Serum specificity

Serum specificity is much more difficult to establish than method specificity and it is generally agreed that it is impossible to establish specificity of staining using only immunocytochemical controls. This is because all tests of serum specificity are tests of the antibody and show what the antibody is *capable* of binding to and not what the antibody in an immunocytochemical study is *actually* binding to. Thus, cautious workers normally refer to 'antigen X-like' immunoreactivity. This is a particularly difficult problem in the field of neuropeptide research where there are large families of related peptides with sequences of homologous amino acids. There are examples in the literature of well controlled studies describing the distribution of a peptide immunoreactivity which subsequently proves to be due to a different molecule. In some situations the identity of the antigen giving rise to immunostaining is still unknown and disputed. Thus all immunocytochemical studies need to be complemented by biochemical isolation and characterization studies of the relevant tissues. In addition, physiological controls should be performed to establish that patterns of staining reflect the known distribution and relative levels of antigen. However, there are a number of commonly used immunocytochemical controls for serum specificity and these are listed below.

(i) As with developing antibodies [point (i) above], any I° antiserum consists of a mixed population of antibodies of different affinity and specificity. Antisera should be used at the highest possible dilution in order to select high affinity antibodies.

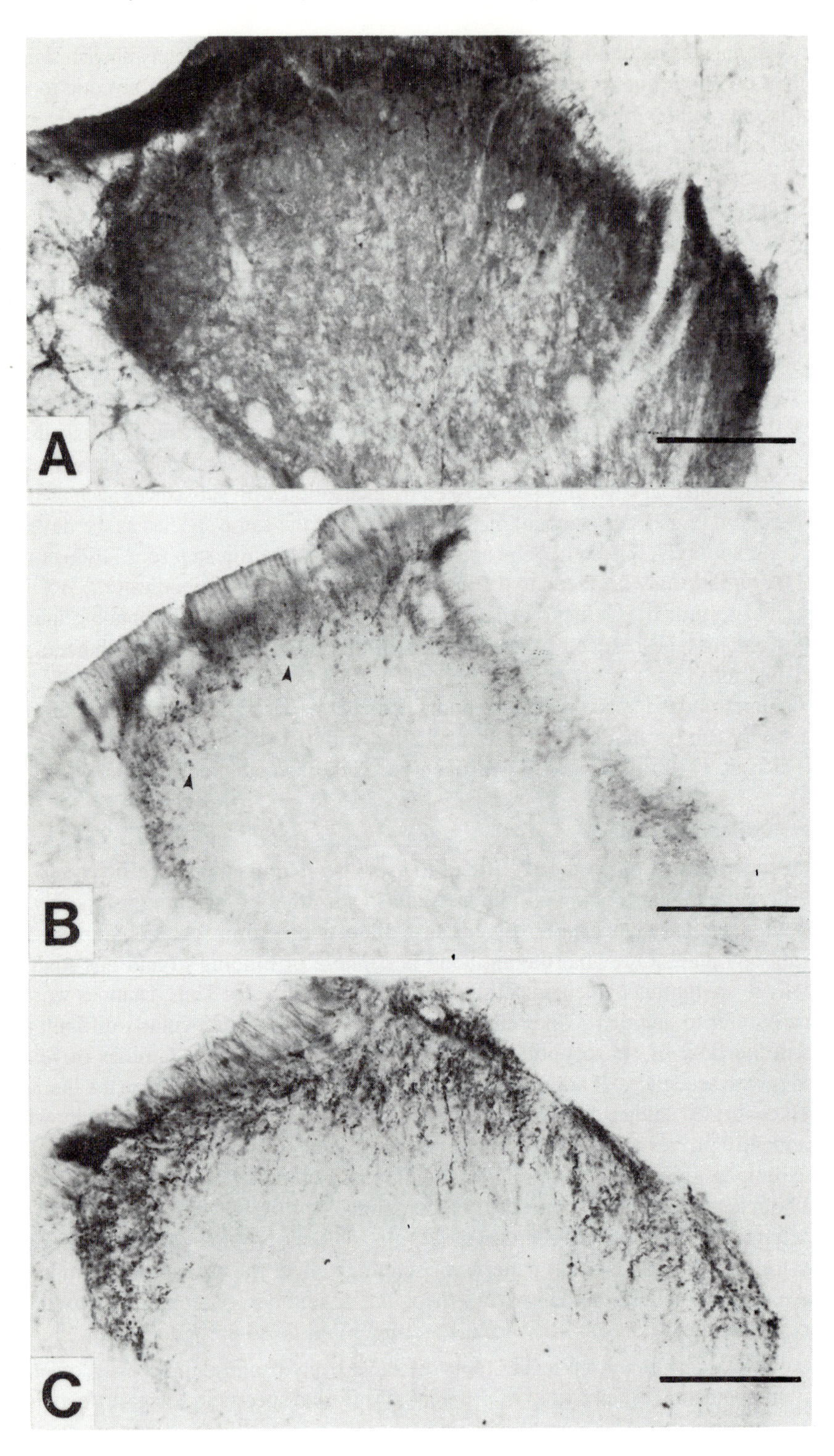
A
B
C

If necessary, the staining obtained may then be intensified in various ways (see e.g. Sections 3.1.4 and 3.3.2). Monoclonal and affinity-purified antibodies should be used and all I° reagents should be pre-adsorbed with possible cross-reacting antigens (Section 3.1.3, *Table 10*). Possible adsorbents include tissue powders (but only if the specific antigen to be localized has first been removed), carrier proteins used in immunization and compounds with sequences or structures related to the antigen to be localized.

(ii) Many workers characterize antisera by RIA. This is useful in terms of indicating the general specificity range of the antisera but the results may not be relevant for immunocytochemistry. Antisera are used at much lower dilution (higher concentration) for immunocytochemistry than for RIA and may therefore bind to antigens to which they actually have rather low affinity. In addition, tissue fixation alters molecular structure so that antibody recognition of fixed antigens may be quite different to that of antigens in liquid phase (i.e. RIA) (see Section 2.1.1). Some of the drawbacks of RIA can be overcome by the use of model tissue systems. Antigens may be coupled to supports such as Sepharose beads, gelatin gels, filter paper or nitrocellulose paper and these model systems can then be immunostained (*Table 15*). Antibodies may then be characterized to show which antigens they will or will not stain and how such staining is modified by different fixation schedules.

(iii) The most commonly used control is to establish that staining is abolished following pre-adsorption of the antibody with the antigen to be localized (*Table 10*). Staining should not be affected by pre-adsorption with other related compounds. This should be carried out for each tissue under study.

The three controls above provide a good basis for assessing serum specificity in an immunocytochemical experiment. However none of the tests are conclusive. The adsorption test, for example, can be misleading. Staining by an antibody which cross-reacts with two compounds (A and B) will be abolished by pre-incubation with either substance (A or B). The test does not indicate which of the two compounds the antibody actually binds to in the tissue. Ultimately all controls must also be supported by chromatographic analysis of tissue extracts in order to establish the identity of particular immunoreactive tissue components.

5. DOUBLE IMMUNOCYTOCHEMICAL LABELLING

A large number of procedures are available for staining two different antigens in light microscopic sections. Such double staining procedures have various applications in

Figure 9. Effect of antibody purity and dilution on specific and non-specific staining. Substance P immunostaining in the dorsal horn of the rat spinal cord using a monoclonal substance P antibody. PAP procedure. (**A**) 1:100 dilution substance P antibody, 1:500 dilution commercial rat PAP antiserum. (**B**) 1:500 dilution substance P antibody, 1:500 dilution commercial rat PAP antiserum. (**C**) 1:500 dilution substance P antibody, 1:5 dilution monoclonal rat PAP. Increasing the dilution of the I° antibody from 1:100 to 1:500 (**A** and **B**) decreases the background staining and allows individual immunostained fibres to be seen (arrow heads in **B**). However the specific staining is not very strong. The specific staining is much stronger in **C** because a monoclonal PAP has been used instead of normal PAP produced from polyclonal antiserum. The monoclonal reagent may be used at low dilution without markedly increasing the background non-specific staining. Scale bars = 100 μm.

Table 15. Immunocytochemical model tissue systems.

(a) *Filter paper (42)*
1. Dissolve antigen at a range of concentrations (100–0.01 μM) in double distilled water (or 5 mM HCl for basic peptides) and spot onto strips of Whatman No. 1 filter paper (10 μl per spot).
2. Following drying at room temperature, place the papers in a 1 litre jar containing 5–7 g of paraformaldehyde powder at the bottom.
3. Incubate the closed jar for 1 h at 80°C.
4. Remove the fixed papers and stain using any standard immunocytochemical protocol.

(b) *Gelatin gel (43)*
1. Dissolve gelatin in distilled water (100 mg/ml) by gentle heating to 60°C.
2. Dissolve antigen in distilled water at a range of concentrations (2000–0.2 μM) and mix with equal volumes of the gelatin solution.
3. Immediately transfer 300 μl aliquots to chilled plastic cups (1 h at 4°C).
4. After gelation, remove the gels from the cups and incubate in 6 ml of fixative (2 h at 4°C).
5. After fixation wash the gels in phosphate buffer for at least 18 h.
6. Cut and stain using any standard immunocytochemical protocol.

chemical neuroanatomy but the most widely used ones are for the demonstration of co-existence and for analysis of interactions between fibres stained for one of the antigens and cell bodies stained for the other. Co-existence in cell bodies (but not terminals) may be shown by staining serial sections for different antigens but it may be more convenient to carry out double staining on single sections. Interactions between different fibres and cell bodies may be postulated on the basis of separate single staining studies but may only be conclusively shown by double staining. To show synaptic interactions such double staining must be carried out at EM level but a double staining LM study is an important and necessary first step.

Many double immunocytochemical staining procedures involve the differential localization of two different I° antibodies by using developing antibodies which have different and distinguishable labels. The two commonest approaches are to use indirect immunofluorescence with two different fluorochromes (usually FITC and TRITC) or to use peroxidase-labelled developing antibodies localized using different chromogens. These two approaches are described in Sections 5.1 and 5.2 and then some remaining double labelling techniques are listed in Section 5.3. The first double staining protocols to be developed involved the combination of two direct labelled procedures and this approach may still be used successfully (see Section 5.3). However, most workers are not in a position to produce directly labelled I° antibodies and so indirect procedures are mainly covered below.

With all indirect double immunocytochemical staining protocols very careful controls have to be carried out to ensure that the antibodies used to stain the two different antigens do not cross-react with one another. With directly labelled I° antibodies this is not a problem but if indirect techniques are being used cross-reactivity is likely to be a major difficulty. Possible solutions to this problem include the use of I° antibodies raised in different species and elution or masking of the first set of antibodies before addition of further antibodies. All these different approaches are used in the various protocols listed below and they are referred to in the appropriate sections. However, it should be noted that none of the double staining techniques are ideal and all have to be performed under carefully controlled and defined conditions.

Table 16. Filter combinations for dual colour immunofluorescence.

Haaijman (44) recommends the following combinations for selective viewing of FITC and TRITC.

FITC	Excitation: 2 mm GG 450 or GG 475 + 2 × KP 490
	Additional: 2–4 mm BG 38
	Beam splitting mirror: TK 490
	Emission: LP 520 + KP 560 (equivalent to BP 525)
TRITC	Excitation: LP 520 + KP 560
	Additional: 1–2 mm BG 36
	Beam splitting mirror: TK 580
	Emission: LP 580

On a Leitz epi-fluorescence illuminator we (45) have successfully used filter blocks L2 (FITC) and N2 (TRITC):

FITC	Excitation: BP 450–490
	Beam splitting mirror: RKP 510
	Emission: BP 525/20
TRITC	Excitation: BP 530–560
	Beam splitting mirror: RKP 580
	Emission: LP 580

5.1 **Dual colour indirect immunofluorescence**

Dual colour indirect immunofluorescence may be carried out if two I° antibodies are available which have been raised in different species. Then a staining sequence such as the following may be applied:

rabbit anti-A (appropriate dilution)
1:20 FITC-labelled goat anti-rabbit IgG
rat anti-B (appropriate dilution)
TRITC-labelled goat anti-rat IgG

FITC and TRITC may be viewed selectively by using special combinations of filters (*Table 16*). However, whatever filter combination is used it is important to test that it is selective under the experimental conditions being used. Generally fluorescence yield decreases with increasing selectivity of filter combinations. Thus the combinations listed in *Table 16* are each a compromise between yield and selectivity and should not be regarded as being completely specific for a particular fluorochrome. The following points should also be noted.

(i) For colour photography a fast daylight film should be used such as Kodak Ektachrome 400 push processed to ASA 800 or 1600.
(ii) Commercially available second antibodies may not be completely species specific and should be tested for unwanted cross-reactivities such as between rabbit and rat. Species-specific antibodies may be produced by affinity purification (*Table 9*).

5.2 **Double immunoperoxidase labelling** (46)

5.2.1 *Double staining with antibody removal*

The indirect labelling peroxidase or PAP methods give an insoluble precipitate which marks the antibody binding site. Thus, for double staining, the first sequence of an-

Table 17. Antibody elution procedures.

1. Acid elution
   Wash sections for 1 h with several changes of glycine-HCl buffer, 0.2 M, pH 2.2. The length of time in the acid buffer that is required for complete elution has to be determined empirically for each antibody. Best results are obtained by staining the first set of antibodies with DAB and the second set with 4-Cl-naphthol or α-napthol pyronin.
2. Acid permanganate elution (48)
   Wash section for 1 min in 0.15 M $KMnO_4$, 0.01 M H_2SO_4, pH 1.8. If necessary the sections may be decolourized in 0.5% $Na_2S_2O_5$. DAB and 4-Cl-naphthol may be used as chromogens. If the DAB is used first, the $KMnO_4-H_2SO_4$ oxidation and subsequent $Na_2S_2SO_5$ reduction turn the ODAB precipitate from brown to yellow. Acid permanganate provides more complete elution than glycine HCl, but may produce more loss of antigenicity.
3. Elution by electrophoresis (46)
   Antibodies may be removed by electrophoresis at 20 V/cm in 0.05 M glycine buffer (pH 2.2) containing 30% dimethyl formamide. This may most conveniently be done in a horizontal electrophoresis apparatus. If complete elution is not obtained a higher voltage may be used but then cooling is necessary.

tibodies may be removed completely before commencing the second set of staining. This avoids the problems of cross-reactivity referred to above. Developing antibodies and certain low affinity I° antibodies may be removed using acid buffer but this treatment may not remove high affinity antibodies and, in addition, the acid denatures certain tissue antigens leading to a loss of immunoreactivity. Therefore alternative elution procedures should be tested, including acid permanganate oxidation and electrophoresis (*Table 17*).

5.2.2 *Double staining without antibody removal*

Sternberger and Joseph (49) have reported that dual colour staining may be carried out without antibody elution using two PAP sequences. However, this is only possible if the antibodies and peroxidase activity of the first set of reagents are blocked before addition of the second set of antibodies. This appears to occur if DAB at quite high concentration (at least 0.05%) is used as first chromogen. The ODAB precipitate appears to prevent access by the second set of antibodies to the first set of reagents. The following sequences of chromogens have been used successfully by various authors:

DAB followed by 4-Cl-naphthol (*Table 14*)
DAB followed by α-naphthol pyronin
DAB + Co and/or Ni followed by DAB (*Table 13*)

The absence of cross-reactivity also depends on using the first I° antibody at high concentration in order to give a very intense reaction deposit at the first antigen site. Thus the first sequence I° antibody may have to be used at lower dilution for double staining than for single staining and the specificity of the double staining may vary with the concentration of tissue antigen.

5.3 **Other double labelling procedures**

The protocols listed in Sections 5.1 and 5.2 are the most widely used procedures for LM double immunolabelling. In addition, PAP may be combined with immunofluorescence (50) and with an alkaline phosphatase anti-alkaline phosphatase pro-

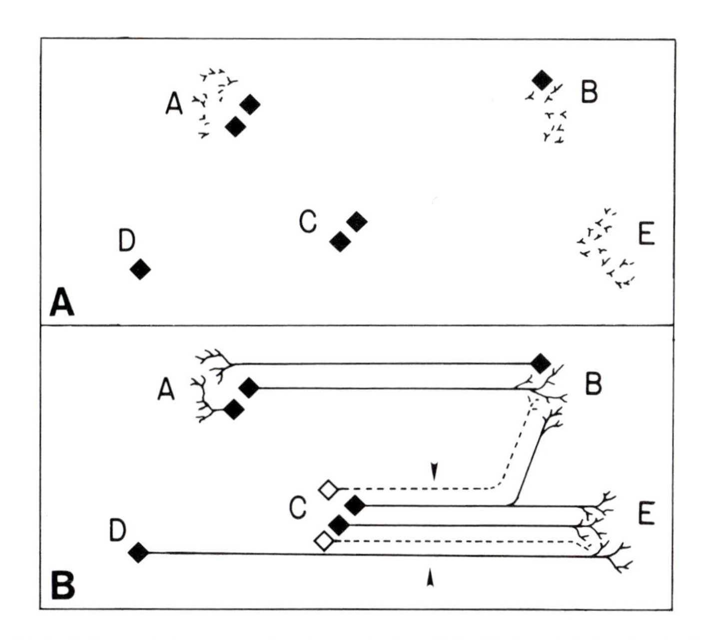

Figure 10. Methods for exploring neuronal pathways in the CNS. (**A**) A model situation is shown of five CNS nuclei (A–E) containing cell bodies (diamonds) and/or nerve terminals. Some antigens (e.g. most neuropeptides) are concentrated in terminals and only appear in cell bodies after axonal transport has been inhibited (e.g. with colchicine). A large number of different pathways could connect the different nuclei. (**B**) One possible arrangement of pathways is shown. Nuclei A and B are connected reciprocally and in addition A contains immunoreactive interneurons. C contains both immunoreactive and non-immunoreactive cells (hollow diamonds) which project via a fibre bundle (arrow heads) to B and E. D projects to E via the same bundle. To establish that such a set of pathways exist, the following approaches could be used: (a) The fibre bundle could be cut with a small stereotaxically-positioned knife (arrow heads). A loss of terminal staining in E might be detected but it would be difficult to establish whether the terminals were derived from C or D. A loss of terminals in B might not be detected because of the intact contribution from A. Specific neurotoxins could be used to make lesions but their use is open to similar interpretative problems. (b) A retrogradely transported tracer could be injected into E and this would label cell bodies in C and D. Combining the retrograde transport with immunocytochemistry would establish that C contains both immunoreactive and non-immunoreactive projection neurons. If the tracer spread by diffusion into B, cells in A would be retrogradely labelled and falsely interpreted as having a projection to E. An anterogradely transported tracer injected into B would label terminals in A. With existing techniques it would be very difficult unequivocally to establish that some of the cells in A are interneurones.

cedure (51). For EM double labelling a signal is required which can be easily distinguished from the peroxidase-based ODAB precipitate and one solution is to use particulate labels. Suitable labels include ferritin, radioactivity (visualized using autoradiography) and colloidal gold (45, 52). It is also possible to combine immunocytochemistry with other specific staining methods such as cholinesterase or catecholamine histochemistry, receptor autoradiography and transmitter uptake autoradiography. Original references to these various procedures can be found in reviews 53–56.

6. IMMUNOCYTOCHEMISTRY AND TRACT-TRACING PROCEDURES

Routine light microscopic immunocytochemistry can be used to map the distribution of substances in the nervous system but in order to localize a substance in a specific pathway more complex staining procedures are required. For example, in the simple situation of a pathway with cell bodies located in area A and a projection to area B, it is necessary to demonstrate an anatomical link between immunostained somata in A and terminals in B (*Figure 10*). Identification of collateral projections and the axonal arborization of local circuit neurones raises additional more complex problems. In rare situations immunostaining may be obtained throughout the cell body, dendrites and axon of a neurone. This occurs with certain cell surface antigens and certain diffusable intracellular antigens. In such cases it may be possible to trace a group of axons from a nucleus to their target areas and this has been achieved in the case of some catecholamine-containing pathways. However this is difficult even for simple pathways and is not possible at all if the antigen can only be localized in restricted parts of neurones. For example, many transmitters and transmitter markers are highly concentrated in terminals but present in very low quantities in pre-terminal axons. In such situations more complex mapping procedures must be used. It may be necessary first to build up concentrations in cell bodies in order to identify the possible cells of origin of the terminals. For many transmitter markers this can be done by blocking axonal transport with colchicine.

It is necessary to establish an anatomical link between cell bodies and terminals. There are a large number of ways in which this can be achieved but they fall into two broad categories. In the first category, the putative pathway is lesioned and loss of terminal staining indicates that the terminals were part of the destroyed pathway. Stereotaxically placed knife cuts can be used to lesion certain pathways, or cells of origin can be destroyed using electrolytic lesions or specific neurotoxins (57). Suitable toxins include kainic acid (cell bodies), 6-OH dopamine (catecholamine cells), 5,7-dihydroxytryptamine (indoleamine cells) and capsaicin (certain I° afferent neurones). However it is often difficult to control the extent and specificity of such lesions or to detect proportionately small losses of terminals. Thus these lesioning approaches have generally been replaced by the second category of tracing methods which involve the injection and subsequent localization of anterogradely or retrogradely transported markers. For example, a retrogradely transported marker injected into a terminal area will be transported to and label the cells of origin.

In the following sections (6.1, 6.2, 6.3) three immunocytochemically based tract-tracing procedures are described. Many of the most important tract-tracing procedures are not immunocytochemically based although they may be combined with immunocytochemistry (Sections 6.2 and 6.3). For details of these other techniques the reader is referred to other recent reviews (39, 55).

6.1 **Immunocytochemical localization of lectins**

Lectins are plant proteins which bind with very high affinity to specific glycoproteins and glycolipids. Certain lectins bind to neurones and are then taken up by receptor-mediated (adsorptive) endocytosis. They are then transported retrogradely and/or anterogradely within the cell. The two most widely used lectins for tract-tracing are *Triticum vulgaris* agglutinin (wheat germ agglutinin, WGA) and *Phaseolus vulgaris* leucoagglutinin (red kidney bean, also referred to as phytohaemagglutinin-L, PHA-L). WGA is used for both retrograde and anterograde labelling and is superior to unconjugated HRP as a tracer because labelling can be obtained with smaller injections and

Table 18. Immunocytochemical localization of WGA and PHA-L.

WGA

0.1−0.5 μl injections of 0.1−0.5% WGA in 0.1 M phosphate buffer may be made by pressure injection through glass micropipettes (40 μm tip) or directly with a Hamilton microsyringe. Animals should be left for 4−36 h survival time to allow for transport of the tracer. Fixation with 4% paraformaldehyde or Bouin's solution (*Table 4*, fixative #4) is recommended.

PHA-L

PHA-L must be delivered iontophoretically. 2.5% PHA-L in 0.1 M PBS may be injected through 10 μm micropipettes using 5 μA positive current delivered over 15 min in 7 sec pulses every 14 sec. Survival times of up to 17 days may be used. Double pH paraformaldehyde fixative (*Table 4*, #6) with 0.05% glutaraldehyde added to the second fixative is recommended.

because it appears to be taken up only slightly by fibres-of-passage. A full discussion of these points can be found in Mesulam (39). This review also covers many of the complex interpretative problems that must be considered in any tract-tracing study. PHA-L is used for anterograde tracing and is unusual amongst tracers in that it appears to fill all parts of labelled cells. Thus cell bodies and dendrites are labelled at the injection site and both the pre-terminal and terminal axonal arborization of labelled cells can be visualized (58). WGA is normally used conjugated to HRP and is then visualized histochemically but unconjugated WGA may be localized immunocytochemically. PHA-L is normally localized immunocytochemically.

Any of the protocols given in Section 3 may be used for their immunocytochemical visualization and double immunostaining (Section 5) allows combined staining of the tracer and an endogenous tissue antigen. *Table 18* lists any special requirements in the use of WGA and PHA-L. Sources of lectins and antibodies are included in Appendix 1.

6.2 **Immunocytochemistry combined with retrograde transport of fluorescent dyes**

A large number of fluorescent dyes are available which are taken up by nerve terminals and subsequently retrogradely transported to label cell bodies. The dyes can be visualized in tissue sections with a fluorescent microscope and used singly or multiply for tract-tracing and may also be applied in combination with techniques such as immunocytochemistry, aldehyde-induced monoamine fluorescence and cholinesterase histochemistry (55). The dyes differ from one another in terms of factors such as excitation and emission characteristics, subcellular localization, solubility, rate of transport and degree of trans-neuronal labelling. These differences allow great flexibility in the use of the dyes but mean also that great care must be taken in choosing and working with a particular dye. One interesting application has been their use in combination with immunofluorescence to allow the identification of transmitter characterized neuronal pathways. The protocol given below is for such an application and is modified from reference 55. That review also describes the characteristics of different dyes. Fast blue, true blue and propidium iodide appear to be the most suitable dyes for combination with immunocytochemistry but the use of only one of these dyes (propidium iodide) is covered below.

(i) Inject by pressure 0.1−0.5 μl of 3% propidium iodide in water.

(ii) After 48−72 h survival perfuse with 4% paraformaldehyde/0.1 M phosphate (pH 7.4). For some transmitter antigens it is necessary to treat first with col-

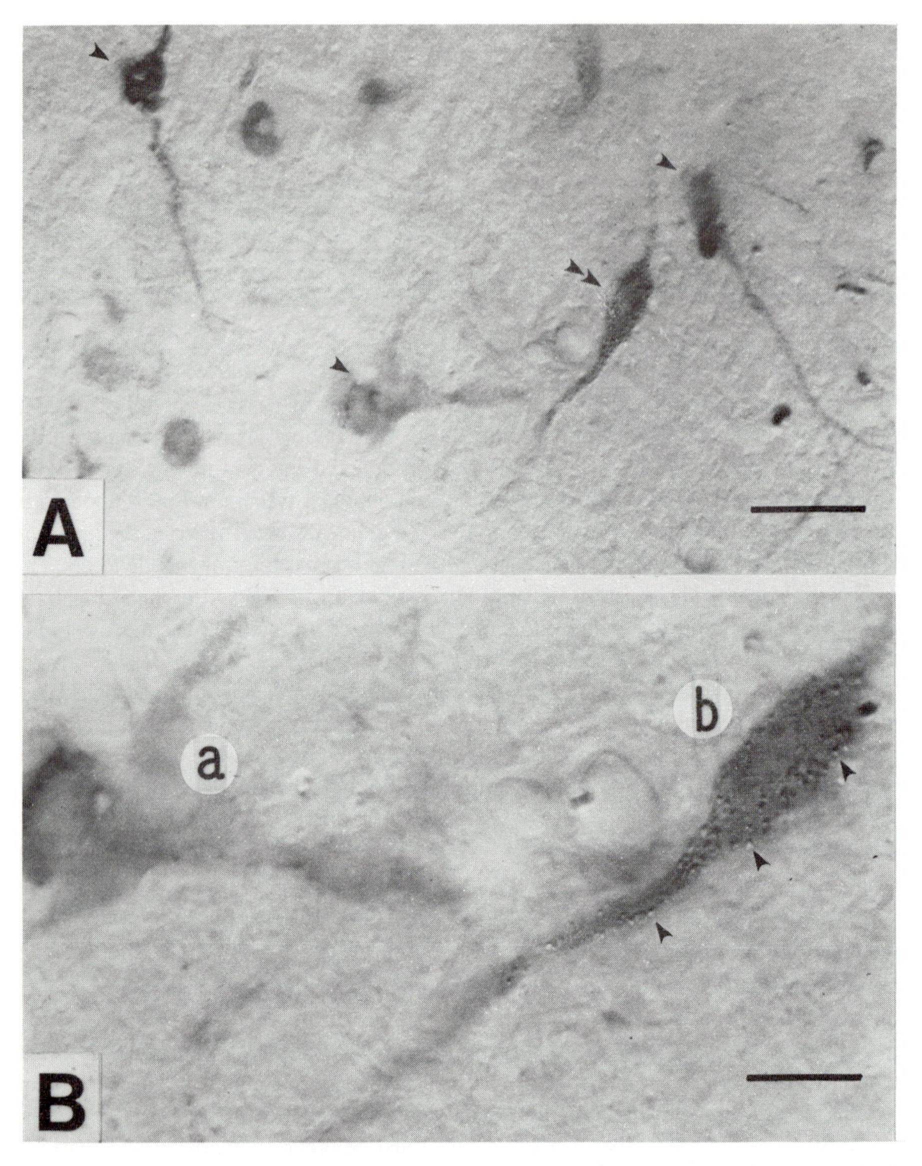

Figure 11. PAP immunocytochemistry combined with retrograde HRP transport. 5-HT immunostaining in raphe dorsalis following an injection of HRP into the striatum (59). Interference contrast illumination. (**A**) Four 5-HT immunostained cells are shown (arrow heads). One of the cells (double arrow head) is also retrogradely labelled. Scale bar = 25 μm. (**B**) High magnification micrograph of two of the cells shown in **A**. Cell (a) is immunostained, cell (b) is both retrogradely labelled and immunostained. The retrogradely transported HRP is packaged into lysosomes and these appear as prominent granules (arrow heads). Scale bar = 10 μm.

chicine (60–120 μg in 20 μl for rats, injected intraventricularly) in order to build up antigen levels in cell bodies. In such cases the animal should be left for an additional 24 h before carrying out the perfusion.

(iii) Following fixation, process the tissue according to the immunofluorescence protocol (Section 3.2.2) using an FITC-labelled second antibody.

(iv) Observe the FITC using a KP 500 excitation filter and LP 520 stop filter. Propidium iodide is viewed using Schott BP 546 primary and LP 590 secondary filter.
(v) With double-stained sections an additional LP 560 stop filter must be used with the FITC combination in order to exclude propidium iodide fluorescence.

Propidium iodide is not good for examining long pathways. With long survival times the injection site is difficult to delineate because of dye diffusion and, in addition, staining of glial cells in the area of retrograde labelling may occur.

6.3 **PAP immunocytochemistry combined with retrograde transport of HRP**

In Section 6.2 a method is described for combining retrograde fluorescent dye transport with fluorescence immunocytochemistry in order to map transmitter characterized neuronal pathways. In an analogous way, retrograde HRP transport may be combined with PAP immunocytochemistry (59, *Figure 11*). The advantage of the peroxidase-based method is that it gives a preparation which is permanent and which may be extended to the electron microscopic level (see Section 7.4). The disadvantage of the method is that it can sometimes be difficult to distinguish the two different peroxidase labels and quite large HRP injections must be used in order to get good double labelling.

(i) Inject by pressure 0.1 μl of 25–50% HRP or 0.1% WGA–HRP.
(ii) Allow 24–48 h survival and then perfuse with 4% paraformaldehyde/0.2% glutaraldehyde/0.1 M phosphate, pH 7.4. If necessary, inject colchicine and allow an additional 24 h survival before perfusion.
(iii) Following perfusion dissect out the brain, leave in fixative for a further 2 h and then transfer to phosphate buffer containing 20% sucrose.
(iv) Leave the tissue until infiltrated (e.g. overnight). Alternatively, perfuse with the phosphate-sucrose after completion of the fixative perfusion. The sucrose perfusion method gives better HRP retrograde labelling but poorer immunocytochemistry.
(v) Cut 40 μm frozen sections (Section 3.1.2), wash in 0.1 M phosphate buffer pH 7.4 and then stain for peroxidase using the Co and Ni intensified DAB procedure (*Table 13*).
(vi) Following staining, wash the sections thoroughly and then carry out PAP immunostaining (Section 3.3) using DAB as chromogen (*Table 13*).
(vii) Wash, dehydrate and coverslip the specimens. Retrograde HRP labelling appears as black granules, immunostaining is brown (*Figure 11*).

7. ELECTRON MICROSCOPIC (EM) IMMUNOCYTOCHEMISTRY

7.1 **General introduction**

Immunocytochemistry must be carried out at EM level if subcellular localization is required and, in the field of neuroscience, EM analysis is necessary in any study designed to establish the existence of synaptic contacts between profiles. Any immunocytochemical technique which employs an electron-dense marker can be extended to the EM level but in practice only two labels are widely used today. Peroxidase with DAB as chromogen is used for both pre- and post-embedding staining (see below) while colloidal gold is used for post-embedding studies.

Pre-embedding and post-embedding staining are the two different methods that are

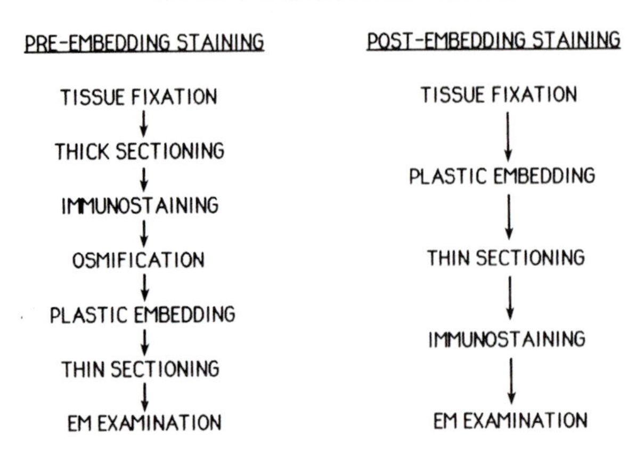

Figure 12. Stages in pre- and post-embedding EM immunocytochemistry.

used for EM immunocytochemistry (*Figure 12*). In pre-embedding immunocytochemistry the staining is carried out on thick tissue sections (e.g. 40 μm) which are then embedded in plastic, sectioned and analysed using relatively routine EM procedures. In post-embedding immunocytochemistry the staining is carried out on plastic-embedded ultrathin EM sections. Post-embedding staining is only suitable for localizing antigens which survive dehydration and plastic embedding and has therefore been used mainly for analysis of cell surface antigens and of intracellular antigens which are present in very high concentrations. In addition, osmium post-fixation normally cannot be included in post-embedding staining and so the method does not give very good membrane preservation and contrast. Thus pre-embedding staining has been the method of choice for localizing CNS transmitters and transmitter markers. Pre-embedding staining is therefore covered in more detail in this chapter but a basic protocol for post-embedding staining is included in Section 7.3.

In pre-embedding staining all the factors relevant to LM immunostaining apply but with the added complication that conditions must be adjusted in order to maintain adequate ultrastructure. Thus long incubations, detergents and frozen sectioning procedures must all be avoided. Fixatives with at least a small proportion of glutaraldehyde normally have to be used. Unfortunately these modifications generally also decrease the amount of immunostaining and so a compromise between retention of staining and preservation of structure has to be adopted. Full methodological details can be found in two recent and comprehensive reviews (56, 60) but the protocol below gives typical conditions suitable for localizing CNS intracellular antigens.

7.2 Pre-embedding staining protocol

(i) Day 1. Perfuse through the ascending aorta with 4% paraformaldehyde/0.2% glutaraldehyde in 0.1 M phosphate buffer (pH 7.4). Dissect out the tissue of interest and leave in fixative for a further 3 h. Cut the tissue into blocks and then transfer them to phosphate buffer containing 30% sucrose and leave until fully infiltrated (6–8 h). Rapidly freeze the blocks by plunging them into liquid nitrogen. Thaw the blocks by returning them to 30% sucrose and then transfer them to PBS. Cut 40 μm thick sections on a Vibratome® . Incubate the sections in 10% normal serum (e.g. goat) for 30 min, wash in PBS, incubate in primary antibody (e.g. rabbit anti-x) overnight. All antisera are diluted in PBS and incubations are carried out at 4°C.

(ii) Day 2. Wash the sections in PBS, incubate in link antibody for 1 h (e.g. 1:10 goat anti-rabbit IgG), wash, incubate in PAP reagent for 1.5 h (e.g. 1:50 rabbit PAP), wash. Stain the sections using DAB (10 min in 0.6% DAB in 0.05 M Tris-HCl pH 7.6, room temperature. Add H_2O_2 to give 0.01% H_2O_2). Terminate the reaction (after 5–10 min) by replacing the DAB solution with PBS. Wash the sections thoroughly, stain for 1.5 h with 1% OsO_4 in 0.1 M phosphate buffer, dehydrate and bring the sections to resin. For flat embedding, Durcupan is recommended because the polymerized resin will easily lift off glass microscope slides (see below).

(iii) Day 3. Place the sections in a drop of Durcupan on a warmed glass microscope slide. Cover with a plastic coverslip. Polymerize the resin (56°C for 2 days).

The slide-mounted sections can be examined in a light microscope. Subsequently the slide is warmed on a hot plate and areas of interest cut from the resin-embedded tissue. Such areas can be mounted in an EM embedding capsule using fresh resin or can simply be glued onto a polymerized blank capsule using a rapid setting epoxy adhesive. Thin sections are cut for EM using standard procedures. Analysis is facilitated by using single slot formvar-coated grids and by preparing grids both with and without lead citrate contrasting. Uncontrasted material is examined using low voltage EM settings (40 kV).

7.3 **Post-embedding staining protocol**

The following protocol is for indirect immunogold labelling of epoxy resin-embedded material and is modified from reference 31. Other protocols may be found in references 40, 52, 56.

Fix the material by perfusion with 4% paraformaldehyde/0.2% glutaraldehyde, post-fix for 2 h, wash and then dehydrate and embed in epoxy resin. Do not include an OsO_4 post-fixation stage. Cut 60–100 nm sections onto nickel or gold grids and allow to dry overnight.

Staining may be carried out in multi-well microtest plates. All solutions should be made up from double-distilled water and should be Millipore filtered (0.45 nm pore).

(i) Etch sections for 1–3 min in 10% H_2O_2. Wash in distilled water.

(ii) Incubate in 1:30 normal serum for 30 min. Drain and then incubate in I° antibody at 4°C for 24 h.

(iii) Wash thoroughly in 0.05 M Tris, pH 7.2, followed by three changes of the same buffer containing 0.2% BSA.

(iv) Place the grids in 0.05 M Tris, pH 8.2, 1% BSA.

(v) Dilute gold-labelled second antibody (e.g. goat anti-rabbit IgG) in 0.05 M Tris, pH 8.2, 1% BSA and centrifuge at 2000 *g* for 20 min to remove micro-aggregates of gold particles.

(vi) Incubate the grids in supernatant for 1 h at room temperature.

(vii) Wash thoroughly in 0.05 M Tris, pH 7.2, 0.2% BSA followed by Tris without BSA and then by water.
(viii) Dry the grids with fibre-free absorbent paper. Counter-stain with uranyl acetate and lead citrate.

7.4 Studies of synaptic circuitry

One interesting and expanding field in which EM immunocytochemistry is applied to neuroscience is in studies of neuronal circuitry. This area is too specialized for detailed coverage in this chapter but it should be pointed out that pre-embedding staining (Section 7.2) can be used in ways which allow double immunolabelling (45) and immunostaining in combination with retrograde HRP tract-tracing (60). These techniques are variants of the protocols described in Sections 5 and 6.3. In addition pre-embedding staining can be combined with anterograde labelling techniques, transmitter uptake autoradiography, Golgi electron microscopy and intracellular labelling. Applied in combination these techniques allow analysis of the synaptic input to neurochemically, morphologically and physiologically characterized cell types. The combined techniques also allow the identification of different components in a neuronal chain. Original references on these procedures can be found in reference 4.

8. CONCLUSION

This chapter has described in detail many of the experimental variables in immunocytochemistry and has presented some specific protocols which are widely used in neuroscience. Immunocytochemistry itself is a rapidly developing field and new variations are always being introduced in an attempt to improve both sensitivity and specificity. Inevitably therefore this chapter has not been able to include all the latest developments and some procedures which have been emphasized here may well have been superseded in a few years. However, I hope that this chapter provides enough information to enable the reader to use effectively the existing techniques and to understand many of the variations which are used in published immunocytochemical studies.

9. ACKNOWLEDGEMENTS

Antibodies were generously provided by Dr T.Görcs (Belle Chasse) and Professor A.C.Cuello (Montreal). I would like to extend special thanks to Miss K.Rudge for expert technical, secretarial and photographic assistance and Mr F.Alvarez (London and Madrid) for help with *Figures 3–5* and for very helpful and critical discussion. Financial support from the Wellcome Trust and the Medical Research Council (UK) is gratefully acknowledged.

10. REFERENCES

1. Emson,P.C. (1983) *Chemical Neuroanatomy*. Raven Press, New York.
2. Björklund,A. and Hökfelt,T. (1985) *Handbook of Chemical Neuroanatomy*. Vols. 1–4. Elsevier Science Publishers, Amsterdam.
3. Cuello,A.C. (1978) In *Handbook of Psychopharmacology*. Iversen,L.L., Iversen,S.D. and Snyder,S.H. (eds), Plenum Publishing Corporation, New York, Vol. 9, p. 69.
4. Cuello,A.C., Priestley,J.V. and Sofroniew,M.V. (1983) *Q. J. Exp. Physiol.*, **68**, 545.
5. Galfre,G. and Milstein,C. (1981) In *Methods in Enzymology*. Lagone,J.J. and Van Vunakis,H. (eds), Academic Press Inc., London and New York, Vol. 73, p. 3.

6. Sofroniew,M.V., Couture,R. and Cuello,A.C. (1983) In *Handbook of Chemical Neuroanatomy.* Bjorklund,A. and Hokfelt,T. (eds), Elsevier Science Publishers, Amsterdam, Vol. 1, p. 210.
7. Cuello,A.C. (1983) *Immunohistochemistry.* John Wiley and Sons, Chichester.
8. Geffard,M., Rock,A.M., Dulluc,J. and Seguela,P. (1985) *Neurochem. Int.*, **7**, 403.
9. Milstein,C., Wright,B. and Cuello,A.C. (1983) *Mol. Immunol.*, **20**, 113.
10. Eckenstein,F., Barde,Y.A. and Thoenen,H. (1981) *Neuroscience*, **6**, 993.
11. Rossier,J. (1981) *Neuroscience*, **6**, 989.
12. Levey,A.I., Armstrong,D.M., Atweh,S.F., Terry,R.D. and Wainer,B.H. (1983) *J. Neurosci.*, **3**, 1.
13. Zipser,B. and McKay,R. (1981) *Nature*, **289**, 549.
14. Sternberger,L.A., Harwell,L.W. and Sternberger,N.H. (1982) *Proc. Natl. Acad. Sci. USA*, **79**, 1326.
15. Garson,J.A., Beverley,P.C.L., Coakham,H.B. and Harper,E.I. (1982) *Nature*, **298**, 375.
16. Valentino,K.L., Winter,J. and Reichardt,L.F. (1985) *Annu. Rev. Neurosci.*, **8**, 199.
17. Ouimet,C.C., Miller,P.E., Hemmings,H.C., Walaas,S.I. and Greengard,P. (1984) *J. Neurosci.*, **4**, 111.
18. Monaghan,P.L., Beitz,A.J., Larson,A.A., Altschuler,R.A., Madl,J.E. and Mullett,M.A. (1986) *Brain Res.*, **363**, 364.
19. Chan-Palay,V., Wu,J.-Y. and Paley,S.L. (1979) *Proc. Natl. Acad. Sci. USA*, **76**, 2067.
20. Matsas,R., Kenny,A.J. and Turner,A.J. (1986) *Neuroscience,* **18**, 991 – 10121.
21. Patel,B.T. and Tudball,N. (1986) *Brain Res.*, **370**, 250.
22. Chan-Palay,V. and Palay,S.L. (1979) *Proc. Natl. Acad. Sci. USA*, **76**, 1485.
23. Schachner,M. (1983) In *Immunohistochemistry.* Cuello,A.C. (ed.), John Wiley and Sons, Chichester, p. 399.
24. Jessell,T.M. and Dodd,J. (1985) *Phil. Trans. R. Soc. Lond. Ser. B*, **308**, 271.
25. Wood,J.N., Hudson,L., Jessell,T.M. and Yamamoto,M. (1982) *Nature*, **296**, 34.
26. Sutcliffe,J.G. and Milner,R.G. (1984) *Trends Biochem Sci.*, 95.
27. Solter,D. and Knowles,B.B. (1979) *Curr. Top. Dev. Biol.*, **13**, 139.
28. Sternberger,L.A. (1979) *Immunocytochemistry*, John Wiley, New York, 2nd edn.
29. Cuello,A.C., Milstein,C. and Galfre,G. (1983) In *Immunohistochemistry.* Cuello,A.C. (ed.), John Wiley and Sons, Chichester, p. 215.
30. Van Den Pol,A.N. (1984) *Q. J. Exp. Physiol.*, **69**, 1.
31. Polak,J.M. and Van Noorden,S. (1984) *Royal Microscopical Society, Microscopy Handbooks Vol. 11*, Oxford University Press.
32. Vaughn,J.E., Barber,R.P., Ribak,C.E. and Houser,C.R. (1981) In *Current Trends in Morphological Techniques Vol III.* Johnson,J.E. (ed.), CRC Press, Florida, p. 33.
33. Pickel,V.M. (1981) In *Neuroanatomical Tract-Tracing Methods.* Heimer,L. and Robards,M.J. (eds), Plenum Press, New York, p. 483.
34. Sofroniew,M.V. and Schrell,V. (1982) *J. Histochem. Cytochem.*, **30**, 504.
35. Causton,B.E. (1984) In *Immunolabelling for Electron Microscopy.* Polak,J.M. and Varndell,I.M. (eds), Elsevier, Amsterdam, p. 29.
36. De May,J.R. (1983) In *Immunohistochemistry.* Cuello,A.C. (ed.), John Wiley and Sons, Chichester, p. 347.
37. Adams,J.C. (1981) *J. Histochem. Cytochem.*, **29**, 775.
38. Hanker,J.S., Yates,P.E., Metz,C.B. and Rustioni,A. (1977) *Histochem. J.*, **9**, 789.
39. Mesulam,M.-M. (1982) *Tracing Neural Connections with Horseradish Peroxidase.* John Wiley and Sons, Chichester.
40. Pool,Chr.W., Buijs,R.M., Swaab,D.F., Boer,G.J. and van Leeuwen,F.W. (1983) In *Immunohistochemistry.* Cuello,A.C. (ed.), John Wiley and Sons, Chichester, p. 1.
41. Petrusz,P., Ordronneau,P. and Finley,J.C.W. (1980) *Histochem. J.*, **12**, 333.
42. Larsson,L.-I. (1981) *J. Histochem. Cytochem.*, **29**, 408.
43. Schipper,J. and Tilders,F.J.H. (1983) *J. Histochem. Cytochem.*, **31**, 12.
44. Haaijman,J.J. (1983) In *Immunohistochemistry.* Cuello,A.C. (ed.), John Wiley and Sons, Chichester, p. 47.
45. Priestley,J.V. and Cuello,A.C. (1982) In *Co-transmission.* Cuello,A.C. (ed.), Macmillan Press Ltd, London, p. 165.
46. Vandescande,F. (1983) In *Immunohistochemistry.* Cuello,A.C. (ed.), John Wiley and Sons, Chichester, p. 101.
47. Nakane,P.K. (1968) *J. Histochem. Cytochem.*, **16**, 557.
48. Tramu,G., Pillez,A. and Leonardelli,J. (1978) *J. Histochem. Cytochem.*, **26**, 322.
49. Sternberger,L.G. and Joseph,S.A. (1979) *J. Histochem. Cytochem.*, **27**, 1424.
50. Lechago,J., Sun,N.C.J. and Weinstein,W.M. (1978) *J. Histochem. Cytochem.*, **27**, 1221.
51. Mason,D.Y. and Sammons,R. (1978) *J. Clin. Pathol.*, **31**, 454.
52. Varndell,I.M. and Polak,J.M. (1984) In *Immunolabelling for Electron Microscopy.* Polak,J.M. and

Varndell,I.M. (eds), Elsevier, Amsterdam, p. 155.
53. Sladek,J.R. and McNeill,T.H. (1980) *Cell Tissue Res.*, **210**, 181.
54. Chan-Palay,V. (1981) In *Current Trends in Morphological Techniques Vol 11.* Johnson,J.E. (ed.), CRC Press, Florida, p. 53.
55. Hökfelt,R., Skagerberg,G., Skirboll,L. and Björklund,A. (1983) In *Handbook of Chemical Neuroanatomy, Vol 1.* Björklund,A. and Hökfelt,T. (eds), Elsevier Science Publishers, Amsterdam, p. 228.
56. Priestley,J.V. and Cuello,A.C. (1983) In *Immunohistochemistry.* Cuello,A.C. (ed.), John Wiley and Sons, Chichester, p. 273.
57. Cuello,A.C., Del Fiacco-Lampis,M. and Paxinos,G. (1983) In *Immunohistochemistry.* Cuello,A.C. (ed.) John Wiley and Sons, Chichester, p. 477.
58. Gerfen,C.R. and Sawchenko,P.E. (1984) *Brain Res.*, **290**, 219.
59. Priestley,J.V., Somogyi,P. and Cuello,A.C. (1981) *Brain Res.*, **220**, 231.
60. Priestley,J.V. (1984) In *Immunolabelling for Electron Microscopy.* Polak,J.M. and Varndell,I.M. (eds), Elsevier, Amsterdam, p. 37.

APPENDIX 1

Suppliers of antibodies

Amersham International plc

Monoclonal antibodies against cytoskeletal proteins. Various developing antibodies including biotin-streptavidin reagents.

White Lion Road,
Amersham,
Buckinghamshire
HP7 9LL, UK

Boehringer Mannheim Biochemicals

Various antibodies including ChAT monoclonal

7941 Castelway Dr,
PO Box 50816,
IN 46250,
USA

Calbiochem

Monoclonal and polyclonal antisera to human and animal proteins, certain peptides and hormones. Various developing antibodies.

PO Box 12087,
San Diego,
CA 92112,
USA

UK distributor:
Cambridge Bioscience,
Newton House,
42 Devonshire Road,
Cambridge CB1 2BL

Cambridge Research Biochemicals Ltd

Peptide antisera.

Button End,
Harston,
Cambridge,
CB2 5NX, UK

PO Box 58,
1887 Park Street,
Atlantic Beach,
NY 11509,
USA

Cappel Laboratories

Monoclonal and polyclonal antisera to human and animal serum proteins. Wide range of developing antibodies.

Cappel Scientific Division,
Cooper Biomedical, Inc.
One Technology Court,
Malvern, PA 19355,
USA

UK distributor:
Lorne Diagnostics Ltd,
PO Box 6,
Twyford,
Reading RG10 9NL

DAKO (Denmark)

Antisera to certain plasma proteins and hormones. Developing antibodies.

Mercia Brocades Ltd,
Brocades House,
Pyrford Road,
West Byfleet,
Weybridge,
KT14 6RA, UK

DAKD Corporation,
22F North Milpas Street,
Santa Barbara,
CA 93103,
USA

E-Y Laboratories Inc

Antisera to human and animal serum proteins, lectins and certain peptides and hormones. Various developing antibodies including colloidal gold.

127 North Amphlett Blvd,
San Mateo,
CA 94401,
USA

UK distributor:
Bioprocessing Ltd,
Consett No. 1 Industrial Estate,
Medomsley Rd,
Consett,
Co Durham DH8 6TJ

Ferring

Peptide antisera

Ferring GMBH,
Wittland 11,
Postfach 21 45,
D-2300 Kiel 1,
FRG

International Laboratory Services

Antisera against various drugs and hormones

14-15 Newbury Street,
London,
EC1A 7HU,
UK

Janssen (Belgium)

Colloidal gold reagents

Janssen Life Sciences Products,
Grove,
Wantage,
OX12 0DQ, UK

Janssen Life Sciences Products,
40 Kingsbridge Road,
Piscataway,
NJ 08854, USA

Miles Scientific

Monoclonal and polyclonal antibodies to human and animal serum proteins, cytoskeletal proteins and certain peptides and hormones. Wide range of developing antibodies.

ICN Biomedicals Inc,
PO Box 19536,
Irvine,
CA 92713, USA

ICN Biomedicals Ltd,
Free Press House,
Castle Street,
High Wycombe, HP13 6RN, UK

Monosan

Monoclonal antibodies to plasma proteins

Sanbio bv,
Heinsbergenstraat 50,
5402 EG UDEN,
PO Box 540, 5400 AM Uden
The Netherlands

UK distributor:
Cambridge Bioscience,
(see Calbiochem)

New England Nuclear

Monoclonal antibodies to plasma proteins. Developing antibodies.

Du Pont, NEN Products,
549 Albany Street,
Boston,
MA 02118, USA

Du Pont (UK) Ltd,
Biotechnology Systems Division,
NEN Research Products,
Stevenage
SG1 4QN, UK

Peninsula Laboratories Inc

Peptide antisera

611 Taylor Way,
Belmont,
CA 94002,
USA

Box 62,
17K Westside Ind. Estate,
Jackson Street, St. Helens,
WA9 3AJ, UK

Pharmacia Inc

Developing antibodies

800 Centennial Avenue,
Piscataway,
NJ 08854,
USA

Pharmacia House,
Midsummer Boulevard,
Central Milton Keynes,
MK9 3HP, UK

Polysciences Inc.

Antisera to plasma proteins, certain cytoskeletal proteins, hormones and lectins. Developing antibodies including biotin avidin and colloidal gold.

400 Valley Road,
Warrington,
PA 18976,
USA

24 Low Farm Park,
Moulton Park,
Northampton,
NN3 1HY, UK

Sigma

Monoclonal and polyclonal antisera to blood cells, plasma proteins and certain lectins. Wide range of developing antibodies.

PO Box 14508,
St Louis,
MO 63178, USA

Fancy Road,
Poole,
BH17 7NH, UK

Sternberger Meyer Immunocytochemicals Inc

PAP and other developing antibodies.

3739 Jarrettsville Pike,
Jarrettsville,
MD 21084, USA

Vector Laboratories

Various peptide and lectin antisera. Developing antibodies.

1429 Rollins Road,
Burlingame,
CA 94010, USA

UK distributor for lectin antibodies:
BDH Chemicals Ltd,
Freshwater Road,
Dagenham RM8 1RZ

Zymed Laboratories Ltd

Monoclonal and polyclonal antibodies to plasma proteins and certain hormones. Developing antibodies including biotin avidin and streptavidin.

52 South Linden Avenue #4,
South San Francisco,
CA 94080, USA

UK distributors:
Cambridge Bioscience
see Calbiochem

APPENDIX 2

Suppliers of Equipment

Bright

Microtomes and cryostats

Bright Instrument Company Ltd,
St Margaret Way,
Stukeley Meadows Industrial Estate,
Huntingdon PE18 6EB, UK

Hacker

Microtomes

Hacker Instruments Inc,
Box 657,
Fairfield,
NJ 07007, USA

Leitz (FRG)

Microscopes and microtomes

E Leitz Inc.,
Instrument Division,
24 Link Drive,
Rockleigh,
NJ 07647, USA

E Leitz (Instruments) Ltd,
48 Park Street,
Luton LU1 3HP,
UK

LKB (Sweden)

Microtomes

LKB Instruments Inc,
9319 Gaither Road,
Gaithersburg,
MD 20877,
USA

LKB Instruments,
232 Addington Road,
Selsdon,
Surrey CR2 8YD
UK

Olympus (Japan)

Microscopes

UK distributor:
Gallenkamp,
Belton Road West,
Loughborough,
LE11 0TR

Olympus Corporation,
4 Nevada Drive,
Lake Success,
NY 11042,
USA

Polysciences Inc

Vibratome (previously Oxford Labs, previously Lancer/Sherwood Medical) Bio-Microslicer (vibratome type sectioner) (Address:Appendix 1)

Reichart-Jung

Microscopes and microtomes

Reichart Scientific Instruments,
Warner-Lambert Technologies Inc,
PO Box 123,
Buffalo,
NY 14240, USA

Reichert-Jung Ltd,
820 Yeovil Road,
Slough SL1 4JB
UK

Zeiss (FRG)

Microscopes

Carl Zeiss Inc,
One Zeiss Drive,
Thornwood,
NY 10594, USA

PO Box 78,
Woodfield Road,
Welwyn Garden City,
AL7 1LU, UK

CHAPTER 4

The use of bioluminescence techniques in neurobiology, with emphasis on the cholinergic system

MAURICE ISRAEL and BERNARD LESBATS

1. INTRODUCTION

The essential work of Ikuta *et al.* (1) who purified from a bacterium (*Arthrobacter globiformis*) an enzyme (choline oxidase) catalysing the oxidation of choline with oxygen consumption, and the useful phospholipid assay subsequently developed by Takayama *et al.* (2), have led to our own contribution on the determination of acetylcholine (ACh). Our objective was to couple the oxidation of the choline resulting from ACh hydrolysis to the detection of hydrogen peroxide (H_2O_2) with one of the very sensitive reactions known to detect this compound. An electrometric procedure would have certainly been possible, but we preferred to use a sensitive and rapid chemiluminescent reaction which would permit us to monitor continuously the amount of ACh present. Several reactions were possible, chemiluminescent ones using luminol or isoluminol plus a catalyst, or lucigenine, or a reaction using chlorophenyloxalate plus a fluorescent probe. Other H_2O_2 detection methods use bioluminescent reactions (pholas luciferin, or earthworm bioluminescence). All these reactions are described in ref. 3: see for example, chapters by Michelson (4), Seitz (5) and Mulkerrin *et al.* (6); for the pholas system, see Henry *et al.* (7,8). The chlorophenyloxalate system and other detectors of H_2O_2 are also compared in Van Dyke's book on bioluminescence and chemiluminescence (9) (see for example, ref. 10). We immediately obtained excellent detection of H_2O_2 with the luminol–peroxidase reaction and therefore did not examine the other possible reactions available.

In the ACh assay described, the enzyme acetylcholinesterase hydrolyses ACh to give choline, which is oxidized to betaine by choline oxidase with H_2O_2 production. In the presence of luminol and horseradish peroxidase (HRP), light emission is generated and recorded with a photomultiplier (see refs. 11–15).

This assay permits us to measure ACh even in samples containing choline without the necessity of separating these compounds by chromatographic or electrophoretic procedures. This simplifies the task of biochemists or pharmacologists working in the cholinergic field. One of the major applications of this procedure is the continuous measurement of ACh release from tissues incubated in a physiological solution containing the assay mixture. This is possible because the reactions are very rapid, the release process being rate limiting, and in addition the compounds employed do not alter the tissues.

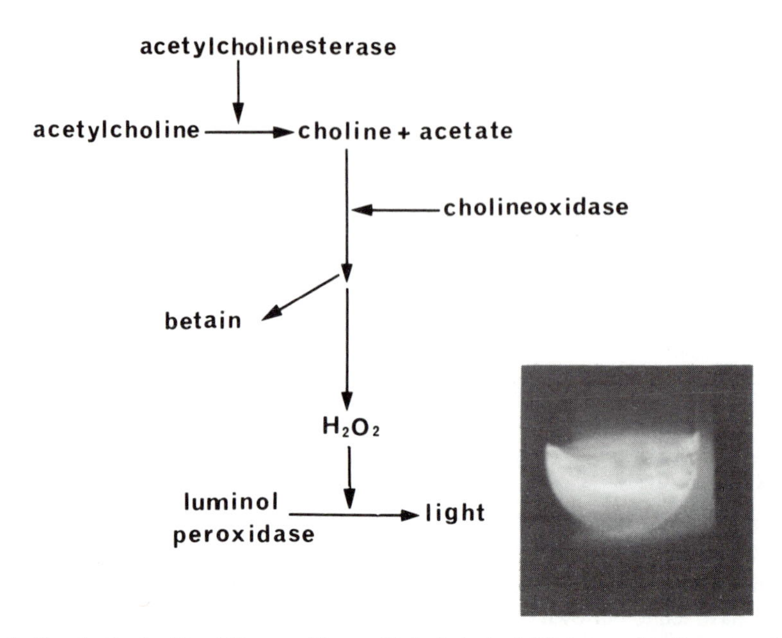

Figure 1. Chemical basis of the ACh assay. The specific hydrolysis of ACh by acetylcholinesterase, followed by the specific oxidation of choline to betaine and H_2O_2 by choline oxidase, leads to a chemiluminescent reaction of luminol with H_2O_2 proportional to the concentration of ACh; HRP is used as a catalyst in this last step. A blue light could be photographed (inset) for a large amount of ACh.

We first describe the assay and its limitations, and then give the procedures for measuring choline acetyltransferase or acetylcholinesterase. Then we show examples in which the compartmentation of ACh and its release can be measured; examples in which the uptake and efflux of ACh were studied are also given.

Finally we describe some further applications of bioluminescence techniques of interest to neurobiologists.

2. THE CHEMILUMINESCENT ACETYLCHOLINE ASSAY

Figure 1 describes the reactions which permit the detection of ACh. When a sample is added to a reaction mixture containing choline oxidase, luminol and HRP only the presence of choline or peroxides will give rise to a light emission which will return to its baseline when the oxidation is terminated. Then if the sample contains ACh, the addition of acetylcholinesterase will trigger a new light emission resulting from the hydrolysis of ACh and the production of choline. Hence it is possible to measure a few pmoles of ACh even in the presence of large amounts of choline which is first oxidized and therefore eliminated by the reaction mixture. No separation of choline and ACh is necessary and this is an essential advantage of the procedure.

2.1 **Stock solutions**

2.1.1 *Choline oxidase (EC 1.1.3.17)*

The enzyme purchased from Wako Pure Chemicals (Osaka, Japan) is in powder form with a specific activity of about 12 units per mg of powder. Stock solutions of 250 units/ml are prepared in distilled water and stored as 100 μl aliquots at −40°C. Choline oxidase is also available from Sigma in powder form (10 units per mg). The stock solution is prepared and stored as for the Wako preparation. It is less convenient to use the enzyme offered by Boehringer which is sold as a solution of 50 units/ml in 4 M NaCl plus 10 mM EDTA.

2.1.2 *Peroxidase (EC 1.11.1.7)*

Several preparations were tested; we have essentially worked with HRP type II from Sigma (150−200 units/mg solid material). The stock solution of 2 mg/ml is prepared in distilled water and stored in 100 μl aliquots at −40°C.

It is also possible to use a lactoperoxidase preparation sold by Boehringer, which has to be diluted until the blank is negligible, or a microperoxidase (MPII) from Sigma.

2.1.3 *Acetylcholinesterase (EC 3.1.1.7)*

To get a very low blank value the stock solution must be filtered on Sephadex G50 gels. We used the acetylcholinesterase from *Electrophorus electricus* sold by Boehringer. A solution of 1000 units/ml in glass-distilled water is first prepared, 250 μl of it are passed through a 5 ml Sephadex G50 coarse column equilibrated in distilled water. The elution requires a further addition of 1.45 ml of water before collecting the enzyme in 0.7 ml. Aliquots of 100 μl are stored at −40°C.

2.1.4 *Luminol (5-amino-1,2,3,4-tetrahydrophthalazindione-1,4)*

This compound was obtained from Merck. A 1 mM stock solution is prepared by dissolving 18 mg in a few drops of 1 M NaOH, and the volume is brought to 100 ml with 0.2 M Tris buffer [Tris (hydroxymethyl) aminomethane] at a final pH of 8.6. Aliquots of 1 ml are stored at −40°C.

2.2 **Chemiluminescent assay mixture (M)**

Just before an ACh determination the stock solutions of choline oxidase, HRP, luminol and acetylcholinesterase are thawed and the following mixture (M) is prepared:

100 μl of choline oxidase
50 μl of HRP
100 μl of luminol

This mixture permits the determination of 16 ACh samples.

2.3 **Determination of acetylcholine**

Several buffers can be used; good results are obtained in 200 mM sodium phosphate buffer, pH 8.6, or 50 mM sodium phosphate buffer, pH 8.6, containing 50 mM NaCl. A volume of 0.25 ml or 0.5 ml of buffer is in general sufficient (it can be increased to 1 ml if it is necessary to buffer more acidic samples). Add the buffer in a glass disposable haemolysis tube which should be extremely clean. A small Teflon-coated magnet is used to stir the solution. Add 15 μl of the assay mixture (M), and a sample (1−100 μl) containing ACh (0.5−50 pmol). As noted above, if the sample contains choline or peroxides these will immediately be eliminated by the reaction mixture. When

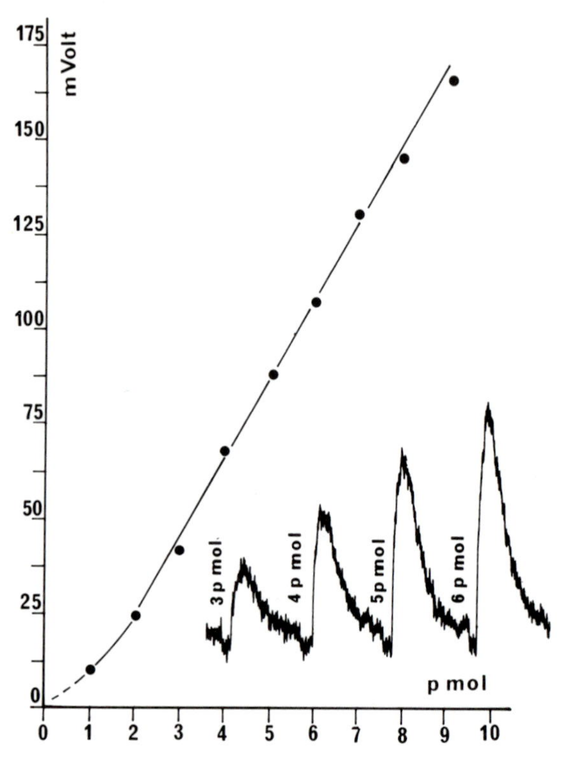

Figure 2. Curve showing the light emission for increasing amounts of ACh. The inset shows the records obtained in the pmole range.

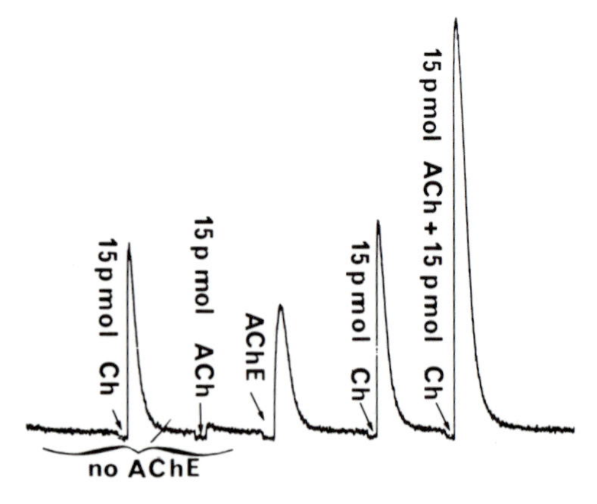

Figure 3. Distinction between choline and ACh. In the absence of acetylcholinesterase the reaction system expands only to choline (15 pmol), whereas the addition of ACh gave no response. However if acetylcholinesterase is added an immediate response for the ACh is elicited. The system is still able to react with choline (15 pmol), and a mixture of choline plus ACh gave the expected summation.

the light emission returns to baseline, add 3–5 μl of acetylcholinesterase to trigger the hydrolysis of ACh and light is emitted proportional to the concentration of ACh, as shown in *Figure 2*.

Figure 3 shows that it is possible to distinguish choline from ACh; in the absence of acetylcholinesterase the addition of 15 pmol of choline gives light emission while the addition of 15 pmol of ACh gives no response; when acetylcholinesterase is injected the hydrolysis of ACh takes place and gives the expected light emission. The figure also shows that a fresh choline injection again triggers light emission, and that the injection of ACh plus choline gives the expected summation.

2.4 **Determination of ACh in biological samples**

2.4.1 *Extraction of ACh*

We obtained good results with the classical trichloroacetic acid (TCA) extraction procedure.

(i) Mince the tissue (5–100 mg) in 0.5 ml of 5% TCA (w/v), and after 60 min centrifuge the denatured proteins.
(ii) Shake the extract (0.5 ml) with ether (5–10 ml) and discard the ether phase. After 5–6 ether washings the pH of the water phase is 4, indicating that most of the TCA has been removed by the ether.
(iii) Evaporate the residual ether.

The entire procedure is carried out with ice-cold solutions. The ether is saturated with water before use.

2.4.2 *Oxidation of the sample and ACh determination*

In several mammalian tissue extracts, and also for other species, the presence of an interfering reducing substance (perhaps ascorbate) blocks the chemiluminescent reaction so it is essential to oxidize these samples before the assay. To 100 μl of the sample add 10 μl of 0.5% sodium metaperiodate (w/v) or 10 μl of 0.5% potassium iodate (w/v). *Figure 4* shows that the addition of 50 μl of the oxidized sample to the assay reagents triggers a light emission due to the presence of the oxidant. This light should decay in about 10 min; if not, the sample has to be diluted. Then acetylcholinesterase is added and a new light emission proportional to the ACh content of the sample is recorded, and calibrated with ACh standards. It has also been possible to determine ACh in a mouse sterno-mastoid muscle: six neural zones were dissected and extracted in 0.5 ml of ice-cold TCA, the sample was ether-washed and treated as above, and the oxidant used was 5 μl of potassium iodate (0.5% w/v) in 50 μl samples.

The method was also used recently (16) to determine the topographical distribution of ACh in the cat brain. Animals were sacrificed after fluothane anaesthesia. After freezing the brain on dry ice, serial (400 μm thick) slices were cut using a refrigerated cryostat. Frontal slices corresponding to stereotaxic planes P2.5, P2, P1 according to Bergman's Atlas (17) were analysed.

An example is shown in *Figure 5*. The slice was punched with a stainless steel tube with a sharpened edge (diameter 1.8 mm) (see ref. 18). The areas taken were as indicated. Each punch gave about 1 mg of tissue, which was extracted with 100 μl of 5% TCA, and ether-washed after centrifuging the denatured proteins which were kept for analysis. The determination of ACh was performed as described above, taking into account the smaller volume to be washed with ether and to be oxidized. The protein pellets were then dissolved in 1 M NaOH, the solution neutralized and buffered. Proteins

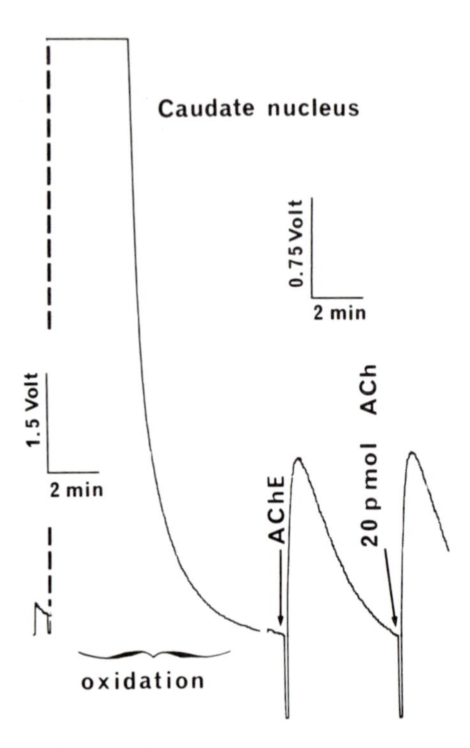

Figure 4. Acetylcholine assay in mammalian tissue extracts. The presence of an interfering substance in several biological tissue extracts quenches the reaction; oxidation of the biological extract overcomes this difficulty. After treating a tricholoroacetic acid, ether-washed extract of caudate nucleus with sodium meta-periodate, the interfering substance is neutralized. The tissue sample is then added to the chemiluminescent reaction mixture and the oxidant triggers a first light emission which should decay in several minutes. The addition of acetylcholinesterase then permits the determination of the ACh content of the sample, calibrated by injecting a 20 pmol standard (see text for details).

were then assayed to compare the samples. The values of ACh in pmol/mg protein are given (*Figure 5*) for three of the areas analysed [cuneiform nucleus (CNF), central tegmental field (FCT), paralemniscal tegmental field (FTP)].

2.4.3 *Comments*

The oxidation step is crucial if the sample is quenched: this is the first thing to test when a new biological tissue is analysed. Also it is essential to check the esterase blank since the lower sensitivity of the method depends on how low this blank is. If the blank is above 0.5−2 pmol after cleaning the enzyme on the Sephadex column it can be decreased further by increasing the proportion of HRP in the reaction mixture — this reduces the amplitude of the ACh response but brings the blank to an extremely low level. We do not know the reason for this empirical observation; probably the esterase blank is partly due to the presence of peroxides. Hence the correct esterase−peroxidase ratio is essential when the assay is performed on samples very low in ACh. Several other difficulties are easy to avoid — traces of ether or organic solvents must be com-

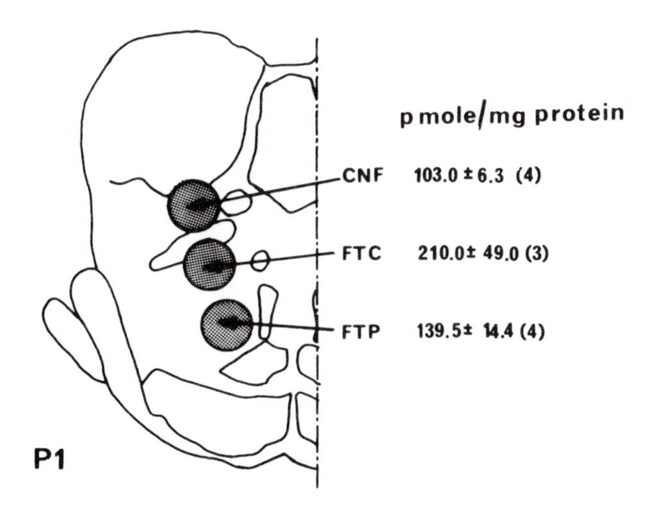

Figure 5. Topographical distribution of ACh in cat brain sections. A typical example is shown, giving the ACh content of the cuneiform nucleus (CNF), the central tegmental field (FTC) and the paralemniscal tegmental field (FTP) (ref. 16).

pletely removed. The buffers must not be prepared by adding concentrated HCl to Tris base or to disodium phosphate, but by mixing the corresponding salts; concentrated HCl brings oxidizing agents which reduce the sensitivity and give an unstable baseline.

3. INSTRUMENTS FOR MEASURING THE CHEMILUMINESCENCE

Several devices for measuring ATP using the firefly–luciferin–luciferase procedure, or calcium with the aequorine method, are commercially available. It is also possible to use scintillation counters to evaluate the chemiluminescence but several difficulties have in this case to be overcome: first the initial flash of light is lost before counting starts (slowing down the esterase by diluting it is therefore useful). It should also be realized that the counter windows have to be closed since the light produced is relatively intense, and may slow down the counting device or even block it. To give an example, in an Intertechnique scintillation counter, after cutting the coincidence, the windows were closed until the blank of the reaction mixture gave 700 counts in a time set at 0.4 min (the mixture was in 200 μl buffer, pH 8.6). The addition of 3 pmol of ACh increased the counts to 1000 showing that some choline was present; after adding 5 μl of acetylcholinesterase, 100 000 counts were obtained. It is therefore possible to use scintillation counters to measure ACh but it is not very convenient. A comparison of different luminometers can be found in ref. 9.

We obtained good results with a 'luminometer' built in the laboratory (12).

4. CONTINUOUS DETERMINATION OF CHOLINE ACETYLTRANSFERASE

In order to measure choline acetyltransferase, it is always possible to extract and measure as a function of time the amount of ACh synthesized by the enzyme. Such a procedure would be more time consuming than the classical radioactive assay described by Fonnum

(19). More useful would be to develop a continuous assay. This was performed in collaboration with Vigo Fosse from Dr Fonnum's laboratory. Since choline acetyltransferase is a reversible enzyme, it is possible to assay it by measuring the choline generated by the reaction:

$$\text{ACh} + \text{CoA} \rightarrow \text{choline} + \text{acetyl-CoA}$$

After inhibiting acetylcholinesterase completely by an organophosphorus inhibitor (ecothiopate iodide) it was possible to show that the addition of CoA triggered the hydrolysis of ACh by choline acetyltransferase, the continuous production of choline being measured by the chemiluminescent reaction. The assay described was developed using a purified choline acetyltransferase preparation or with a supernatant fraction rich in the enzyme obtained from electric organ.

4.1 **Stock solutions**

4.1.1 *Choline acetyltransferase*

(i) Assay the enzyme in tissue extracts or purchase it from Sigma.
(ii) Prepare a solution of 16 units/ml of water and deep freeze.
(iii) Before the assay, add 10 μl of phospholine (ecothiopate iodide; 4.7×10^{-4} M) to inhibit all traces of acetylcholinesterase.

4.1.2 *Coenzyme A (CoA)*

The stock solution is 0.8 mg/ml in distilled water.

4.1.3 *Acetylcholine chloride*

(i) Use the pre-weighed powder (Sigma) as aliquots of 150 mg per vial.
(ii) Prepare a 1 mM stock solution in distilled water.

4.1.4 *Buffer*

Prepare a 50 mM sodium phosphate buffer containing 300 mM NaCl at a pH of exactly 7.7; above that pH the non-enzymatic reaction becomes too high, below that pH the chemiluminescent reaction is not sensitive enough.

4.1.5 *Chemiluminescent reaction mixture for choline acetyltransferase assay (A)*

Use the same stock solutions as described for the ACh assay. A mixture containing 50 μl of choline oxidase, 10 μl of HRP, 50 μl of luminol and 10 ml of buffer pH 7.7 is sufficient for 20 determinations.

4.2 **The continuous choline acetyltransferase assay**

(i) In a glass haemolysis tube containing a small Teflon-coated magnet add 0.5 ml of the chemiluminescent reaction mixture (A), 20 μl of ACh and 5–20 μl of the choline acetyltransferase solution which contains the esterase inhibitor.
(ii) Start the choline acetyltransferase reverse reaction by adding 10 μl of the CoA solution. *Figure 6* shows that the light emission continuously recorded is proportional to the amount of enzyme.

In this case a biological extract was used; a supernatant fraction (S_2) was prepared

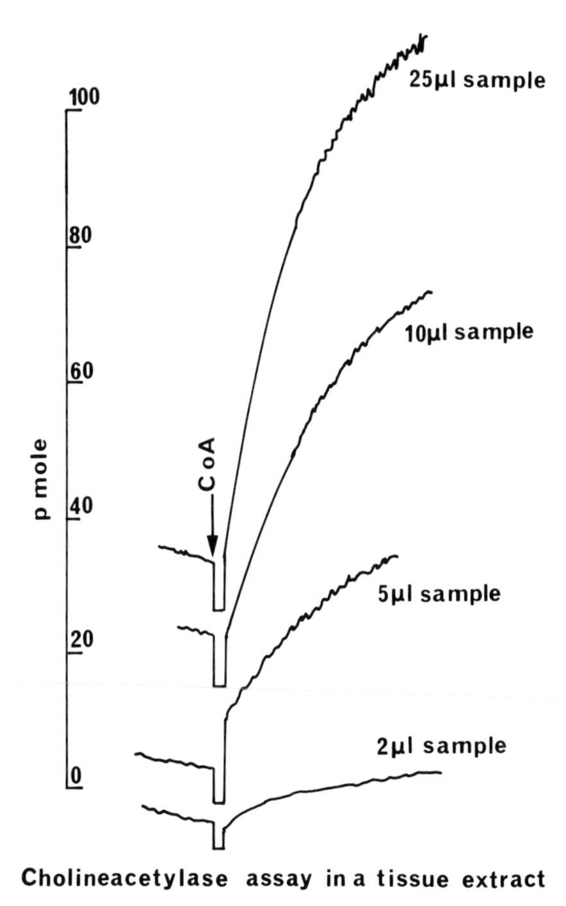

Figure 6. Continuous choline acetyltransferase assay in a biological extract. The reverse reaction (CoA-dependent hydrolysis of ACh) generates choline in the chemiluminescent reaction mixture. The slope of the light emission is proportional to the enzymatic activity. All esterase activities were inhibited before starting the assay (see text).

from an electric organ homogenate (0.5 g/ml) in *Torpedo* physiological saline. After centrifugation (6000 *g*, 20 min) the supernatant (1 ml) was incubated for 30 min in 5×10^{-5} M phospholine to block all esterase activity. Then the sample (0.5 ml) was filtered through a 5 ml Sephadex G50 coarse column equilibrated in distilled water. The void volume was collected after 1.2 ml in 0.5 ml. This filtration step removes choline and CoA from the protein peak collected in the void volume. In *Figure 6* the amount of S_2 was varied from 2 to 25 µl showing that it is possible to measure the enzyme in a few milligrams of tissue with a linear response within minutes. In order to convert the light emission into moles of choline it is preferable to inject a standard before starting the reaction, since the acetyl CoA produced by the enzyme might re-acetylate the choline of the standard.

5. ACETYLCHOLINESTERASE DETERMINATION

It is sometimes useful to measure acetylcholinesterase with its natural substrate ACh, rather than with the very convenient Ellman procedure (20) which is based on the hydro-

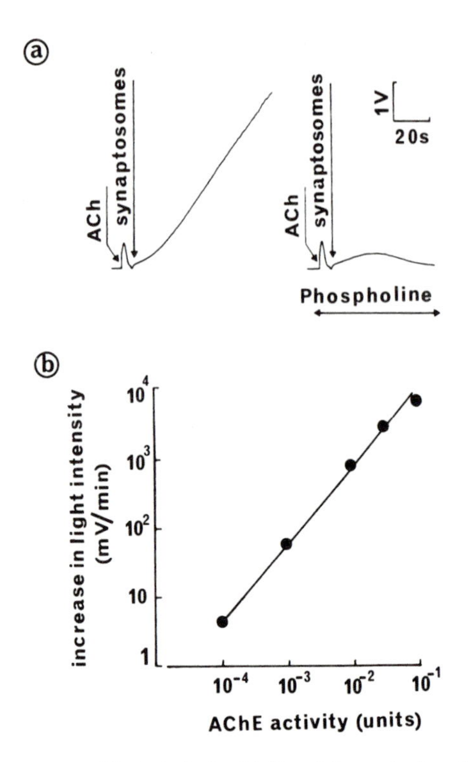

Figure 7. Acetylcholinesterase assay. The hydrolysis of ACh and the production of choline in the chemiluminescent reaction mixture is followed by the constantly rising light emission. In (**a**) the assay was performed on a suspension of synaptosomes showing that it was possible to measure the enzyme on the external face of the membrane. The addition of phospholine inhibited the esterase activity. The response curve for increasing amounts of the enzyme acetylcholinesterase is shown in (**b**).

lysis of the analogue, acetyl thiocholine. The chemiluminescent ACh assay was adapted for such a purpose by Birman (21). In the presence of excess ACh the esterase activity is followed by the production of choline which is continuously oxidized by the chemiluminescent reaction mixture giving a constantly rising light emission.

(i) Place the following into each tube:
sodium phosphate buffer, 0.1 M pH 8 (0.5 ml);
choline oxidase (2.5 μl of stock solution of 250 units/ml);
HRP (5 μl of a stock solution of 2 mg/ml);
luminol (15 μl of a 1 mM stock solution in 0.1 M sodium phosphate buffer, pH 8);
ACh, 0.5 mM final concentration.

(ii) Stir the solution constantly with a small magnet. The addition of ACh always shows a small light emission due to some spontaneous hydrolysis of the stock solution.

(iii) Start the enzymatic reaction by adding the enzyme. As for the choline acetyltransferase procedure it is better to remove any free choline from the enzyme mixture by passing it through a 5 ml Sephadex G50 column.

Figure 7 shows the light emission resulting from the acetylcholinesterase activity of a few microlitres of synaptosomal suspension and its complete inhibition by the esterase inhibitor, phospholine (50 μM). This figure shows the response curve for different acetylcholinesterase concentrations, one unit corresponding to 1 μmol of substrate hydrolysed per min. Absolute activities could be easily determined in the same conditions by using a discontinuous method. Portions of the enzyme were added to the reaction mixture in the absence of choline oxidase. After 1 min, the reaction was terminated by injection of the cholinesterase inhibitor, phospholine (ecothiopate iodide, 100 μM), and choline oxidase was added. The choline generated led then to a light emission, which was immediately calibrated by injecting known amounts of choline into the mixture. A blank without enzyme allowed an evaluation of the amount of spontaneous hydrolysis, which was subtracted. When high sensitivity is required it is possible to assay the enzyme discontinuously: in this case acetylcholinesterase is added to a 0.5 mM ACh solution in sodium phosphate buffer, pH 7, and the reaction stopped after 3 h, for example by adding phospholine (200 μM). It is then as easy to determine the amount of choline produced by the chemiluminescent method as for the ACh assay.

6. THE RELEASE OF ACETYLCHOLINE FROM TISSUES, MONITORED BY THE CHEMILUMINESCENT ASSAY

In our first work (12) it was found that the release of ACh could be monitored from slices of electric organ depolarized in a physiological solution containing calcium and the chemiluminescent reaction mixture. It was also shown that the release of ACh could be induced by electrical stimulation of the nerve. In this latter case it was essential to overcome an artefact resulting from the decomposition of luminol by the electrical field and anodal oxidation. In order to illustrate the possibilities of the chemiluminescent detection of ACh release from tissues we describe three examples in which it was used, the first is the release of ACh from synaptosomes, the next is the release of ACh from mouse caudate nucleus slices and finally the experimental conditions used to assay the transmitter *in vivo* in push–pull cannulae perfusates (18).

6.1 **Acetylcholine release from Torpedo electric organ synaptosomes**

The essential breakthrough resulting from the choline oxidase–chemiluminescence procedure was to record, in line and in real time, the release of ACh from nerve terminals by a biochemical procedure. Several technical problems had to be overcome. First, the assay had to be sensitive and rapid so that the rate-limiting step would be the release of the transmitter rather than the enzymatic reactions. Second, the synaptosomes prepared from the electric organ as previously described (22,23), had to be obtained in conditions which did not alter their physiological properties. This was achieved by isolating them in physiological solutions in which minimum amounts of sucrose were added for the purpose of separation. Osmolarity and salt composition were maintained throughout the procedure. The solution in which the release of ACh is measured is very similar to the gradient layer in which synaptosomes (*Figure 8*) are collected: 0.4 M sucrose, 280 mM NaCl, 3 mM KCl, 1.8 mM $MgCl_2$, 5.6 mM glucose and 50 mM Tris buffer, pH 8.6.

(i) In a tube containing a magnetic stirrer add:
 230 μl of this physiological solution;
 5 μl of the choline oxidase stock solution;

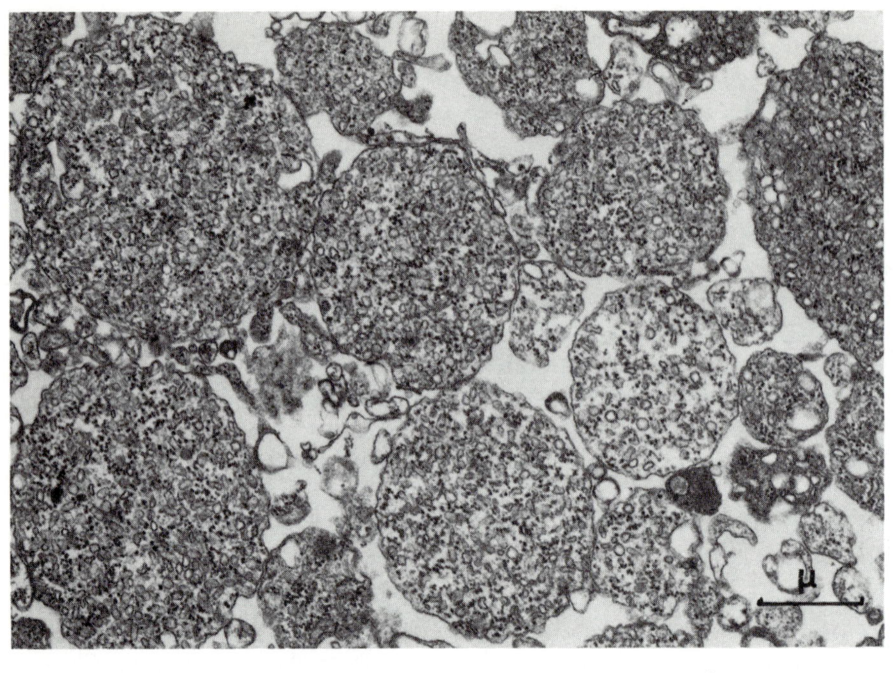

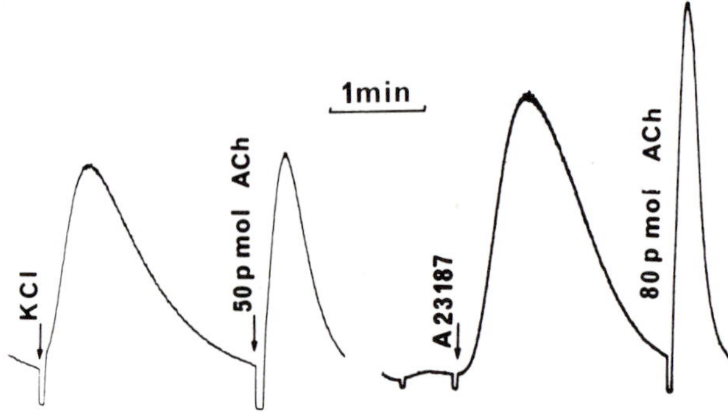

Figure 8. The release of ACh from synaptosomes measured continuously with the chemiluminescent procedure. **Top:** *Torpedo* electric organ synaptosomes specifically prepared in physiological solutions as described in the text. **Bottom:** the suspension of synaptosomes was added to the chemiluminescent reaction mixture and the release of ACh triggered either by KCl depolarization (left) or by the calcium ionophore, A23187 (right). In both cases the injection of ACh standards permits a calibration.

5 μl of HRP stock solution;
10 μl of luminol stock solution.
(These four solutions can be mixed in advance for several samples.)
50 μl of synaptosomal suspension;
$CaCl_2$ (1 – 10 mM final concentration)

The synaptosomes contain enough acetylcholinesterase when we perform the release on an aliquot representing about 1 nmol of occluded ACh. It is possible but often not necessary to add acetylcholinesterase (3 – 5 μl of the gel-filtered stock solution).

(ii) Trigger the release of ACh by the addition of KCl (50 mM) to depolarize the synaptosomes or by the calcium ionophore (A23187, 4–14 μM) or by any other releasing agent (*Figure 8*).

(iii) Calibrate the release curves by injecting an ACh standard (*Figure 8*).

Note that the rising phase of the release curve is less steep than for the ACh standard showing that the release rate is the limiting factor. This was also found with all the releasing agents tested (11–15).

6.2 Evaluation of ACh compartments in synaptosomes

It had earlier been suggested that when synaptosomes were frozen and thawed in the reaction mixture, the first light emission recorded corresponded most probably to the cytoplasmic ACh compartment which was liberated during the thawing process. If the preparation was frozen and thawed again several times, no significant amounts of ACh leaked out from the preparation. This is in agreement with the observation that the ACh associated with the vesicular compartment remains bound even after freezing and thawing the tissue, from which synaptic vesicles full of ACh are prepared. In order to evaluate the vesicular ACh compartment (60–70% of the total) it is necessary to add a detergent (Triton X-100) to liberate that pool. *Figure 9* shows that freezing and thawing liberated some ACh and that two more freezings had no further effect. The vesicular pool of ACh was then liberated by the addition of Triton X-100. Hence it is possible to use the chemiluminescent detection of ACh to measure also its compartmentation before or after inducing the release of ACh.

6.3 Acetylcholine release from mouse caudate nucleus slices

The protocol described is simple: however any modification of the tissue washing phase inevitably leads to difficulties, due to an interfering reducing compound eliminated by the washing phase, which is therefore crucial.

(i) Anaesthetize a mouse with ether and dissect the brain on a filter paper moistened with physiological solution. Take the two caudate nuclei and cut each to produce eight slices of equal thickness.

(ii) Incubate the 16 slices in 200 ml of a solution consisting of 136 mM NaCl, 5.6 mM KCl, 1.2 mM $MgCl_2$ and 10 mM Tris buffer, pH 8.6; gas the solution with O_2 which constantly stirs the slices. After 30 min renew the solution and wash the slices for 15 more minutes.

(iii) ACh release experiment: prepare a mixture for 10 release tests consisting of 7.5 ml of the same physiological solution at pH 8.6, but not gassed with O_2, 100 μl of choline oxidase (stock solution of 250 units/ml), 50 μl of HRP (stock solution of 2 mg/ml) and 100 μl of luminol (stock solution, 1 mM). In a glass haemolysis tube add 0.75 ml of this mixture, 5 μl of gel-filtered acetylcholinesterase (stock solution of 1000 units/ml) and 5 μl of 1 M $CaCl_2$ giving a final calcium concentration of 6.4 mM. Add the slice to the tube and stir constantly and gently with a small magnet. Add KCl (20 μl of 3 M solution giving a final concentration of ~77 M) to evoke the release of ACh.

Figure 10 shows a typical ACh release curve calibrated by the injection of an ACh standard. The release can be controlled in several ways: the omission of calcium abolishes

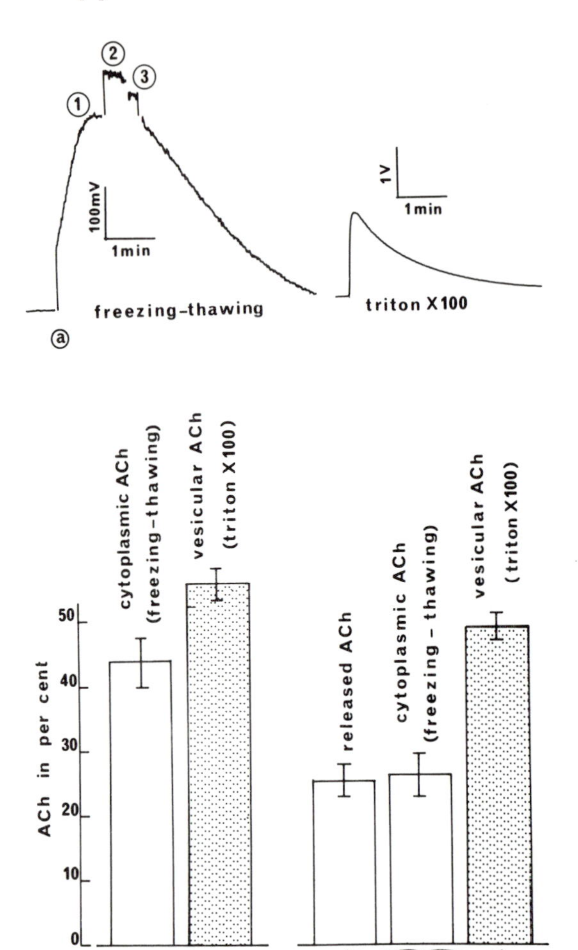

Figure 9. Use of the chemiluminescent method to evaluate the compartmentation of ACh in synaptosomes. After freezing and thawing, the synaptosomes in the chemiluminescent reaction mixture, the first light emission elicited (1) measures the leakage of the cytoplasmic pool; a second or third freezing and thawing does not liberate more ACh (**a**). To measure the vesicular compartment it is necessary to add a detergent (Triton X-100). In (**b**) the two compartments were measured at rest (left) or after triggering the release of ACh (by A23187 or gramicidine). The blocks on the right show: the evoked release of ACh (~50% of the cytoplasmic pool), the remaining cytoplasmic compartment liberated by freeze–thawing, and the vesicular compartment. These compartments are evaluated by measuring the area below the corresponding light emission curves.

it (*Figure 10*) and the inhibition of acetylcholinesterase by phospholine diminishes it considerably, showing that it is indeed ACh and not choline which is released. The omission of choline oxidase abolishes the reaction. If Mg is increased the release is strongly reduced. In conclusion, it is possible to monitor the release of transmitter from a slice of caudate nucleus with the chemiluminescent procedure.

6.4 Assay of ACh in perfusates collected by the push–pull cannula method

Another example in which the assay can be used was found in experiments by Morot-Gaudry *et al.* (24), who implanted three push–pull cannulae in the brain regions to

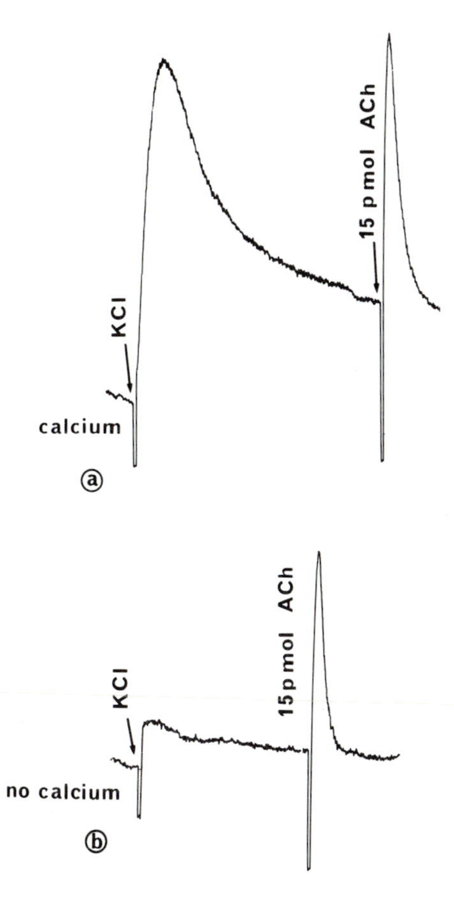

Figure 10. Release of ACh from mouse caudate nucleus slices. In **(a)** the release is triggered by KCl depolarization in the presence of calcium. In **(b)** the omission of calcium abolished the release of ACh (the slices had been previously washed as described in the text to remove any interfering substance). A standard of 15 pmol of ACh is injected to calibrate the amount released.

be explored. The experiments were performed on cats anaesthetized with halothane, maintained under artificial ventilation, and the blood pressure and body temperature were controlled.

The Horsley–Clarke stereotaxic frame and Gaddum's original push–pull cannula, as modified by Nioulon *et al.* (25), are used. The perfusate is an 'artificial cerebrospinal fluid' consisting of 126.5 mM NaCl, 0.5 mM Na_2SO_4, 2.4 mM KCl, 1.1 mM $CaCl_2$, 0.83 mM $MgCl_2$, 0.5 mM KH_2PO_4, 27.5 mM $NaHCO_3$, 5.9 mM glucose, the pH is brought to 7.3 with O_2/CO_2 (95/5 v/v). The medium contains 10^{-4} M phospholine during the first 90 min of perfusion and then 10^{-6} M for the rest of the experiment. The perfusion rate is 10 μl/min. Perfusates are collected only after 120 min for successive 10 min periods. As far as the assay is concerned, the fractions are collected in ice-cold tubes and kept frozen at -80°C. The determination is performed by the standard procedure described above on aliquots of 20–40 μl, the amounts detected are in the pmole range per minute of perfusion. The esterase required to trigger the reaction is sufficient to overcome the presence of the small amount of phospholine (0.6×10^{-7} M) present. No oxidation is necessary to assay ACh in these perfusates, the interfering substance probably being washed out by the perfusion.

6.5 Comments

It is often desired to test the effect of biological extracts or toxins on the release of ACh; these extracts might contain reducing agents which can be removed by gel filtration but if this is not done, it is still possible to unquench the sample by adding small amounts of a diluted H_2O_2 solution until a standard gives the expected unquenched response. It is also useful to know that some drugs (ionophores for example) are dissolved in organic solvents: ethanol will be oxidized at high concentrations, so it is essential to check the blank. Other solvents such as dimethyl sulphoxide do not interfere with the assay.

7. ACETYLCHOLINE FLUXES FROM SYNAPTIC VESICLES DETERMINED WITH THE CHEMILUMINESCENT ASSAY

Although pure cholinergic synaptic vesicles have been available from the electric organ since 1968 (26), *in vitro* experiments dealing with ACh vesicular exchanges have been limited by the absence of techniques allowing a continuous recording of the ACh released. The chemiluminescent method offers a technical improvement which enables a study of the processes which can drive ACh efflux from isolated vesicles (27). An

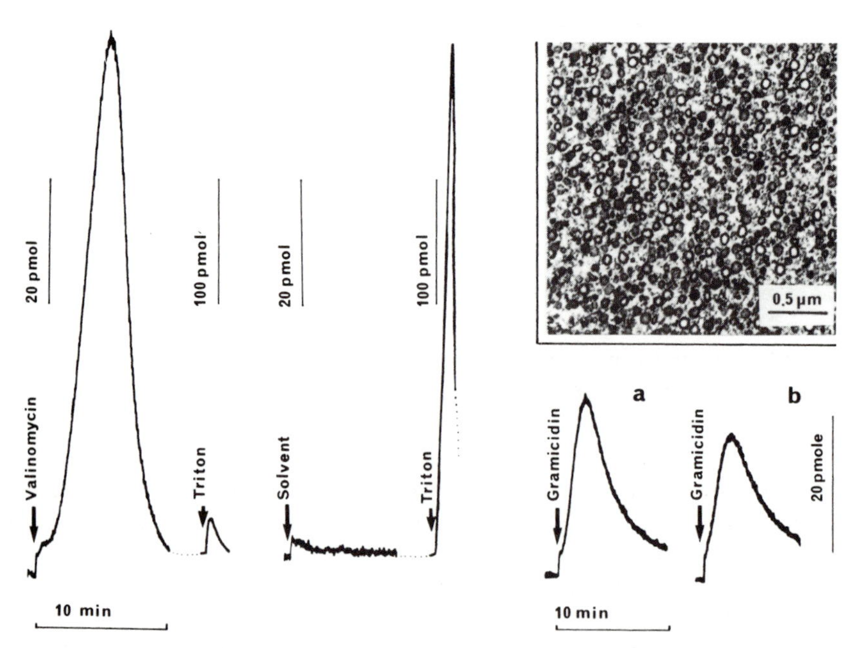

Figure 11. Measurement of ACh efflux from synaptic vesicles. The inset shows a preparation of synaptic vesicles isolated from *Torpedo* electric organ. The suspension of synaptic vesicles is incubated in the chemiluminescent reaction mixture to detect the efflux of ACh elicited by various ionophores on the left; the efflux is first elicited by valinomycin, then the detergent (Triton X-100) liberates the ACh remaining in the vesicles. This is to be compared with a control in which solvent replaces the ionophore while the addition of detergent gives the content of the vesicles. In (**a**) and (**b**) the ionophore gramicidin was used to trigger the efflux of ACh from the vesicles.

adaptation of the experimental protocol provides for studies of the process of ACh incorporation into vesicles.

7.1 Isolation of Torpedo synaptic vesicles

The basic procedure for isolating synaptic vesicles from the *Torpedo* electric organ is the one previously described (28,29). The vesicles are purified in a discontinuous sucrose−KCl iso-osmotic gradient and collected at the interface, 0.25−0.38 M sucrose−KCl. It is important to note that the fractionation procedure described here maintains the synaptic vesicles in a K^+-rich environment as in the normal intracellular fluid.

The homogeneity of the vesicle fraction by electron microscopic examination is illustrated in the inset of *Figure 11*. Cholinergic synaptic vesicles are characterized by their high ACh and ATP content, in a ratio close to 4:1. The standard isolation procedure described above produces a vesicular fraction which routinely contains 75 nmol of ACh, and 20 nmol of ATP, per gram of wet weight of original tissue. Vesicles can be further purified to remove soluble proteins diffusing from the supernatant.

7.2 Endogenous ACh efflux from isolated synaptic vesicles

The chemiluminescent method for ACh measurement permits a continuous detection of the efflux of ACh from isolated vesicles. The efflux itself can be induced by various agents. The illustration is limited to two examples, where ionophores have been used as releasing agents.

7.2.1 *Reagents*

(i) Standard incubation medium: 0.15 M glycine; 60 mM Tris buffer, pH 8, and 360 mM KCl or NaCl. Store at room temperature.
(ii) Mixture for chemiluminescent detection: 100 μl of choline oxidase, 20 μl of HRP, 100 μl of luminol. Keep at 4°C (same stock solutions as in previous sections).
(iii) Vesicular fraction F_3: keep at 4°C.
(iv) Ionophore solutions: dilute the ethanolic stocks of valinomycin (2.2 mM) and gramicidin (5 mM) in the standard mixture and keep at 4°C. Valinomycin (1:60, v/v); gramicidine (1:20, v/v).

7.2.2 *Procedure*

(i) Add the vesicle fraction (5−10 μl; 200−400 pmol ACh) to 400 μl of the standard incubation medium supplemented with acetylcholinesterase and 15−20 μl of the mixture of choline oxidase−peroxidase and luminol.
(ii) Allow the sample to equilibrate at room temperature for 5 min. After that time, the choline present in the vesicular fraction as a soluble component diffusing from the lighter fractions of the gradient has been destroyed and a stable background light is obtained.
(iii) Initiate the ACh efflux by addition of a few microlitres of ionophore or diluted solvent as control (*Figure 11*).
(iv) Determine the ACh content of the vesicle samples before and after the efflux test in parallel runs by the addition of 2 μl of 5% Triton X-100 buffered at pH 8.6.

(v) Inject internal ACh standards into each test tube to estimate the amount of ACh released and the residual ACh content.

The total amount of ACh liberated is converted to molar units using the area under the curve. This is generally done by the weighing method. Other calibration procedures can be used. For the determination of content, the detergent effect of Triton is so rapid that it is sufficient to measure the amplitude of the response.

7.2.3 *Comment*

The composition of the incubation medium can be adapted according to the experimental purpose. However, one must take into account some fundamental principles:

(i) the iso-osmolarity of the medium must be maintained (850 mOsm);
(ii) a high concentration of glycine affects the activity of the acetylcholinesterase;
(iii) each assay must be calibrated by an internal standard.

8. ACETYLCHOLINE UPTAKE BY ISOLATED SYNAPTIC VESICLES

ACh uptake by isolated vesicles is followed as an increase in the vesicular content after incubation in the presence of exogenous ACh. This procedure requires preliminary experimental steps:

(i) depletion of the vesicles of their endogenous ACh content;
(ii) concentration of the vesicular material;
(iii) inhibition of any acetylcholinesterase activity in the vesicular preparation.

8.1 **Starting material**

(i) Submit the freshly prepared vesicular fraction F_3 (see above) to hypo-osmotic shock by diluting it 3.8-fold with 10 mM Tris buffer, pH 7.1.
(ii) After 15 min at room temperature, centrifuge the mixture at 60 000 r.p.m. for 60 min in a Beckman Ti 60 rotor.
(iii) Resuspend the pellet in a small volume of 0.9 M sucrose, 10 mM Tris buffer, pH 7.1, to a final concentration of about 40 g initial tissue/ml.
(iv) Partition this concentrated preparation in sealed vials and keep frozen at −80°C until use within weeks.

8.2 **Experimental procedure**

Thaw part of the concentrated vesicular preparation (300 μl) and treat for 15 min at room temperature with 5 μl of 47 $\times$ 10^{-4} M phospholine iodide as an acetylcholinesterase inhibitor. This pre-treatment is an obligatory prerequisite.

The subsequent operations aim to get a controlled composition of both the internal and external media. Hence, for each problem, the conditions should be chosen according to one's specific purpose. A typical rate-type experiment is described below (27).

(i) Add an equal volume of 2 mM Tris buffer, pH 6.2, to the vesicular preparation pre-treated with phospholine.
(ii) Allow the mixture to equilibrate at room temperature for 15 min and at 4°C for a further 90 min.
(iii) Pass the suspension through a column of Sephadex G50 coarse (1 cm $\times$ 5 cm) equilibrated with 0.9 M sucrose, 2 mM Tris buffer, pH 7.1.

(iv) Elute the column with the same buffered solution and recover the vesicles in 600 μl.
(v) Rapidly start the incubation with ACh.
(vi) Divide the vesicular effluent into aliquots (240 μl) and inject ACh (10 μl of a 375 mM solution) at $t = 0$.
(vii) After t minutes, remove a 60 μl sample and apply it to a column of Sephadex G50 coarse (5 mm × 56 mm) equilibrated with 0.9 M sucrose, 10 mM Tris buffer, pH 7.1.
(viii) Elute with the same buffer and recover the vesicles in 140 μl. The filtration process is achieved within 1–2 min.
(ix) Keep effluents on ice until the ACh assay (within 1 h).

For the determination of the initial ACh content, a sample without added ACh is similarly treated.

8.3 Chemiluminescent detection of the ACh uptake

Measurement of the ACh incorporated into the vesicles is carried out as fast as possible after the sampling.

(i) The standard medium for the ACh detection consists of 400 mM KCl, 10 mM phosphate buffer, pH 8.6.
(ii) The chemiluminescent mixture contains 100 μl of choline oxidase, 50 μl of peroxidase and 100 μl of luminol, prepared as above.
(iii) Acetylcholinesterase is purified as described above.

8.3.1 *Procedure*

The vesicular sample is divided into two portions: one is designed for the chemiluminescent assay, the other is used for the protein determination.

(i) For the chemiluminescent assay transfer a sample (80 μl) containing 1–3 μg of protein to a test tube and mix with 3 μl of the acetylcholinesterase and 6 μl of the chemiluminescent assay mixture. After 1–2 min at 20°C the traces of free ACh which have eluted with the vesicular material have been efficiently removed by conversion to betaine.
(ii) Transfer the treated sample to a test tube containing 250 μl of standard medium, 15 μl of fresh chemiluminescent mixture and 3 μl of acetylcholinesterase. A stable low background light is obtained.
(iii) Determine the vesicular ACh content by addition of 2 μl of 10% Triton X-100, pH 8.6.
(iv) Add an internal standard to calibrate the light amplitude response.

The raw data obtained using the experimental procedure described above are shown in *Figure 12*. The values are normalized using the protein concentration in each sample and a graph as shown in the inset is then obtained.

8.3.2 *Comments*

The experimental procedure involves two filtration steps through Sephadex G50 columns. The first one serves several purposes:

(i) the vesicular sample is cleaned from the aggregates remaining after the pellet resuspension;

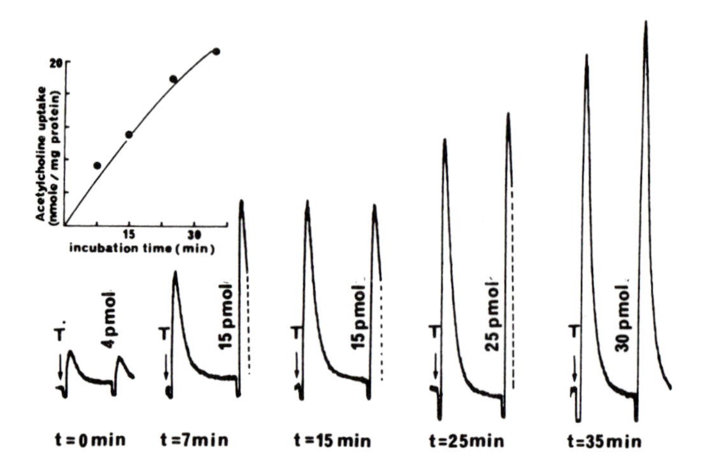

Figure 12. Uptake of ACh by synaptic vesicles measured with the chemiluminescent procedure. After incubating the vesicles with ACh, the suspension is gel-filtered through Sephadex. Then the residual extra-vesicular ACh is destroyed by a small amount of the chemiluminescent reaction mixture which converts it to betaine. The vesicles are then lysed by Triton X-100, in a new solution containing the reaction mixture, to measure the occluded ACh. The figure shows as a function of time the increase of occluded ACh, each sample being followed by the injection of a standard. The inset shows the uptake expressed per mg protein (see text).

(ii) excess of unreacted phospholine is eliminated.

(iii) the elution medium, which corresponds to the incubation external medium, has a controlled composition.

The second filtration is aimed to stop the incubation. Although other procedures may be used, this method has proved to be accurate and reliable. The strength of the buffer used for elution is reinforced to allow a sufficient buffering of the H^+ liberated during the hydrolysis of the free ACh by the acetylcholinesterase. Indeed, acidification of the external medium has been shown to provoke the release of endogenous ACh from isolated synpatic vesicles.

The chemiluminescent assay for measuring the uptake of ACh into the vesicles offers two main advantages, besides its rapidity and easiness. Only the ACh incorporated inside the vesicles is measured. Adsorbed ACh is destroyed by the esterase and the resulting choline is oxidized during the pre-treatment step of the sample.

Drawbacks resulting from the use of radioactive tracers and variable specific activities are avoided.

9. FURTHER APPLICATIONS OF CHEMILUMINESCENT TECHNIQUES

9.1 Measurement of energy metabolites

The firefly luciferin–luciferase reaction (30) has been adapted to monitor the release of ATP from synaptosomes (31–33), and so can be directly related to release of ACh as described above. Furthermore, ATP, ADP and creatine phosphate in synaptosome extracts can be measured by this technique via the appropriate enzyme links. The following method, based on the ATP assay of Lundin and Thorne (34), is in routine use (Herman Bachelard, personal communication; 41). Freshly prepared or incubated synaptosomes can be used directly but it is often advantageous to separate them from incubation media by centrifugation or filtration. Very rapid filtration is achieved by suction through a combination of Whatman GFF and GFB filters in a Swinney holder.

(i) Prepare synaptosome extracts by homogenization in ice-cold 0.6 M perchloric acid (usually 1 ml per 1.5 mg synaptosomal protein).
(ii) After centrifugation at 10 000 *g* for 10 min, neutralize the supernatant with the calculated volume of 1 M $KHCO_3$ at 0°C.
(iii) Keep at 0°C for 10 min and centrifuge to remove precipitated $KClO_4$.

This synaptosome extract is usually diluted 10-fold with the appropriate assay medium immediately before use.

ATP and ADP are measured consecutively on the same sample.

(i) Place 200 μl of diluted synaptosome extract (containing 2 mM $MgSO_4$, 5 mM K_2SO_4, 100 μM phosphoenolpyruvate, 2 mM Tris/H_2SO_4, pH 7.75) in a cuvette in the luminometer, and add 40 μl of ATP monitoring reagent (LKB).
(ii) Follow the light output on a flat-bed recorder; at the light output proportional to the ATP concentration, a plateau is observed.
(iii) Now add 20 μl of a 5 mg/ml solution of pyruvate kinase (type III) in 20 mM Tris/H_2SO_4, pH 7.75.
(iv) After total conversion of ADP to ATP the light output reaches a new plateau, the height of which is proportional to the total ADP plus ATP concentration. Add an internal standard (40 pmol ATP) and record the further plateau in light output.
(v) Run ATP and ADP standards (0–50 pmol) with each assay.

Creatine phosphate is measured following complete transfer of its phosphate group to ADP with the formation of ATP by a pre-incubation with creatine kinase.

(i) Incubate 200 μl of the diluted synaptosome extract (containing 2 mM $MgSO_4$, 5 mM K_2SO_4, 10 μM ADP, 2 μg of creatine kinase in Tris/H_2SO_4, pH 7.7) for 30 min at 30°C in the luminometer cuvette.
(ii) Add 40 μl of ATP monitoring reagent (LKB) and record the increase in light output.
(iii) At the plateau, add an internal standard of 60 pmol ATP and record again.

Creatine phosphate standards (0–200 pmol) are run with each assay.

An internal standard is necessary to allow correction if residual concentrations of potassium perchlorate, inhibitory to the luciferin/luciferase system, are present in the sample. The initial 10-fold sample dilution helps reduce this problem.

After correction for perchlorate inhibition, and comparison with standard curves, ATP and ADP values are derived directly from the first assay. Creatine phosphate values are calculated by subtracting the ATP value from the first assay, from creatine phosphate plus ATP from the second.

Analogous methods are available for measurement of coenzymes such as nicotinamide nucleotides and flavin nucleotides (see ref. 3) and therefore can be adapted to measurement of activities of enzymes for which these act as coenzymes. Recently, a chemiluminescent assay for glutamate was published by Fosse *et al.* (40). Glutamate is oxidized by glutamate dehydrogenase; the NADH produced is measured using the luciferase of *Photobacterium fischerei* coupled to an FMN-oxido reductase. This gives light in the presence of decaldehyde (see ref. 3). Such an approach is likely to produce assay methods for other neurotransmitters.

9.2 **Calcium**

A chemiluminescent assay for calcium has been described which uses the photoprotein, aequorine, from the jelly fish, *Aequora aequora* (35). Injection of aequorine into a cell or medium containing minute amounts of calcium produces a flash of light which can be directly related to the concentration of calcium present (36,37).

9.3 **Monoamine oxidase**

This chemiluminescent assay (Tenne's method, 38) is both sensitive and rapid and depends on the bioluminescence generated by certain dark mutants of luminous bacteria (*Vibrio harveyi* M42) in response to long-chain aliphatic aldehydes generated in the enzymic reaction. The particular mutant strain, originally described by Ulitzur and Hastings (39), is an aldehyde-requiring mutant that increases its luminescence upon addition of long-chain aliphatic aldehydes to the cell culture. Thus, any enzyme activity that generates such aldehydes could be detected readily in the bioluminescence system. The intensity of the luminescence is proportional to the aldehyde concentration. In the case of monoamine oxidase (MAO), the reaction sequence is:

(i) $RCH_2NH_2 + O_2 + H_2O \xrightarrow{MAO} RCHO + NH_3 + H_2O_2$

(ii) $RCHO + O_2 + FMNH_2 \xrightarrow{luciferase} RCOOH + FMN + H_2O + light$

The luminescence can be detected as described earlier.

The activity of MAO can be measured either continuously, or in a discontinuous assay using, for example, octylamine or decylamine as substrate.

10. COMMENT

The above descriptions give a brief account of the ways in which bioluminescent techniques are being applied to the sensitive and accurate estimation of a variety of compounds and processes of interest to neurobiologists. Future developments can be envisaged where such phenomena as proton gradients and membrane potentials, as well as various neurotransmitters, could be continuously monitored if the appropriate biological chemiluminescent systems can be discovered. There is a wide source of light-emitting reactions in biology which could provide the neurobiologist with new possibilities of studying the rapid processes which participate in neuronal communication.

11. REFERENCES

1. Ikuta,S., Shigeyuki,I., Misaki,H. and Horiuti,Y. (1977) *J. Biochem.*, **82**, 1741.
2. Takayama,M., Itoh,S., Nagasaki,T. and Tanihizu,T. (1977) *Clin. Chim. Acta*, **79**, 93.
3. De Luca,A., ed. (1978) *Methods in Enzymology. Vol. 57.* Academic Press, New York.
4. Michelson,A.M. (1978) In *Methods in Enzymology.* De Luca,A. (ed.), Academic Press, New York, Vol. 57, p. 385.
5. Sietz,R.W. (1978) In *Methods in Enzymology.* De Luca,A. (ed.), Academic Press, New York, Vol. 57, p. 445.
6. Mulkerrin,M.G. and Wampler,J.E. (1978) In *Methods in Enzymology.* De Luca,A. (ed.), Academic Press, New York, Vol. 57, p. 375.
7. Henry,J.P., Isambert,M.F. and Michelson,A.M. (1979) *Biochim. Biophys. Acta*, **205**, 437.
8. Henry,J.P. and Monny,C. (1977) *Biochemistry*, **16**, 2517.

9. Van Dyke,K., ed. (1985) *Bioluminescence and Chemiluminescence: Instruments and Applications. Vol. 1*, C.R.C. Press, Boca Raton, Florida.
10. Tsuji,A., Masaka,M. and Hidetoski,A. (1985) In *Bioluminescence and Chemiluminescence: Instruments and Applications.* Van Dyke,K. (ed.), C.R.C. Press, Boca Raton, Florida, Vol. 1, p. 185.
11. Israël,M. and Lesbats,B. (1980) *C.R. Acad. Sci. Ser. D (Paris)*, **291**, 713.
12. Israël.M. and Lesbats,B. (1981a) *Neurochem. Int.*, **3**, 81.
13. Israël,M. and Lesbats,B. (1981b) *J. Neurochem.*, **37**, 1475.
14. Israël,M. and Lesbats,B. (1982) *J. Neurochem.*, **39**, 248.
15. Israël,M. and Lesbats,B. (1985) In *Bioluminescence and Chemiluminescence: Instruments and Applications.* Van Dyke,K. (ed.), C.R.C. Press, Boca Raton, Florida, Vol. 2, p. 1.
16. Cheramy,A., Romo,R., Morot Gaudry-Talarmin,Y., Israël,M. and Glowinski,J. (1987) in preparation.
17. Bergman,A. (1968) *The Brain Stem of the Rat.* The University of Wisconsin Press.
18. Tassin,J.P., Chéramy,A., Blanc,G., Thiery,A.M. and Glowinski,J. (1977) *Brain Res.*, **107**, 291.
19. Fonnum,F. (1975) *J. Neurochem.*, **24**, 407.
20. Ellman,G.L., Courtney,K.D., Andres,V., Jr. and Featherstone,R.M. (1961) *Biochem. Pharmacol.*, **7**, 88.
21. Birman,S. (1985) *Biochem. J.*, **225**, 825.
22. Israël,M., Manaranche,R., Mastour-Frachon,P. and Morel,N. (1976) *Biochem. J.*, **160**, 113.
23. Morel,N., Israël,M., Manaranche,R. and Mastour,P. (1966) *J. Cell Biol.*, **75**, 43.
24. Morot-Gaudry,Y., Romo,R., Lesbats,B., Chéramy,A., Godeheu,G., Glowinski,J. and Israël,M. (1985) *Eur. J. Pharmacol.*, **110**, 81.
25. Nieoullon,A., Chéramy,A. and Glowinski,J. (1977) *J. Neurochem.*, **28**, 819.
26. Israël,M., Gautron,J. and Lesbats,B. (1968) *C.R. Hebd. Séances Acad. Sci. Ser. D*, **266**, 273.
27. Diebler,M.F. (1982) *J. Neurochem.*, **39**, 1405; and unpublished experiments.
28. Israël,M., Gautron,J. and Lesbats,B. (1970) *J. Neurochem.*, **17**, 1441.
29. Israël,M., Manaranche,R., Marsal,J., Meunier,F.M., Morel,N., Frachon,P. and Lesbats,B. (1980) *J. Membr. Biol.*, **54**, 90.
30. McElroy,W.D. (1947) *Proc. Natl. Acad. Sci. USA*, **33**, 342.
31. Israël,M., Lesbats,B., Meunier,F.M. and Stinnakre,J. (1976) *Proc. R. Soc. Lond.*, **193**, 461.
32. Israël,M. and Meunier,F.M. (1978) *J. Physiol. Paris*, **74**, 485.
33. White,T.D. (1977) *J. Neurochem.*, **30**, 329.
34. Lundin,A. and Thorn,A. (1975) *Anal. Biochem.*, **66**, 47.
35. Johnson,F.H. and Shimomura,O. (1972) *Nature*, **237**, 287.
36. Stinnakre,J. and Tauc,L. (1973) *Nature*, **242**, 113.
37. Llinas,R., Blinks,J.R. and Nicholson,C. (1972) *Science*, **176**, 1127.
38. Tenne,M., Finberg,J.P.M., Youdim,M.B.H. and Ulitzur,S. (1985) *J. Neurochem.*, **44**, 1378.
39. Ulitzur,S. and Hastings,J.W. (1978) *Proc. Natl. Acad. Sci. USA*, **76**, 265.
40. Fosse,V.M., Kolstad,J. and Fonnum,F. (1986) *J. Neurochem.*, **47**, 340.
41. Park,I., Thorn,M.B. and Bachelard,H.S. (1987) *J. Neurochem.*, **49**, in press.

CHAPTER 5

Neuroreceptors

G.G. LUNT

1. INTRODUCTION

The study of neurotransmitter receptors has grown rapidly and has almost achieved the status of a separate discipline within the field of neurobiology. In spite, however, of this rapid growth of activity, the nicotinic acetylcholine receptor (nAChR) is still the only receptor protein that is completely characterized at the molecular level. Receptor research can be divided into two distinct classes. First there are the research programmes that seek to isolate and to characterize the receptor proteins. In general, progress in this area is slow and although the experience gained with the nAChR has been of great benefit, it is increasingly apparent that different receptors may require quite different isolation strategies.

The second type of receptor research is fundamentally different in that the prime concern is to obtain a measure of the presence and pharmacological specificity of the recognition site of the receptor. The receptor need not be purified for such studies. The key to success in this type of experiment is a sensitive and specific assay for the neurotransmitter binding site of the receptor. By far the most widely used general method for this involves allowing radiolabelled ligand to interact with the receptor and then separating the receptor-bound ligand and quantifying it. This inherently simple procedure, first described by Paton and Rang in 1965 (1) has developed into a major activity in both basic research and in the pharmaceutical industry where such assays offer the prospect of a simple rapid screen for evaluation of receptor drugs. In this chapter I will attempt to steer the would-be 'binder' a safe course through the minefield of neuroreceptor binding assays.

2. GENERAL RECEPTOR PURIFICATION STRATEGIES

Neuroreceptors are integral proteins of the plasma membrane and therefore a preliminary separation of soluble tissue constituents from insoluble material gives an initial enrichment in binding activity. However, dissolution of the structure of the plasma membrane is a necessary prerequisite to any further purification.

Receptor purification can be considered in three basic stages. First the tissue is homogenized and the homogenate processed by differential centrifugation to obtain a membrane fraction. Second, the membranes are treated with detergent in order to solubilize the receptor which is then purified in a third, affinity-chromatography step. The successful purifications of the vertebrate nAChR and the γ-aminobutyric acid receptor complex (GABAR) are based on this general procedure. However it should be emphasized that receptor purification is a relatively young science and there are no guarantees that these methods can be successfully applied to all neuroreceptors.

2.1 Preparation of membrane fractions

Most receptor preparations start with the formation of a total homogenate of the whole tissue. The composition of the buffer(s) used is in most cases taken directly from procedures for the preparation of subcellular fractions. Thus 0.32 M sucrose buffered at pH 7.4 with Tris is a widely used starting medium. In the case of vertebrate brain a motor-driven Potter–Elvejhem type of homogenizer is most often used to produce a homogenate of between 10 and 20% (tissue:buffer, w/v). In general, subsequent handling of the homogenate closely follows the steps used to produce a crude synaptosomal pellet (see Chapter 1). There is little to be gained in proceeding with further purification of the membrane fraction because the relatively modest enrichments in receptor activity of such purified membranes are outweighed by the losses of material and the extra processing time.

It is often considered desirable to introduce more extensive washing steps than is usual in subcellular fractionation procedures. The rationale for this is that disruption of the tissue may release the entire synaptic pool of endogenous neurotransmitter, thus the receptor binding site is likely to be occupied making assay of the binding activity more difficult. In certain cases this does not occur, an example being the preparation of nAChR from electric organ. In this case the extremely high activity of acetylcholinesterase present in the tissue homogenates makes it unlikely that any significant amounts of free acetylcholine (ACh) will be retained. In the case of isolation of the GABAR from mammalian brain it has been found to be advantageous to expose the membrane pellet to osmotic shock by resuspension in ice-cold water and in some cases to combine this with a series of rapid freeze–thaw cycles. Such procedures are developed on an empirical basis and it cannot be stated with any certainty that it is necessary to adopt them for other receptor types.

A major problem in receptor purification is proteolysis. The endogenous proteases of neural tissues are not, in general, well characterized and it is consequently difficult to adopt procedures that will reliably reduce their activity. A number of anti-protease 'cocktails' has been introduced into the literature but their composition is rarely the result of systematic investigation. Several such cocktails contain, in addition to true anti-proteases, agents that should more properly be termed bacteriocides.

There is no doubt that in most cases receptor integrity is better preserved if protease activity is minimized. It is therefore worthwhile checking the effectiveness of a number of agents in improving receptor yield and activity. *Table 1* lists the composition of a typical cocktail.

Many of the agents listed in *Table 1* are extremely expensive and their use is often confined to the buffers used in the first homogenization and early centrifugation steps. They should be added to the buffer immediately before use; for example, phenylmethylsulphonyl fluoride (PMSF) is rapidly hydrolysed in aqueous solution and cannot be incorporated into stock buffer solutions. The choice of which agents to use can only be made empirically, for example in my own laboratory in studies on the muscarinic AChR of insect brain, EDTA and EGTA were found to be the most effective for ensuring high levels of binding activity; the other agents in *Table 1* had no significant effects. In most studies on vertebrate receptors PMSF seems to have become *de rigueur*! The two thiol reagents, iodoacetamide and N-ethylmaleimide, have been reported to inhibit the binding activity of the nAChR from vertebrate muscle; therefore, in this

Table I. Commonly used protease inhibitors.

Compound	*Concentration*	*Possible site of activity*
EDTA or EGTA	10 mM	Ca^{2+}-activated proteases
Benzamidine hydrochloride	2 mM	Prevents procollagen hydrolysis
Benzethonium chloride	0.1 mM	Bacteriocidal
Pepstatin	5 g/ml	Lysosomal proteases
Leupeptin	10 g/ml	Lysosomal proteases
Phenylmethylsulphonyl fluoride	5 mM	Serine proteases
Bacitracin	0.1 mg/ml	Prevents glucagon degradation; non-specific
Iodoacetamide	10 mM	SH groups of proteases
N-Ethylmaleimide	10 mM	SH groups of proteases
Soybean trypsin inhibitor	0.5 mg/ml	Trypsin

particular case, they would not be included in the anti-protease cocktail. Similar effects may well be found on other receptors, thus it is emphasized that the possible pharmacological effects of the anti-proteases should be checked prior to inclusion in the preparation buffers.

In addition to the anti-proteases described above, temperature and speed of processing are two factors to which attention should be given. All tissue homogenates and subsequent membrane fractions should be processed at 0–4°C and efforts should be made to reduce the total preparation time. In my laboratory it was found that when purifying nAChR from human muscle, reducing the overall preparation time from 36 to 24 h gave valuable increases in the activity of receptor preparations and in the reproducibility of binding characteristics and subunit composition.

2.2 Membrane solubilization

Solubilization of the receptor protein is perhaps the step in the overall purification scheme that can most affect the binding properties of the receptor. It is necessary to completely disrupt the structure of the membrane and, in effect, replace the physiological lipid environment of the receptor with a detergent coat. There is now a greater awareness of the importance of protein–lipid interactions than was the case in the early days of receptor purification and increasingly detergent–phospholipid mixtures are being employed in solubilization procedures.

The detergents used in receptor solubilization may be either ionic or non-ionic. In the latter class the most widely used are the polyoxyethylene ethers of which Triton X-100 is the most common. A similar type of detergent is Lubrol which is a mixture of condensates of ethylene oxide and long chain fatty alcohols. Less common, but worthy of consideration, are digitonin and octylglucoside. Among the ionic detergents, cholate and deoxycholate have been widely used and, more recently, the zwitterionic detergent, 3-(3-cholamidopropyl)-dimethylammoniopropane sulphonate (CHAPS), has been found to be particularly useful in some cases.

The choice of detergent is difficult and to a very great extent is done on an empirical basis. There are, however, some basic guidelines. The critical micelle concentration (CMC) value of the detergent is a factor to be borne in mind because there is an indication that detergents are more effective in solubilizing receptors when used at concentrations below their CMC. The detergent:membrane protein ratio can also be considered and detergent:protein ratios should be in the range of 10:1 to 0.1:1. It should also be borne in mind that the interaction of the detergent with the membrane proteins and lipids is an equilibrium state in which some detergent is associated as mixed micelles with membrane constituents and some detergent is free in solution. Thus if some subsequent manipulations of the solubilized protein solution involve, for example, dilution of samples, then the equilibrium will be changed. Similarly, in order to reproduce the degree of solubilization of a receptor it is necessary to ensure that both membrane and detergent concentrations are known. The time taken to reach equilibrium can be quite long, particularly if the operation is carried out at 4°C to minimize protease activity; thus again it is important that extraction times are carefully adhered to in order to ensure reproducibility of solubilization.

A further consideration is the choice of buffer in which to carry out the solubilization. Phosphate generally is beneficial in solubilization procedures; the reason for this is not well understood but, for example, it is known that phosphate ions cause extensive restructuring of water molecules. Thus it may be advantageous to change from a buffer such as Tris, which is widely used in membrane preparative procedures, to a phosphate buffer for the solubilization step.

The decision as to whether or not to use phospholipids can again only be based on trial and error. Thus the nAChR from electric organ has since the mid 1970s been solubilized by using Triton X-100 at a concentration of 1%. Such preparations show good ligand binding activities and have a pharmacological profile reminiscent of the native receptor *in situ*. However receptor protein purified in this way could not be successfully reconstituted into phospholipid bilayers. Lindstrom and his colleagues (2) devised a procedure in which solubilization of the receptor was achieved by using 2% cholate and 5 mg/ml phosphatidylcholine and this detergent:phospholipid mix was used in all subsequent manipulations. Such preparations still retained all the pharmacological properties of the native receptor but additionally the protein could be incorporated into phospholipid bilayers with full retention of biological activity.

A mixture of soybean lipids known as azolectin is often employed in conjunction with other detergents as, for example, in the case of solubilization of the GABAR from mammalian brain. In this particular case a combination of 1.5% CHAPS with 0.15% azolectin is found to give greatly improved binding activity compared with preparations made using either deoxycholate or Triton X-100. However, it is important to realize that successful solubilization procedures are likely to be developed in an empirical manner. Thus, as briefly mentioned above, the use of CHAPS in solubilizing the GABAR is beneficial but in the case of nAChR from mammalian brain this detergent markedly reduced the binding activity of the solubilized receptor (S.Wonnacott, unpublished experiments). Similarly we have found in my own laboratory that CHAPS reduces the binding activity of both nicotinic and muscarinic AChR from insect brain.

A potential problem with non-ionic detergents of the substituted polyoxyethylene type such as Triton X-100, Lubrol and Brij 35 is the presence of impurities that have strong oxidizing properties. Ashani and Catravas (3) carried out a detailed analysis of several such detergents and found that they could oxidize SH groups of many proteins. Such activity may well affect receptor activity. Most suppliers offer several grades of detergent and it is a wise precaution to use the highest quality grade for receptor solubilization; additionally Ashani and Catravas (3) describe procedures for removing the impurities.

Reynolds (4) has written an excellent short review on some of the fundamental aspects of the interaction of detergent with membrane proteins.

It should be realized that the solubilization step in the overall receptor purification scheme will not necessarily give any substantial purification of binding activity. That is, the detergent treatment is likely to solubilize the majority of the membrane proteins. The solubilized mixture of proteins does however constitute the starting material for the next stage in the purification, namely the affinity chromatography step.

2.3 **Affinity chromatography of receptors**

Conventional protein purification procedures do not in general provide a useful means of purifying neuroreceptors. This is because the receptor proteins are not significantly different in terms of their physicochemical properties from other integral membrane proteins; additionally receptor proteins constitute a very small proportion of the total membrane proteins and proportions of less than 1% are common. The receptor proteins do however possess one unique characteristic, that is, they carry a recognition site that binds very specifically and often with high affinity to particular ligands. It is this property that can be exploited in affinity chromatography purification procedures. Thus, in simple outline, a ligand that binds to the receptor with high affinity and specificity is covalently coupled to an inert support. The solubilized membrane proteins are allowed to interact with the coupled ligand and non-interacting proteins are removed by washing; the receptor is then specifically displaced, most commonly with a competing ligand.

Needless to say the practice is less straightforward than the theory. There are four important considerations in the design of such procedures. These are the support material, the ligand, possibly a spacer arm and finally a method of covalently linking the ligand to the support.

2.3.1 *Choice of support*

A matrix is required that is insoluble and has good flow characteristics. Frequently the preparation of the affinity adsorbent is an expensive and time-consuming process. Therefore, the material should have a long working life. Cross-linked agarose derivatives such as Sepharose CL are possibly the most satisfactory support media. The cross-linking gives considerable additional rigidity to the Sepharose beads and, as such, higher flow-rates can be used without the danger of the beads collapsing. Similarly, more vigorous washing procedures can be employed than is the case with native Sepharose beads. Sepharose also has a large, open pore structure compared with some other possible supports (such as cellulose or polyacrylamide) and this permits easy access of large proteins to the coupled ligand.

There are a variety of coupling methods available but the most generally used is the cyanogen bromide method first described by Axen *et al.* (5). Cyanogen bromide is thought to react with the hydroxyl groups of Sepharose to form imidocarbonates that will then readily react with primary amines to give an isourea derivative. Thus there is one important restraint on the use of the CNBr activation, namely that the ligand to be coupled must contain an amino group. An alternative coupling procedure utilizes carbodiimides which are used to form a peptide bond; thus by using a Sepharose that has available either an amino or carboxyl group, available ligands with carboxyl or amino groups can be coupled.

In the first instance it is convenient to use preparations of Sepharose that are pre-activated. Pharmacia list a number of such derivatives. CNBr-activated Sepharose can be used to immobilize proteins, peptides and ligands with a free amino group. Two further products from Pharmacia, AH-Sepharose and CH-Sepharose, both have a spacer arm of six carbon atoms terminating in an amino or carboxyl group, respectively. Ligands with carboxyl and amino groups can then be coupled using the carbodiimide procedure. The use of a spacer arm may be advantageous in that it can reduce steric hindrance, thus the ligand is held well clear of the Sepharose surface. The spacer also allows considerable movement and both factors can facilitate interaction of the ligand with the binding site on large multi-subunit receptor proteins.

The precise choice of support and the coupling conditions can only be determined by experimentation. Pharmacia Fine Chemicals produce a handbook '*Affinity Chromatography Principles and Methods*', which forms an excellent starting point for preliminary experiments; thereafter procedures can be modified in the light of experience.

2.3.2 *Adsorption of the receptor protein*

Having prepared a satisfactory affinity support the next stage is the adsorption of the receptor protein. Such details as buffer strength, pH and detergent concentration can be taken from the conditions employed in the assay of the solubilized receptor. There is a choice of whether to carry out the adsorption step as a column procedure or to use a batch treatment method. In the case of purification of the GABAR and nAChR both column and batch procedures have been used by different groups.

It is difficult to generalize about these procedures but the preference in my own laboratory is to use a batch method for the adsorption stage. In the case of nAChR from human muscle we gently stir the detergent-solubilized receptor extract with Sepharose affinity beads (the immobilized ligand is the α-toxin from *Naja* venom) overnight at 4°C. In contrast, Sigel *et al.* (6) describe a purification of the GABAR complex from bovine brain in which adsorption of the solubilized receptor to a benzodiazepine–agarose affinity gel column is achieved in about 70 min. As with other aspects of receptor purification it is not possible to give any hard and fast rules about how adsorption should be carried out. However, bearing in mind earlier comments about proteolysis it is probably advantageous to carry out the procedure at 4°C.

2.3.3 *Washing the gel*

Once adsorption is complete, which can be monitored by assaying either the column eluate or the soluble phase of a stirred batch for receptor activity, washing the affinity gel can proceed. Extensive washings are needed to remove non-receptor protein; such proteins could well represent 90% or more of the total solubilized protein. It is frequently found that the ligand–receptor complex is more stable than would be predicted from a consideration of the K_D for the same interaction in solution. This phenomenon can often be exploited in that washing conditions can be employed that will quantitatively elute proteins bound in a non-specific manner to the affinity gel through either ionic or hydrophobic interactions. Thus it may be advantageous to wash with several different buffers. Taking the nAChR of vertebrate muscle once more as an example, the affinity beads are washed alternately with 0.5 M NaCl in 0.01 M phosphate containing 1 mM EDTA and with 0.1 M phosphate buffer alone. Generally some 10–100 gel volumes of washings are used. As a general rule washings commence with high ionic strength buffers that will remove proteins that are bound by ionic interactions and these are followed by buffers of very low ionic strength which will elute proteins that are hydrophobically associated with the gel.

In the case of batch treatments the washings are conveniently done using a sintered glass funnel and gentle vacuum, care being taken not to allow the beads to become completely dry between washes. The amount of washing required can be determined simply by monitoring the protein content of the washes and when this falls to zero specific elution of the receptor can commence.

2.3.4 *Removal of the bound receptor*

In principle, specific removal of the bound receptor is achieved by introducing a competing soluble ligand for the receptor binding site. Thus, as the receptor spontaneously dissociates from the immobilized ligand the soluble ligand binds and the now mobile receptor–ligand complex is eluted. Clearly there are several important factors to be considered. The competing, soluble ligand should have a high affinity for the receptor and must be present at a much higher concentration than the effective concentration of the immobilized ligand. Additionally the receptor may, in addition to the specific interaction with the immobilized ligand, bind non-specifically to the gel matrix as would any other protein. Therefore it may be necessary to combine, for example, high ionic strength buffers and a specific displacing ligand. As mentioned above, it is sometimes found that binding of receptor to immobilized ligand is much tighter than if the same ligand is in solution. In some circumstances it may be necessary to resort to chaotropic agents, such as lithium salts or thiosulphate in order to release the bound receptor. Such treatments carry a risk of disruption of conformation of the receptor protein that may lead to an altered binding activity but this possibility can be checked by examining the pharmacological specificity of the eluted receptor. It is desirable to avoid such procedures if possible and if there is a choice of ligand with which to prepare the affinity gel it is worth carrying out preliminary experiments to determine the ease with which bound receptor can be eluted. For example in early studies of the nAChR from *Torpedo* electric organ it was found that the receptor could be quantitatively adsorbed onto an affinity gel of α-bungarotoxin coupled to Sepharose; however, it was found to be impossible to elute the receptor protein. The α-toxin from *Naja* venom, which has a K_D about three orders of magnitude higher than α-bungarotoxin, was found to be suitable and is now generally used for the purification of nAChR from vertebrate muscle.

The elution of receptor protein with a competing ligand does of course raise the problem of removing the ligand from the eluted protein. This can often be achieved by arranging an ion-exchange column in tandem with the affinity column. A further refinement is to recycle the free ligand back through the affinity column. In the case of nAChR from human muscle we have employed such a procedure; the receptor is eluted from the *Naja* toxin–Sepharose with either 0.5 M carbamoylcholine (carbachol) or 3 mM benzoquinonium chloride directly onto a DEAE–cellulose column, the eluate from this column is then recycled back through the two columns. This procedure is carried out for 16 h at 4°C and a good recovery of the receptor is obtained. After extensive washing of the DEAE–cellulose (50 column volumes) the receptor protein is eluted with 0.5 M NaCl. This procedure can in principle be applied to any receptor protein where there is differential binding of the protein and the ligand to an ion-exchange material.

An alternative procedure for removing the free ligand from the receptor is by dialysis; this however is more time consuming than the twin-column system described above and there is a danger that the receptor protein may adsorb onto the dialysis membrane thereby giving very low recoveries. Even after dialysis there is likely to be a requirement for an ion-exchange step to remove the last traces of the free ligand.

Generally, long elution times are necessary to recover the receptor quantitatively from the affinity column and during this time the receptor is exposed to very high ligand concentrations. Thus, in the case of the nAChR the receptor is in the presence of the potent agonist carbachol at a concentration of up to 1 M for 16–20 h. It has been found that such conditions can result in altered pharmacological specificity of the purified receptor presumably as a consequence of agonist-induced changes in conformation. It is for this reason that many laboratories prefer to use the ligand benzoquinonium chloride which gives effective displacement of bound receptor at low concentrations. As a general rule a displacing ligand of high affinity ($K_D < 10^{-5}$ M) should be sought to avoid problems of this sort.

It is emphasized that the procedures outlined above are necessarily rather general and lacking in detail. Receptor isolation is still rather an unpredictable exercise. We simply do not understand enough about precisely how membrane proteins interact with membrane lipids and possibly with other membrane proteins to be able to predict how they will react with detergents and how the detergent–receptor complexes will interact with pharmacological agents. Thus purification strategies must to a large extent be determined empirically.

Some alternatives to the use of a pharmacological ligand in the affinity purification step have been employed and should be considered. Most neuroreceptors are glycoproteins and as such will react with varying degrees of specificity with lectins. Lectin affinity chromatography therefore is a possible intermediate step in the purification process. Thus for example, Concanavalin A–Sepharose has been used as an additional intermediate purification step in the isolation of nAChR. It is unlikely that lectin affinity chromatography will give as high a degree of purification as a system based on a ligand directed at the receptor binding site; many other membrane glycoproteins will share common glycosyl residues with the receptor protein of interest and may therefore co-chromatograph with the receptor. Nevertheless lectin affinity chromatography may give a worthwhile enrichment of the receptor protein.

Antibody affinity chromatography could also be considered, particularly if a monoclonal antibody of high affinity can be generated. Antibodies can be readily coupled to CNBr–Sepharose with virtually no loss of antigen binding activity. Removal of the bound receptor can be a problem however and it may be necessary to use chaotropic agents or extremes of ionic strength and/or pH, all of which may damage the receptor protein. Some success however has been achieved using this approach to isolate the nAChR from human muscle.

In the following two sections details are presented of the purification scheme for two neuroreceptors, the nAChR from mammalian skeletal muscle and the GABAR complex from mammalian brain.

2.4 **Purification of the nAChR from mammalian muscle**

Unless otherwise stated all operations are carried out at 4°C.

(i) Coarsely chop the excised muscle (200–300 g).

(ii) Homogenize the muscle mince in a Sorvall Omnimix or Waring blender at full speed for 2 × 1 min in 10 mM potassium phosphate buffer, pH 7.4 (5 vol) containing 100 mM NaCl, 1 mM EDTA, 0.1 M benzoethonium chloride, 2 mM benzamidine hydrochloride, 0.1 mM PMSF, pepstatin (20 μg/ml), soybean trypsin inhibitor (10 μg/ml) and bacitracin (100 μg/ml).

(iii) Centrifuge the homogenate at 20 000 *g* for 60 min.

(iv) Homogenize the pellet from (iii) at full speed for 2 × 1 min in the same buffer (2 vol) as in (ii) with the addition of 2% (v/v) Triton X-100.

(v) Stir the homogenate for 4 h at room temperature.

(vi) Centrifuge the homogenate at 100 000 *g* for 60 min.

(vii) Gently stir the supernatant from (vi) for 4 h with *Naja naja Siamensis* α-toxin coupled to Sepharose 4B (25 ml beads carrying 12.5 mg of bound α-toxin).

(viii) Filter wash, under light vaccum, the Sepharose beads with 10 mM potassium phosphate, pH 7.4, containing 0.1% Triton X-100 (1 litre) followed by 10 mM potassium phosphate buffer containing 0.1% Triton X-100 and 0.5 M NaCl (1 litre) followed by 10 mM potassium phosphate, pH 7.4, containing 0.1% Triton X-100. All buffers should contain the protease inhibitors given in (ii) above.

(ix) Pack the washed Sepharose beads into a small column (2 × 8 cm) that is closely coupled to a DEAE–cellulose column (2 × 2 cm). The receptor is eluted from the toxin–Sepharose by recycling 30 ml of 3 mM benzoquinonium chloride in 10 mM potassium phosphate, pH 7.4, containing 0.1% Triton X-100 through the columns for 16 h.

(x) Wash the DEAE–cellulose column with a further 100 ml of the phosphate buffer and then elute the receptor with 0.5 M NaCl in 10 mM potassium phosphate, pH 7.4, containing 0.1% Triton X-100 (20 ml). Collect 1 ml fractions, assay for receptor binding activity and pool the peak fractions — these are usually found between fractions 3 and 8.

This procedure, when applied to human skeletal muscle, routinely produces between 10 and 20 pmol of receptor protein, representing a yield of about 15–20% with respect to the amount of receptor present in the initial Triton X-100 extract.

2.5 **Purification of the GABA benzodiazepine receptor complex from mammalian brain**

The GABAR of mammalian brain is a considerably more complex protein than the nAChR. It has at least three separate but interacting binding sites that are capable of affecting the associated Cl channel (6). Some isolation procedures result in a protein that does not show the full range of binding activities that are associated with the native receptor complex. Such procedures however give higher yields of protein than those that retain better the functional activity. The following scheme, kindly provided by Dr F.A.Stephenson (M.R.C. Molecular Neurobiology Unit, Cambridge) encompasses the two options. As was the case with the nAChR preparation, all operations should be carried out at 4°C unless otherwise stated.

(i) Homogenize 100 g of bovine cerebral cortex in 1.4 l of 0.32 M sucrose in 10 mM EDTA, 1 mM benzamidine hydrochloride, 0.2% sodium azide, 0.5 mM PMSF, 1 mg/100 ml soybean trypsin inhibitor, 1 mg/100 ml chicken ovomucoid trypsin inhibitor. Referred to as buffer 1.

(ii) Centrifuge at 900 *g* for 10 min; remove the supernatant and centrifuge it at 20 000 *g* for 40 min.

(iii) Resuspend the pellet from (ii) in 1.2 l of 10 mM Hepes, pH 7.4, containing 0.2% sodium azide, 1 mg/100 ml soybean trypsin inhibitor, 1 mg/100 ml chicken ovomucoid trypsin inhibitor. Referred to as buffer 2.

(iv) Centrifuge at 20 000 *g* for 40 min. Discard the supernatant and resuspend the pellet in 170 ml of buffer 2.

At this point two alternative routes can be followed: route A results in a high yield of protein but with poor binding activity; route B gives a receptor preparation with retention of function but with low yield.

(v) Solubilization

(a) Procedure A

To the pellet suspension from (iv) add 18 mg of bacitracin, 0.36 ml of 150 mM PMSF, 7.9 ml of 3.5 M KCl and 4.5 ml of 20% sodium deoxycholate. This will result in final concentrations of 150 mM KCl, 0.5% sodium deoxycholate, 1 mg/ml bacitracin, 0.5 mM PMSF. Gently stir the suspension for 10 min at 4°C.

(b) Procedure B

Proceed exactly as in procedure A except that sodium deoxycholate is replaced with 6.8 ml of 40% CHAPS containing 4% azolectin — this will result in final concentrations of 1.5% CHAPS, 0.15% azolectin. Gently stir the suspension for 30 min at 4°C.

(vi) Centrifuge the suspensions at 100 000 *g* for 60 min. Collect the supernatant either by aspiration, or if there is a heavy myelin layer, filter through glass wool. The volume obtained should be about 150 ml.

(vii) Apply the supernatant from (vi) at 60 ml/h to an affinity column of the benzodiazepine, R07-1986/1, coupled to adipic dihydrazide agarose (6). Then for route A wash overnight with 600 ml of 100 mM potassium phosphate, pH 7.4, containing 50 mM KCl, 0.1 mM EGTA, 2 mM magnesium acetate, 10% (w/v) sucrose, 0.02% sodium azide, 0.2% Triton X-100. For route B replace the Triton X-100 with 0.5% CHAPS, 0.16% azolectin.

(viii) For route A wash wtih 20 ml of 20 mM potassium phosphate, pH 7.4, containing 10% (w/v) sucrose, 2 mM magnesium acetate, 0.02% sodium azide, 0.2% Triton X-100. For route B increase the potassium phosphate concentration to 200 mM and replace Triton X-100 with 0.5% CHAPS, 0.16% azolectin.

(ix) Specific elution of GABAR from the column is achieved by passing 20 ml of 10 mM chlorazepate in the buffer mixture used in (viii) through the column in a time of 1 h. Follow this with 30 ml of the buffer alone, that is without chlorazepate. Collect fractions of 5 ml. As in (viii), route B differs from route A in the replacement of Triton X-100 by CHAPS/azolectin and the increased potassium phosphate concentration.

(x) Pool the receptor-containing fractions (pre-determined by binding assays). Adjust the pH to 6.5 with 100 mM H_3PO_4. Apply to a DEAE−Sephacel ion-exchange column (1 × 0.5 cm) equilibrated with 20 mM potassium phosphate, pH 6.5, containing 10% (w/v) sucrose, 2 mM magnesium acetate, 0.02% sodium azide, 0.2% Triton X-100. Route B differs in that 0.5% CHAPS and 0.04% azolectin replace the Triton X-100. Application of the sample is done at a flow-rate of 40 ml/h.

(xi) Elute the column with 0.8 M KCl in 1 l of the equilibration buffer used in (x). Use a flow-rate of 10 ml/h and collect 1 ml fractions. As before, for route B replace Triton X-100 with 0.5% CHAPS/azolectin. Assay the fractions for GABAR binding activity and pool the peak fractions.

The above procedure yields about 150 μg of protein that represents about a 2000-fold purification of the benzodiazepine binding activity of the initial detergent extract.

3. MEASUREMENT OF RECEPTOR BINDING ACTIVITY

Binding assays provide direct information about only one aspect of receptor function, that is, the recognition of a ligand by the receptor protein. In terms of physiological function this is only the first step in what may be a complex sequence of events that couples the recognition event to the cellular response. It is necessary therefore to take steps to ensure that the binding activity that is assayed is indeed an integral part of a physiological receptor mechanism.

The literature abounds with binding assays for almost all known receptor types. The methods follow a common pattern in which the receptor preparation is exposed to a radiolabelled ligand until an equilibrium is reached, after which time free ligand is separated from the receptor-bound ligand. Such an apparently simple procedure can unfortunately pose a number of practical problems. Thus the number of neuroreceptor sites with which the radioligand will interact in a specific, physiologically-relevant manner is likely to be a small proportion of the total number of potential binding sites in the tissue preparation. The great majority of these sites may well be physiologically irrelevant but may be defined as 'specific' binding sites within the terms of reference of many binding assays. The problem then is not simply to separate specific binding from non-specific but to distinguish between physiological receptors and 'other' binding sites.

3.1 Specific and non-specific binding sites

As mentioned briefly above, the number of receptor binding sites is small compared with the total number of sites with which the radiolabelled ligand can interact. This latter class of sites encompasses binding to the tissue preparation, to assay tubes, to filters and indeed to any part of the experimental system other than the receptor recognition site. The general procedure for distinguishing between them is to use the 'cold ligand excess' method. In this procedure parallel measurements are made of the binding of radiolabelled ligand in the absence and presence of a large excess of unlabelled ligand. Specific binding is then defined as total binding (in the absence of unlabelled ligand) minus non-specific binding (that remaining in the presence of unlabelled ligand). The method is based on two assumptions, first that the number of non-specific sites is greatly in excess of that of the specific sites and, second, that if the specific radioactivity of the radiolabelled ligand is reduced by addition of a large excess of cold ligand, then specifically-bound radioactivity will become insignificant but the level of non-specific binding is essentially unchanged. The major shortcoming of the method is that it defines 'specificity' in terms of the relationship between the concentration of radiolabelled ligand and the K_D of the binding site [see (7) for discussion]. Thus any site whose K_D is lower than or comparable with the concentration of radioligand used will emerge as a specific site. It is quite possible therefore that if differing concentration ranges of radiolabelled ligand are used then different 'specific' sites will appear. It frequently is the case that as ligands of higher specific radioactivity become available so workers use them at lower concentrations and find new specific binding sites.

Another aspect of the specific versus non-specific problem is the saturability of the sites. It is often stated rather loosely that non-specific sites are unsaturable. Clearly this cannot be entirely true because no matter what the nature of the site there must be a finite number of them in the particular experimental system that is used. However, it is probable that there are many more non-specific sites than there are specific sites such that when using concentrations of radiolabelled ligand in the generally used range of $10^{-9}-10^{-4}$ M, saturation of the non-specific sites is not approached. Putting it a slightly different way, saturability of the various binding sites is operationally defined and 'unsaturable' binding under one set of conditions may well be 'saturable' under different conditions.

It is just over 10 years ago that Cuatrecasas and Hollenberg (8) reported the specific, saturable binding of radiolabelled insulin to talc, similarly radioactive opiates bind to glass fibre filters with K_D values in the nanomolar range and with some degree of pharmacological specificity. More recently the same phenomenon was re-discovered by Bielkiwicz and Cook (9) who reported the specific, high affinity binding of the histamine receptor ligand, mepyramine, to glass fibre filters. Clearly then it may be necessary to take steps to ensure that in any particular binding assay not only can a distinction be made between specific and non-specific sites but also between physiologically relevant and irrelevant sites.

To some extent this can be achieved by a modification of the assay system so that specific binding of the radiolabelled ligand is assessed by displacement not with cold ligand but with a chemically-distinct agonist or antagonist. In this case the probability of competition between the radiolabelled ligand and the chemically-distinct displacing ligand for any common, non-physiological, high affinity site is greatly reduced. Indeed it was recommended some years ago (10) that as a general rule a structurally-distinct ligand should be used for the displacement of specific binding. The extension of this approach is of course to displace the radioligand with a variety of compounds that are known from physiological experiments to act at the appropriate neuroreceptor site.

A further problem of the definition of a binding site may arise from the nature of the receptor preparation. Many assays are performed on whole tissue homogenates or at best crude membrane fractions. Such preparations will inevitably contain large numbers of vesiculated membranes that may actually take up the radiolabel thereby giving apparently high levels of binding. The problem can be minimized by doing the binding assays at 0°C, or perhaps better, by investigating the relationship between binding and temperature; a marked increase in binding at higher temperatures is suggestive of uptake. Additionally because many transport processes show a dependence on Na^+ ions, a comparison of binding in Na^+-free and Na^+-containing buffers should be made.

Finally, steps can be taken to disrupt membrane vesicles in the preparation, for example a series of three or four rapid freeze–thaw cycles or the addition of detergent such as 0.01% Triton X-100, may well reduce considerably the apparent binding.

Bearing these factors in mind we can now consider what is perhaps one of the most critical aspects of the binding assays: namely the procedures used to separate the receptor–ligand complex from free ligand.

3.2 **Separation of free from bound ligand**

The preparations in which neuroreceptor sites are assayed fall into two distinct classes: there are particulate preparations which comprise tissue homogenates and membrane preparations of varying degrees of purity and there are preparations in which the receptor protein has been solubilized. In the former case the choice is between centrifugation and filtration whereas with soluble receptor preparations the choice is much greater and includes differential precipitation of the receptor protein, gel exclusion chromatography, adsorption of free ligand and equilibrium dialysis.

3.2.1 *Separation time*

Whatever the method that is finally chosen, the single most important factor is the time taken to achieve a separation. In the great majority of cases binding measurements are made at equilibrium, therefore as soon as the free ligand is separated from the bound ligand that equilibrium is disturbed and bound ligand will dissociate until a new equilibrium is reached. The equilibrium dissociation constant, K_D, for the binding is of course a function of both the association rate constant, k_{+1} and the dissociation rate constant k_{-1}. In the majority of receptor binding assays, k_{+1} is of the order of 10^6 M/sec. (Association is a diffusion-limited process in most cases but in the case of some large ligands, for example peptides, where multipoint attachment to the receptor may occur, there could be a significant slowing of the association rate.) The variations seen in K_D, which are considered to reflect the affinity of the receptor, are therefore primarily reflections of variation in k_{-1}. The parameter of immediate practical concern is the half-life of the ligand–receptor complex, $T_{1/2}$, and this approximates to $0.7/k_{-1}$. What this means in terms of practical methodology is that in order to limit the loss of bound ligand by dissociation to less than 10%, a separation time of not more than 0.15 $T_{1/2}$ must be employed. Expressed as a function of K_D the permitted separation time increases approximately 10-fold for every 10-fold decrease in K_D and goes from a value of about 0.01 sec at 10^{-6} M up to about 30 h at 10^{-12} M. Most neuroreceptor binding assays are done with ligands with a K_D in the range of 10^{-8} M (permitted separation time 10 sec) to 10^{-10} M (permitted separation time 15 min).

3.2.2 *Vacuum filtration*

Vacuum filtration offers possibly the fastest and most convenient method of achieving separation times of a few seconds. There are several commercial filtration chambers available; the Millipore and BioRad versions are the most widely encountered. With such systems total filtration and washing times of 3–8 sec can be routinely achieved for sample volumes of up to 10 ml using 2.5 cm filters. Depending on the K_D of the receptor–ligand pair it may not be possible to wash the retained material extensively and thus filter assays may suffer from high blank values. Non-specific adsorption of free ligand to the filters may also be a problem. This can often be overcome by a variety or combination of pre-treatments.

(i) The filters can be pre-soaked for a few minutes in a solution of the cold ligand at relatively high concentration (e.g. 100 times the concentration of radiolabelled ligand).

(ii) Pre-soaking the filters in bovine serum albumin, casein or gelatin solutions (1% w/v) often markedly reduces ligand binding, particularly in the case of peptides.

Filter binding is generally lower if the filters are wet at the time the sample is added and in this respect the multiplace filter holders can present a problem in that if the vacuum is present when the samples are added then the later filters will probably dry out during the time that the earlier samples are added. Similar considerations apply to the treatment of the filters between the initial filtration and subsequent washes. In my own laboratory we had an experience of this when assaying the effects of GABA on the binding of the benzodiazepine, [^{3}H]flunitrazepam, to a putative GABAR on membranes from insect ganglia. In several experiments we observed an enhancement of flunitrazepam binding that increased with increasing GABA concentrations but then decreased as the GABA concentration increased further. It was the practice in the laboratory (as in many others) to use a 12-place filter holder and the GABA concentration increased from filters 1 to 12. We noticed that, by varying the order in which samples were filtered, so the degree of enhancement of flunitrazepam binding changed. Clearly the sample treatment on a 12-place holder varies; when we repeated the experiments using a single Millipore filter tower, in which each sample is subjected to an identical washing and drying period, our anomalous results disappeared. Such effects cannot be predicted and in many binding assays no such problems will be encountered. It is nevertheless worthwhile randomizing the order in which samples are filtered on multiplace holders, at least in preliminary experiments, in order to rule out any such effects.

The risk of dissociation of bound ligand during filtration and washing can be further minimized by ensuring that all buffers are ice-cold. However even with all these precautions it is emphasized that satisfactory data will be obtained only if the K_D is 10^{-8} M or less. If the K_D is higher than this, or if there are insurmountable problems of ligand binding to the filters, then the alternative separation procedure of centrifugation must be employed.

3.2.3 *Centrifugation*

Centrifugation has the advantage that until the supernatant is removed the system remains at equilibrium even though one has effectively, by pelleting the particulate material, achieved a separation of free from bound ligand. It is of course not possible to wash the pellet and inevitably there is some physical entrapment of free ligand in the pellet. Some workers superficially rinse the pellet but this does little to reduce the trapped ligand. The problem can be minimized by using very high *g* forces so that small densely-packed pellets are obtained. The small air-driven ultra-centrifuge, the Beckman 'Airfuge', is particularly useful as *g* forces in excess of 100 000 can be generated in less than 1 min. Depending on the nature of the particulate component of the assay it may be possible to obtain satisfactory pellets in relatively low-speed small-capacity bench-top centrifuges such as the MSE Microcentaur. It is not advisable to attempt to remove the pellets from the centrifuge tubes for subsequent counting; instead a small volume (100–200 μl) of a tissue solubilizer, such as Soluene by Packard, is added to the pellet and the tip of the centrifuge tube is then cut off and dropped directly into scintillation vials. These can be left for a period for the pellet to dissolve and scintillant can then be added and counting carried out in the preferred manner.

If one wishes to do binding assays with the natural transmitter it is likely that centrifugation assays will be the only suitable method because most transmitters have a K_D in the micromolar region and it is emphasized again that in this case filter assays will grossly underestimate the bound ligand.

The foregoing discussion applies to preparations in which the receptor is present in its membrane-bound form. If however the receptor has been solubilized by detergent treatment of the membrane preparation then there is a greater choice of methods for separating free ligand from the receptor–ligand complex.

3.2.4 *Equilibrium dialysis*

In principle equilibrium dialysis should be one of the most accurate ways of measuring binding over a very wide range of K_D values because the equilibrium is not disturbed by the sampling procedure. In practice the technique is rarely used.

In its simplest form the receptor-containing solution is tied into small (0.5–1.0 ml) dialysis sacs which are then suspended in a larger volume of buffer containing radiolabelled ligand. The sac contents are then sampled with increasing time until a steady-state value is obtained. At this time the concentration of free ligand inside and outside the sac is the same; thus any radioactivity in the sac over and above this level is the result of receptor binding.

The big disadvantage is that unless very high receptor concentrations are used the amount of radioactivity bound is low compared with the background level; that is, one sees perhaps a 10% greater count rate for the sac contents than for the buffer outside and clearly this greatly reduces sensitivity. The procedure is also tedious and time-consuming although there are some commercial dialysis cells in which small chambers of volume 100–500 μl are separated by a dialysis membrane and multiple assays can be done more conveniently than by tying dialysis sacs. The procedure can also be expensive in that large volumes of radiolabelled ligand are required compared with other assay methods. However if it is required to measure the binding of small molecular weight ligands (<1000) with K_D values in the range $10^{-5}-10^{-7}$ M, then perhaps it is worthwhile making an assessment of equilibrium dialysis.

3.2.5 *Differential precipitation*

Differential precipitation of the receptor–ligand complex has been used for assay of the nAChR. In this case 35% saturated ammonium sulphate promotes precipitation of the detergent-solubilized receptor and bound ligand (in this case α-bungarotoxin) while the free ligand remains in solution. The precipitated receptor–ligand complex can be collected either by vacuum filtration or by centrifugation. An alternative precipitating agent is polyethylene glycol which is widely used to precipitate antibody–antigen complexes but has so far found little application in neuroreceptor assays.

One disadvantage of this general procedure is the relatively long time taken for precipitation (in the case of the nAChR assay about 12 h). However it seems that the equilibrium is not necessarily disturbed; certainly we have found that the binding parameters of the nAChR for α-bungarotoxin are not changed by the ammonium sulphate precipitation. Bearing in mind the speed with which separation of the precipitate can be achieved, the method may be worth considering for receptors with a K_D of greater than 10^{-8} M.

3.2.6 *Ion-exchange*

An alternative assay procedure exploits the predominantly anionic nature of most proteins in the pH region 7–8. Thus if the ligand in question is either uncharged or cationic, a separation of free from bound ligand can be based on ion-exchange media.

Again the nAChR provides a good example where soluble receptor plus bound ligand is retained on a DEAE–cellulose filter whereas the free ligand, α-bungarotoxin, is washed straight through. The method has some limitations in that the filter capacity is rather limited so that if the receptor protein is a small proportion of the total soluble protein in the sample, the filter can effectively be saturated with non-receptor protein with corresponding underestimation of receptor concentrations. One simple way to improve capacity is to use two filters, a procedure adopted in my own laboratory for assaying partially-purified nAChR. Another problem is the relatively low flow-rates through DEAE–cellulose filters compared with the glass fibre GF/B or GF/C filters used in binding assays of particular receptor preparations. Increasing the vacuum can help but there is then a danger of rupturing the filter with a corresponding loss of sample!

An interesting and potentially widely-applicable variation on the ion-exchange filtration method is currently gaining ground. This is based on the coating of GF/B glass fibre filters with the cationic polymer polyethylenimine (11). This procedure effectively transforms the GF/B filter into a robust and fast flowing ion-exchange filter. Filter binding of free ligand, particularly if the ligand is cationic, is particularly low. Preparation of the filters simply involves pre-soaking for 1–24 h in 0.3% polyethylenimine at pH 10 and they are then used directly with no further washing. The system has been successfully used to assay binding to muscarinic and nicotinic acetylcholine receptors, adrenergic receptors, dopaminergic receptors, opiate and other peptide receptors and benzodiazepine receptors. Given the more rapid filtration times compared with DEAE filters it may be possible to extend the range of K_D values to a little above 10^{-8} M.

3.2.7 *Other methods*

There are other filtration methods based on filter membranes of known pore size such that molecules of a particular molecular weight are retained. Such membranes, produced by Millipore or Amicon, tend to have slow flow-rates which of course extend filtration and wash times. Additionally some ligands show a high level of binding to such membranes thereby reducing the sensitivity of the assay.

There are several variations on the basic procedure of adsorption of free ligand onto an insoluble particulate support followed by rapid filtration or centrifugation. Charcoal is the most commonly used adsorbent although talc has also been used. In general these techniques have not been widely used in neuroreceptor binding studies but have been largely confined to the area of hormone receptors. Possibly if unsatisfactory results are obtained with the more common procedures described earlier an investigation of adsorption assays may be worthwhile.

Gel exclusion chromatography has been applied to some receptor-binding assays. In outline, the receptor preparation is incubated with ligand until equilibrium is reached and the mixture is then applied to the gel column, for example, Sepharose G50, the assumption is that receptor is excluded whereas the small molecular weight ligand is included, thus the receptor will appear in the void volume. The time taken for passage through the column is obviously crucial. In practice the columns can be very small (2–3 cm × 0.2–0.3 cm) and the elution can be speeded up by centrifugation. A convenient system can be assembled from disposable pipette tips, which serve as the columns, to which the sample is added followed by a brief (1–2 min) centrifugation in a bench-top centrifuge. The system is unlikely to be suitable for use with receptors of K_D greater than 10^{-8} M, but offers the advantage that 'non-specific binding' (or more correctly co-elution of free ligand with the receptor–ligand complex) is usually very low.

3.3 **Analysis of binding data**

In the majority of cases the parameter of most interest is the affinity of the receptor for the ligand. This is most frequently expressed as K_D, the equilibrium dissociation constant. Assuming that the receptor–ligand interaction obeys simple mass–action rules and that only a single class of non-interacting sites is present then:

$$K_D = \frac{[L_f]\ [R]}{[L_b]} \qquad \text{Equation 1}$$

where L_f and L_b represent free and bound ligand respectively and R is the unoccupied receptor. Thus measurement of free and bound ligand concentrations at constant R will permit the construction of a simple hyperbolic binding curve. A reasonable estimate of K_D can be made by taking the L_f value at which L_b is half-maximum.

In practice, experimental binding data rarely permit determination of K_D by such a simple analysis and a variety of secondary analytical procedures is available. Such analyses are in the main borrowed directly from the field of enzyme kinetics and while this is useful to a point, it is worth remembering that there are some very fundamental differences between receptor–ligand interactions and the catalytic conversion of substrate to product. Nevertheless it is reasonable to see the formation of a receptor–ligand complex as analogous to the formation of an enzyme–substrate complex. Thus receptor–ligand interaction can be presented in the form of a Michaelis–Menten equation:

$$[L_b] = \frac{[L_f]\ [R_t]}{K_D + [L_f]} \qquad \text{Equation 2}$$

where R_t is the total number of receptor sites present. Comparing Equation 2 with the Michaelis–Menten equation:

$$V = \frac{[S]\ V_{max}}{K_m + [S]}$$

it can be seen how tempting it is to equate the parameters K_D with K_m and R_t with V_{max}. This can be misleading however because in our binding experiments L_b and L_f are calculated from a single observation, that is, L_b is directly measured and L_f is determined by subtraction of that value from the total ligand concentration. Thus L_b and L_f are dependent variables and any error in L_b will necessarily occur in L_f also. This is in marked contrast to the case of measurements of enzyme kinetics where S, the counterpart of L_f has a fixed, known value whereas v, the initial velocity, the counterpart of L_b, is the only dependent variable. In this case therefore any errors that occur in the measurement of v will not occur in S. A further point to consider is that in studies of enzyme kinetics, K_m is not identical to K_D because the former parameter can be affected by the rate of formation of product from substrate and clearly this situation cannot occur in ligand–receptor interactions. Thus it will not necessarily follow that the best methods for analysing enzyme kinetic data are appropriate for receptor binding data.

In the majority of cases the two parameters which are of most interest are R_t, the total number of receptor sites, and K_D. Most methods for determining their values involve a transformation of the receptor binding data into a straight line equation. Linear regression analysis of such lines reassures many workers and permits the evaluation of some degree of statistical significance of the data. However it is emphasized that such analyses assume a normal distribution of errors whereas in reality this is not likely to be so in binding experiments. The most widely used analysis is the Scatchard plot which in the context of binding experiments has the form:

$$\frac{[L_b]}{[L_f} = \frac{-1}{K_D}[L_b] + \frac{[R_t]}{K_D}$$

Thus when $\frac{[L_b]}{[L_f]}$ is plotted against $[L_b]$ a straight line of slope $\frac{-1}{K_D}$ is obtained (assuming a single population of non-interacting sites) with an intercept on the L_b axis equal to R_t (because $[R_t] = [L_b]$ when $\frac{[L_b]}{[L_f]} = 0$).

In spite of the very widespread use of the Scatchard plot the literature still abounds with examples of misuses and miscalculations of the procedure. The first important point concerns the range of ligand concentrations used, which should ideally cover the range of 10–90% saturation of the receptor site. In practice this can be extremely difficult and if one can achieve a receptor concentration equivalent to K_D and a ligand concentration range from about 0.1 to 10 times K_D then satisfactory Scatchard plots are likely to be obtained.

Most of the misinterpretations of Scatchard analyses arise when non-linear plots are obtained. These can arise from either heterogeneity of sites or cooperativity within a single class of sites and in most cases the former is the more likely. *Figure 1a* shows the typical appearance of such a plot. The very common error that is made is to try to resolve the curve into two linear components by intuitively drawing tangents to the two ends of the curves and from those tangents calculating K_D and R_t for the presumed two sites.

The first major error in such procedures is to assume that the second intercept on

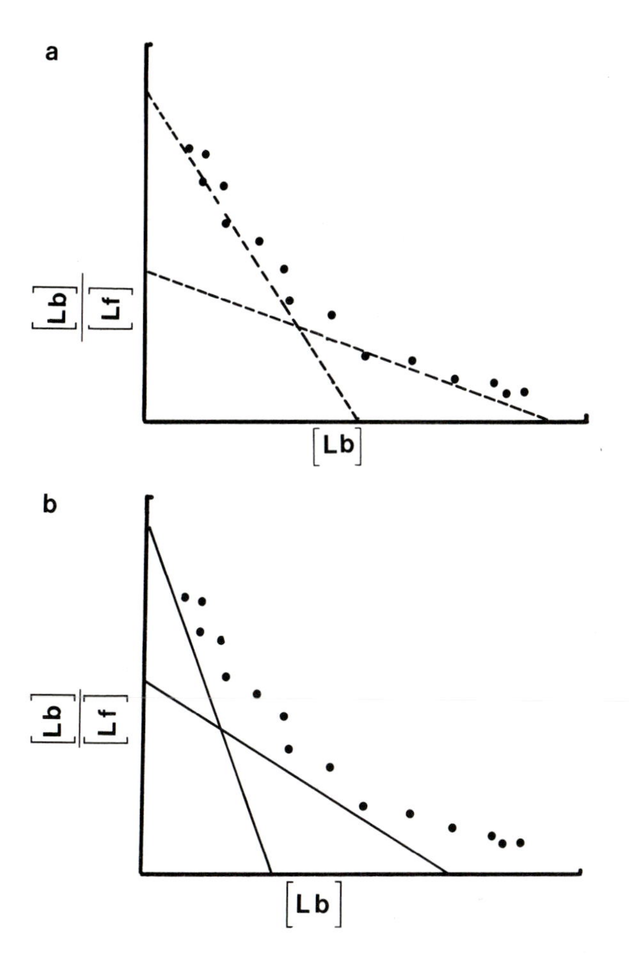

Figure 1. Curvilinear Scatchard plot. (**a**) The experimental points clearly do not fall on a straight line. The two dashed lines represent tangents drawn to the two extremes of the curve; many similar attempts to deal with binding data in this way can be found in the literature. (**b**) In this graph the same experimental points are presented but in this case the curve has been resolved into two linear portions using a computer program based on the method of Rosenthal (12). It can be clearly seen that quite different values for K_D and R_t will be obtained from those resulting from the analysis in (**a**).

the L_b axis corresponds to R_t for the second, lower affinity site; this is simply not true — this intercept represents occupation of all classes of sites. In fact it is not possible to calculate either parameter, that is K_D or R_t, from the tangents shown in *Figure 1a*. At each experimental point the magnitude of L_b and hence of L_f is a function of the amount of ligand bound at both sites. At very low ligand concentrations the higher affinity site may predominate whereas at very high ligand concentrations this site may be saturated and one can effectively measure binding to the low affinity site only. Therefore it is essential, as emphasized previously, to cover as much as possible of the saturation curve. Secondly, when calculating the parameters, the relative contributions of the two sites must be determined. There are several ways in which this can be done: a graphical procedure was described by Rosenthal in 1967 (12) and more recently Norby *et al.* (13) looked again at the problem and found 50 papers published between 1975 and

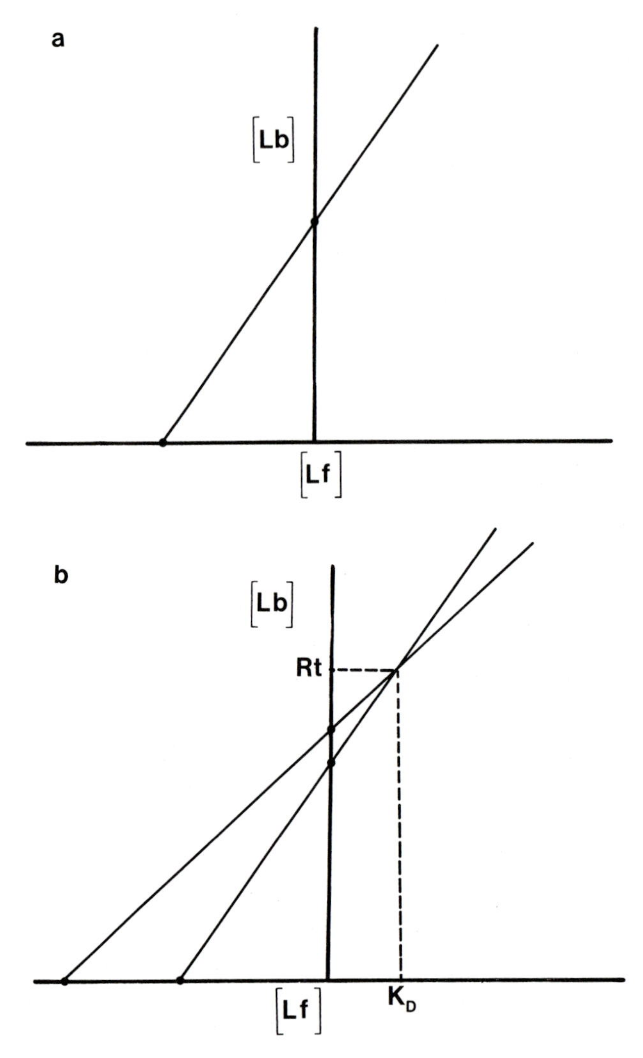

Figure 2. Direct linear plot. In (**a**) a single pair of data points is plotted giving the first 'line', in (**b**) the next pair of observations is plotted and the intercept of the two lines gives an assessment of K_D and R_t.

1980 in which the authors had miscalculated the parameters. Burgisser (14) in 1984 posed the question in an article — 'What's wrong with Scatchard analysis?'

Figure 1b shows the same data as *Figure 1a* but with the calculated lines of best fit drawn in. As a sound working rule, tangents drawn by eye, as in *Figure 1a*, will give reasonable estimates of binding parameters only if the K_Ds of the sites differ by at least two orders of magnitude but if, as is frequently the case, the difference is 5- to 10-fold then errors of up to 100% can easily occur. There are several computer programs that can be used for fitting binding data and frequent discussion and updating of these can be found in the 'Computer Club' columns of '*Trends in Pharmacological Sciences*'.

A procedure that is slowly gaining ground is the direct linear plot first described by Eisenthal and Cornish Bowden (15). The method, originally developed for the analysis of enzyme kinetics data, has much to offer in the field of receptor binding. Axes are set up corresponding to L_b and L_f as shown in *Figure 2*. However, the experimental observations are plotted as lines through the axes rather than as points. The analysis differs fundamentally from almost all others in that the parameters K_D and R_t are seen as variables. Thus for a single pair of values of L_b and L_f there is an infinite variety of corresponding pairs of values of K_D and R_t and this is what is represented in the straight line in *Figure 2a*. When a second observation is made then that second line satisfies all possible values of K_D and R_t that correspond to the measured values of L_b and L_f. However the point of intersection of the two lines uniquely defines a pair of K_D and R_t values that exactly satisfies both sets of experimental observations. Thus with a series of observations a series of intersections is obtained (see *Figure 2b)*. The best estimate of K_D and R_t corresponds to the medial value of the range of intersections and, in this respect, the direct linear plot is advantageous in that it does not depend on assumptions about uniform distribution of error and weightings that are made by all the least squares methods for fitting to straight lines. The statistical advantages of the method are considered in detail by Cornish Bowden and Eisenthal (16).

The method offers the further advantage that it requires very little calculation of raw data before graphical analysis and it is probably the only statistically-sound method for obtaining K_D and R_t estimates by eye. In a comparative study of nine methods for analysing ligand binding data the direct linear plot emerged as the method of choice and was particularly advantageous at ligand concentrations below K_D (R.Eisenthal, personal communication). The direct linear plot is included in a very good software package called 'Enzpack', by P.A.Williams and published by Elsevier Biosoft. Although this material is primarily concerned with the teaching of enzyme kinetics it contains much that is directly relevant to ligand binding experiments.

In the following section detailed directions are presented for some receptor binding assays. As with much of the previous material, it is unlikely that the precise methodology will be applicable to other receptor types but the general procedures should serve as a starting point for the development of particular receptor assays.

3.4 Assay of nAChR binding activity by ammonium sulphate precipitation

This procedure is widely used for the assay of receptor activity in detergent extracts of muscle or electric organ.

(i) Add [^{125}I]α-bungarotoxin ($\sim$600 Ci/mmol; 0.5–1.5 nM) to 100 μl of receptor-containing detergent extract and mix thoroughly.

(ii) Add 75 μl of saturated ammonium sulphate to give a 30% (w/v) final concentration.

(iii) Leave at 4°C for 16 h.

(iv) Filter under vacuum on 2.5 cm diameter Whatman GF/C discs. Wash each disc, under vacuum, with 3 ml of 30% (w/v) ammonium sulphate.

(v) Count the discs in a gamma counter.

The entire assay should be done in the presence and absence of either 0.1 mM d-tubocurarine or 2.5 mM benzoquinonium chloride. Specific binding is then defined as that proportion of the total radioactivity that is displaced by the competing ligand.

3.5 Assay of nAChR by DEAE cellulose ion-exchange

This procedure is used for the assay of receptor activity in partially-purified preparations. It is emphasized that if the receptor is not the major protein in the preparation the method may underestimate R_t.

(i) Incubate the receptor preparation (100 μl) with [^{125}I]α-bungarotoxin (50 μl; 0.1–1.5 nM) for 1.5 h at 23°C.
(ii) Terminate the incubation by the addition of ice-cold 0.01 M potassium phosphate buffer, pH 7.4, containing 1% Triton X-100 and 0.1% (w/v) bovine serum albumin (1 ml).
(iii) Filter under vacuum through two Whatman DE81 cellulose filter discs (2.5 cm) that have been pre-soaked in the buffer.
(iv) Wash the discs under vacuum with 5 ml of the same phosphate buffer.
(v) Remove the discs and count in a gamma counter.

Specific binding is determined exactly as in Section 3.4 above.

3.6 Assay of solubilized nicotinic receptors in brain by polyethyleneimine ion-exchange

This method has been recently developed by Dr S.Wonnacott who has kindly provided the detailed procedure below.

(i) Prepare brain synaptic membranes (P2 membranes) and resuspend them in 50 mM potassium phosphate buffer, pH 7.5, containing 0.5 mM PMSF, 5 mM EDTA and 0.02% (w/v) sodium azide (2 ml/g original wet weight).
(ii) Add Triton X-100 (20% w/v) to give a final concentration of 0.5% (w/v), stir on ice for 60 min.
(iii) Centrifuge at 40 000 *g* for 30 min and remove the supernatant.
(iv) Dilute the supernatant 5-fold with 20 mM Hepes buffer, pH 7.5, containing 118 mM NaCl, 4.8 mM KCl, 2.5 mM $CaCl_2$ and 1.2 mM $MgSO_4$.
(v) Incubate the samples (0.25 ml) of the diluted supernatant with [^{3}H](−)nicotine (10 μl 60 Ci/mmol; 40 nM final concentration) for 40 min at 20°C. Duplicate sets of samples should be prepared and to one set (−)nicotine is added to give a final concentration of 10^{-3} M.
(vi) Transfer the samples to ice and leave for 30 min.
(vii) Filter the samples under vacuum through Whatman GF/B filters (2.5 cm) pre-soaked for 3 h at 40°C in 0.33% (v/v) polyethyleneimine.
(viii) Wash three times with 2 ml of ice-cold 20 mM Hepes buffer. The total filtration and washing time should be less than 20 sec.
(ix) Transfer the filters to scintillation vials, dry and count in the preferred manner.

Specific binding is defined as that component of the total binding displaced by 10^{-3} M (−)nicotine. It is found that in this assay GF/B filters have a higher capacity and give a more quantitative recovery of binding sites than the GF/C filters that are preferred for most filter assays of particulate receptor preparations.

3.7 Availability of radiolabelled ligands

There are two major sources of radiolabelled ligands. Amersham International plc based in Amersham, UK (formerly The Radiochemical Centre) and Du Pont (formerly New

England Nuclear or NEN). Both companies supply, in addition to their main product catalogue, separate listings of receptor ligands. Du Pont produce a large poster, entitled Receptor Site Analysis, that lists about 220 ligands covering both neuroreceptors and some steroid receptors. The poster identifies the pharmacological specificity of the ligands and briefly indicates such properties as reversibility, photoaffinity activity and whether uptake/transport sites are labelled. In addition to this, the company produces an even more extensive Receptor Ligands List which is available on request [from Du Pont (UK) Ltd, Wedgwood Way, Stevenage, Herts SG1 4QN]. The company also produces a regular bulletin, NEN Products News, that highlights new ligands and gives brief summaries of work that has led to their adoption as useful new probes. The bulletin is free on request.

Amersham International also produces specialist listings of neuroreceptor ligands. New technical literature has just been produced under the title of The Receptor Release Programme. This is a scheme whereby interested workers can be informed immediately of the availability of new ligands. Additionally the company has just launched a new bi-monthly information bulletin called Amersham Research News. The bulletin covers the whole product range but highlights products that have potential use in receptor studies. Amersham International also produce ancillary products such as specialist film and equipment for the rapidly-growing area of receptor autoradiography.

One problem that can arise is that in my experience neither company deals adequately with the availability of cold, that is non-radiolabelled, ligand. Thus it is necessary to have available non-labelled ligand in order to proceed with the 'cold ligand excess' method for measuring specific binding. It is also frequently desirable to be able to dilute the radiolabelled ligand, particularly when carrying out a saturation study. Thus at very high ligand concentrations ($>10^{-4}$ M) it may be prohibitively expensive to carry this out using radiolabelled ligand alone. In the case of some ligands it can be extremely difficult to obtain non-labelled material, for example both NEN and Amersham produce radiolabelled cage convulsants (TBPS and TBOB, respectively) that are widely used to label the Cl^- channel component of the GABAR; however the non-labelled compounds are not available commercially. It is my experience that both companies are quite helpful in helping to locate sources of cold ligands but I think it would be of great help to research workers, particularly to those entering the field, if the companies' receptor ligand products covered non-labelled ligands also.

4. REFERENCES

1. Paton,W.D.M. and Rang,H.P. (1965) *Proc. R. Soc. B,* **163**, 1.
2. Lindstrom,J., Anholt,R., Einarson,B., Engel,A., Osame,M. and Montal,M. (1980) *J. Biol. Chem.,* **255**, 8340.
3. Ashani,Y. and Catravas,G.N. (1980) *Anal. Biochem.,* **109**, 55.
4. Reynolds,J.A. (1981) In *Receptors and Recognition, Series B. Volume 11.* Jacobs,S. and Cuatrecasas,P. (eds), Chapman and Hall, London and New York, p. 33.
5. Axen,R., Porath,J. and Ernback,S. (1967) *Nature,* **214**, 1302.
6. Sigel,E., Stephenson,F.A., Mamalaki,C. and Barnard,E.A. (1983) *J. Biol. Chem.,* **258**, 6965.
7. Briley,P.A., Filbin,M.T., Lunt,G.G. and Turner,P.D. (1981) *Mol. Cell Biochem.,* **39**, 347.
8. Cuatrecasas,P. and Hollenberg,M.D. (1975) *Biochem. Biophys. Res. Commun.,* **62**, 31.
9. Bielkiewicz,B. and Cook,D. (1985) *Trends Pharmacol. Sci.,* **6**, 93.
10. Bennett,J.P. (1978) In *Neurotransmitter Receptor Binding.* Yamamura,H.L., Enna,S.J. and Kuhar,M.J. (eds), Raven Press, New York, p. 57.
11. Bruns,R.F., Lawson-Wendling,K. and Pugsley,T.A. (1983) *Anal. Biochem.,* **132**, 74.

12. Rosenthal,H.E. (1967) *Anal. Biochem.*, **20**, 525.
13. Norby,J.G., Ottolenghi,P. and Ercinska,M. (1980) *Anal. Biochem.*, **102**, 318.
14. Burgisser,E. (1984) *Trends Pharmacol. Sci.*, **5**, 142.
15. Eisenthal,R. and Cornish-Bowden,A. (1974) *Biochem. J.*, **139**, 715.
16. Cornish-Bowden,A. and Eisenthal,R. (1974) *Biochem. J.*, **139**, 721.

CHAPTER 6

Second messengers and protein phosphorylation in the nervous system

H. CLIVE PALFREY and PHILIP MOBLEY

1. INTRODUCTION

The role of second messengers in regulating and integrating the activity of the nervous system has become widely appreciated in the last decade. Cyclic nucleotides, Ca^{2+} and the more recently discovered diacylglycerol are elevated in many cells, including neurones, in response to a wide variety of neurotransmitters and hormones, as well as by electrical activity *per se*. While measurement of the levels of second messengers is still important, recent work has also focused on the mechanisms whereby such intracellularly-generated second messengers alter the physiological behaviour of nerve and other cells. In this respect it has become clear that stimulation of specific protein kinases, and alteration of the phosphorylation state of numerous target proteins, is the major mechanism whereby many second messengers exert their final physiological effects.

The problems inherent in the study of the nervous system extend to studies on second messengers: for example, the heterogeneity encountered in the central and peripheral nervous systems (CNS and PNS) hampers biochemical approaches to the study of processes in specific cells. This has led to the development of model systems such as synaptosomes, cultured neurones and other types of cells (e.g. glia) from particular brain regions, or continuous lines of transformed cells that retain some neuronal or glial characteristics (see e.g. Chapters 1 and 2). In addition, the study of single identified neurones in the nervous systems of invertebrates is beginning to prove useful in the elucidation of second messenger action.

The central importance of protein phosphorylation as a key process underlying hormonal action has received particular emphasis in the nervous system (1). Neurones appear to be rich in protein kinases and recent work has demonstrated the existence of several neurone-specific phosphoproteins that may mediate the actions of neurotransmitters in the brain. One class of substrates that deserve special mention are the ion channels, since the activity of several may be modulated by phosphorylation. Another area of significant interest is the nerve terminal, where transmitter release may be regulated by phosphorylation of specific proteins in response to cAMP or Ca^{2+}. Recent reviews may be consulted for overviews of the areas of brain protein kinases and neuronal phosphoproteins (2–4).

In this chapter we cover the practical aspects of studying second messengers and protein phosphorylation in the brain and other elements of the nervous system. Due to space limitations we have not considered the measurement of second messengers other than cyclic nucleotides, for example calcium and diacylglycerol. For further information about recent developments in the area of these second messengers readers are referred to recent reviews (5,6) and, for methodology related to phosphoinositide turnover, to Chapter 7 of this volume.

2. TISSUE PREPARATIONS

2.1 Whole CNS or PNS preparations: individual neurones *in situ*

These types of preparation are the most physiologically relevant in the study of second messengers and their actions. Their limitations lie in the complexity of the tissue and its manipulation in the case of the mammalian nervous system, or in the special technical difficulties of working with identified neurones in the case of the invertebrate nervous system. Several studies have detailed changes in cyclic nucleotide levels in brain *in situ* and intracranial injection of [^{32}P]orthophosphate has been used for the investigation of protein phosphorylation *in vivo* (e.g. 7,8). For instance, rapid labelling of proteins of molecular weight 40 000 (the α subunit of pyruvate dehydrogenase, PDH), 47 000 (protein B-50) and 14 000 and 18 000 (myelin basic protein; see *Table 1* for a listing of some prominent brain phosphoproteins) has been shown using this approach. Recently, techniques for infusion of [^{32}P]orthophosphate into specific brain areas have been refined and coupled to two-dimensional gel electrophoresis (9) offering increased resolution of proteins labelled *in vivo*.

A major drawback of experiments *in vivo* is the difficulty of identifying the neurotransmitter systems that may be involved in mediating any changes in second messenger systems observed with perturbations to the intact animal. Another serious complication in studying second messengers and their effects in brain is the rapidity with which post-mortem changes can take place. For example, decapitation has been shown to increase the level of cAMP several-fold within seconds (10) and it is conceivable that changes in Ca^{2+} and other second messengers could also occur. In order to minimize such changes, the use of methods that fix brain tissue rapidly is essential. A popular method has been rapid freezing, usually by dropping the whole animal into liquid N_2. However, the freezing of brain in intact animals by this method is relatively slow; for example, in the rat, deep brain structures can take over a minute to reach 0°C. In addition, frozen brain is difficult to remove and dissect. For cyclic nucleotide measurements, intact animal studies have benefited from the development of microwave technology. Microwaves lead to rapid heating of tissue and denaturation of enzymes, without affecting small heat-stable molecules such as some neurotransmitters and cyclic nucleotides (11). The rate of heating depends on several factors such as microwave frequency and power output of the instrument as well as the size of the animal. Machines presently approved for animal sacrifice employ a waveguide system which allows the targetting of radiation to the head region. The animal is placed in a restrainer to minimize head movement and inserted in the instrument so that just the head and neck are exposed to the electromagnetic field.

Using currently available 10 kW instruments (e.g. from New Japan Radio Co., Model NJE 2603-10kW, or Palmer Bioscience, Sheerness, UK) it is possible to elevate the brain temperature of a mouse to 90°C in about 300 msec and that of a 200 g rat in about 850 msec. After irradiation animals are cooled in ice water, following which

Table 1. Some identified phosphoproteins of the mammalian CNS.

Protein	*Mol. wt* $\times$ *10^{-3} (SDS)*	*Localization*	*Protein kinase(s)*	*Comments*	*Ref.*[a]
1. MAP2	270	Dendrites	cA-PK, CaM-PKII, PK-C independent PKs	Phosphorylation may regulate microtubule assembly	1
2. Neurofilaments	200 (NF3) 150 (NF2) 70 (NF1)	Widespread in neurones	cA-PK, independent PK(s)	NF2 is principal substrate for cA-PK	2
3. 87 kd	87 (bovine) 80 (rat)	Synaptosomal cytosol. Present in non-neuronal tissues	PK-C	Most acidic phosphoprotein in many cells	3
4. Synapsin I	86 (Ia) 80 (Ib)	Many nerve terminals. Peripheral protein of synaptic vesicles	cA-PK, CaM-PKI, CaM-PKII	Multisite phosphorylation	4
5. Protein III	74 (IIIa) 55 (IIIb)	Many nerve terminals, associated with synaptic vesicles. Adrenal medulla	cA-PK, CaM-PKI, CaM-PKII	Single serine phosphorylation site	5
6. CaM-PKII	60 (β) 51 (α)	Cytosolic and membrane associated, many neurones. Also found in PSDs	CaM-PKII (autophosphorylation)	Subunit ratio differs in cerebrum and cerebellum	6
7. Tyrosine hydroxylase	56	Dopaminergic and noradrenergic nerve terminals	cA-PK, CaM-PKII, PK-C	Multisite phosphorylation	7
8. B-50	47–49	Synaptic membranes of many neurones	PK-C	May be equivalent to GAP-43 (ref. 13)	8
9. Pyruvate dehydrogenase	40 (α)	Mitochondria	PDH-kinase	Phosphorylation inhibits activity	9
10. DARPP-32	32	Cytosolic, dopaminoceptive neurones	cA-PK	Protein phosphatase inhibitor	10
11. G-substrate	23	Purkinje cells	cG-PK	Cytosolic, protein phosphatase inhibitor	11
12. Myelin basic protein	14 (small) 18 (large)	CNS myelin	cA-PK, CaM-PKII, PK-C. independent PK(s)	May be involved in myelin structure/assembly	12

[a]These references are representative and refer to this table only.

(**1**) Schulman (1984) *J. Cell Biol.*, **99**, 11; (**2**) Julien and Mushynski (1982) *J. Biol. Chem.*, **257**, 10467; (**3**) Albert, Wu, Nairn and Greengard (1984) *Proc. Natl. Acad. Sci. USA*, **81**, 3622; (**4**) Huttner, DeGennaro and Greengard (1981) *J. Biol. Chem.*, **256**, 1482; (**5**) Huang, Browning and Greengard (1982) *J. Biol. Chem.*, **257**, 6524; (**6**) McGuinness, Lai and Greengard (1985) *J. Biol. Chem.*, **260**, 1696; (**7**) Albert *et al.* (1984) *Proc. Natl. Acad. Sci. USA*, **84**, 7713; (**8**) Aloyo, Zwiers and Gispen (1983). *J. Neurochem.*, **44**, 649; (**9**) Browning *et al.* (1979) *Science*, **203**, 60; (**10**) Walaas and Greengard (1984) *J. Neurosci.*, **4**, 84; (**11**) Aswad and Greengard (1981) *J. Biol. Chem.*, **256**, 3487; (**12**) Turner *et al.* (1982) *J. Neurochem.*, **39**, 1397; (**13**) Jacobson, Virag and Skene (1986) *J. Neurosci.*, **6**, 1843.

cA-PK, cAMP-dependent protein kinase; CaM-PK, calmodulin-dependent protein kinase; cG-PK, cGMP-dependent protein kinase; PK-C, protein kinase-C.

brains can be removed and regions of interest dissected and prepared for assay of cAMP (see Section 3) or other compounds of interest. Since brain tissue becomes firm after heating, dissection of discrete brain areas is often facilitated. Microwaves may also be useful for estimation of protein phosphorylation in the intact brain. The sodium dodecyl sulphate–polyacrylamide gel electrophoresis (SDS–PAGE) banding pattern of phosphoproteins from brain tissue fixed by microwave radiation has been reported to be similar to that observed with tissues frozen in liquid N_2 (7).

Dissected regions of the PNS have been used extensively for the study of cyclic nucleotides and protein phosphorylation. An advantage with these preparations is that it is often possible to stimulate the tissue electrically in a well-defined manner that mimics physiological activity. An example we consider later is that of mammalian superior cervical ganglion. Pre-ganglionic stimulation results in activation of synapses releasing different transmitters, the post-synaptic consequences of which can be separated pharmacologically. Similarly, identified neurones, such as those from the invertebrate *Aplysia* can be simultaneously monitored electrophysiologically and biochemically. The large size of such cells allows them to be injected with a wide variety of agents including kinases and kinase inhibitors (3). The elegant experiments of Levitan's group (e.g. 12), using injection of [γ-^{32}P]ATP into single neurones to investigate protein phosphorylation, are a pioneering step in this area.

2.2 Brain (and PNS) slices

Slices have been particularly useful in the determination of neurotransmitter effects on cyclic nucleotides and subsequent protein phosphorylation. An early demonstration of cAMP elevation was by Kakiuchi and Rall (13) using slices from rabbit cerebral cortex and cerebellum. A vast amount of data has since accumulated on the effects of putative neurotransmitters on cAMP and cGMP formation in slices from different regions of mammalian brain and PNS (reviewed in 10). Many investigators have used slices to demonstrate neurotransmitter effects on phosphorylation of specific proteins. Because of the usefulness of brain slices their preparation and incubation is described in some detail in *Table 2*.

Table 2. Preparation of brain slices.

1. Dissect brain region of interest from rats sacrificed by decapitation and place in 20 ml vials (20–40 mg tissue/vial) containing 10 ml of ice-cold Krebs–Ringer bicarbonate buffer (KRB: 118 mM NaCl, 5 mM KCl, 2.5 mM $CaCl_2$, 2 mM KH_2PO_4, 2 mM $MgSO_4$, 25 mM $NaHCO_3$, 11.1 mM glucose, pH 7.4), previously saturated with 95% O_2/5% CO_2.
2. Slice tissue at 4°C with a McIlwain tissue chopper (Mickle Labs., UK; Brinkmann Instruments, USA) set for 0.3 mm slices. Return the slices to the vial and briefly vortex in KRB to separate individual slices.
3. Aspirate buffer and add 10 ml of fresh oxygenated KRB at 37°C. Incubate slices with continuous oxygenation in a shaking water bath at 37°C. A convenient way to oxygenate and minimize agitation is by bubbling the gas mixture via 18- or 20-gauge hypodermic needles with a notch filed about 1 cm from the tip; gas flow is adjusted such that it escapes through the notch and not from the tip.
4. Pre-incubate slices for 20–30 min, aspirate medium and replace with fresh KRB at 37°C (or if ^{32}P is to be added use labelled medium: see *Table 3*). Incubate for a further 25 min and then add effectors to the vials for the desired period of time.
5. Terminate incubation by aspirating the buffer and pouring liquid N_2 directly into each vial to freeze the slices. Processing of the slices will then depend on the specific parameter to be measured.

Several problems and variables exist with respect to the use of brain slices.

(i) Slices from 0.25 to 0.5 mm have been most frequently used; in thicker slices anoxia appears to be a problem and in thinner slices too few viable cells will be present [slicing results in tissue damage that extends 50–100 μm beyond the slice surface (14); thus, in a 0.3 mm slice one-third to two-thirds of the slice may consist of dead cells]. Because of tissue damage it is desirable to avoid cross-chopping if possible. Some cells or brain regions may be more sensitive to chopping than others, for example, few cells survive after mechanical chopping of cerebellum; hand-slicing gives better results in this region.

(ii) Oxygenation of the slices is typically accomplished by bubbling a stream of 95% O_2/5% CO_2 into the incubation vessel, but this can cause excessive agitation resulting in the elevation of cAMP levels in the slice (probably due to adenosine release) and possibly affecting other second messengers as well. Minimal agitation is thus advisable.

(iii) Basal levels of cyclic nucleotides: as already indicated, decapitation raises cAMP levels, thus slices must be incubated for an appropriate time to allow such levels to return to normal. This decline to baseline may vary from one region to another: for example, it is relatively rapid in the neocortex but quite slow in cerebellum.

(iv) Termination of the reaction: freezing is both a rapid and versatile method for preparation of the tissue for subsequent assay. If the slices are to be kept frozen and transferred to another vessel for homogenization it is advisable to use incubation vessels made out of Teflon with a smooth surface so that the frozen slices can be easily removed. Frozen slices can be stored prior to processing if the parameter to be measured is stable at the storage temperature, but this must be checked; for example we have found that −20°C is inadequate for cAMP (homogenization in acid is required prior to storage).

2.3 Synaptosomes and subcellular fractions from nervous tissue

Synaptosomes are useful for the study of certain pre-synaptic events, for example transmitter uptake and release. Because they are sealed organelles and contain mitochondria, synaptosomes are viable for extended periods *in vitro* provided that a metabolizable substrate is available. They can also incorporate [^{32}P]orthophosphate into ATP and have been used as a model system for investigating pre-synaptic protein phosphorylation. However, synaptosomes have proved disappointing in the investigation of neurotransmitter effects on second messenger systems. The isolation of synaptosomes is considered in Chapter 1. A recent development is a preparation termed the 'synaptoneurosome' containing re-sealed pre- and post-synaptic elements; the generation of this fraction along with some of its properties is summarized in (15).

Subcellular fractions are useful in the study of enzyme systems and in the identification of substrates for specific enzymes. Some examples are synaptic membranes, post-synaptic densities and myelin. The major disadvantage of many of these preparations is their extreme heterogeneity, since they may contain elements from all of the neurones and some non-neuronal elements in the original preparation. In addition, it is probable that the normal relationship between components in the system is disrupted on tissue homogenization (see Section 4.4). Moreover, it is impossible to study the effects of, for example, neurotransmitters on protein phosphorylation in such preparations. Nevertheless, subcellular fractions are invaluable for identification of such systems as neurotransmitter-sensitive adenylate cyclase and second messenger-sensitive protein kinases and their substrates. Several schemes for the complete fractionation of brain tissue are available (e.g. Chapter 1 and ref. 16).

2.4 Tissue cultures of nervous system elements and continuous cell lines

Isolation of specific cell types *en masse* from the CNS and PNS has become a powerful tool in the analysis of cell function at the single cell level. In second messenger research such preparations can be used for analysis of neurotransmitter effects on second messenger levels and protein phosphorylation. A major advantage is the ability to obtain multiple replicates of a single type of cell, allowing convenient measurement of several parameters. The properties and preparation of such cultures is described in Chapter 2.

Continuous cell lines such as neuroblastomas and gliomas offer yet another level of convenience in that a single homogeneous cell type is studied. Such cells can express 'differentiated' phenotypes and frequently display neurochemically-relevant processes (e.g. neurotransmitter receptors and release). However, it must be remembered that these cells are also transformed and that the expression of certain properties may be altered by this factor. See ref. 17 for some examples.

3. ASSAY OF CYCLIC NUCLEOTIDES

Voluminous amounts have been written about the measurement of cyclic nucleotides in biological materials (see ref. 18 for a compendium on the subject). Protein binding methods have been widely used for cAMP and radioimmunoassay (RIA) kits for both cAMP and cGMP are available from a number of commercial sources. While some of the RIA kits on the market are relatively expensive, the cost of the assay can be reduced by buying the necessary reagents separately from various suppliers (e.g. Research Products International, Mt. Prospect, IL, USA, supplies antibodies to both cAMP and cGMP). The following is a brief account of our experience with some of these methods. We have used the protein kinase binding assays of both Brown (19) and Gilman (20) for cAMP and the RIAs described by Steiner and Brooker (18) for the measurement of cAMP and cGMP with good results. With cAMP, selection of one assay over another depends in part on the amount of nucleotide to be measured and the equipment available. Generally the protein kinase binding assays require more cAMP than the RIAs. RIA is the only satisfactory method for the estimation of cGMP. For higher sensitivity in RIA, acetylation of the cyclic nucleotide is recommended.

3.1 Tissue extraction

Typically, cAMP and cGMP are extracted from tissue with dilute acid, for example 0.3 M perchloric acid, 5% trichloroacetic acid (TCA) or 0.1 M HCl.

(i) Homogenize frozen brain slices, or dissected regions of brain fixed by microwave radiation, in ice-cold 0.3 M perchloric acid (PCA) with a Polytron-type instrument.

(ii) Centrifuge homogenates (10 000 *g* for 10 min), pour the supernatants onto

0.8 × 10 cm Dowex 50 columns (AG50-W-X8, H+ form, 100–200 mesh, Bio-Rad) and elute the cAMP or cGMP with 0.1 M HCl (42).

(iii) Lyophilize the fraction containing cAMP, reconstitute the residue in 1 ml of 0.1 M acetate buffer (pH 4) and assay for cAMP using one of the protein binding assays, or for cGMP by RIA.

For cell cultures, where the limited amount of tissue available may mean assaying small amounts of cyclic nucleotides, RIA is the method of choice. Monolayers are washed twice in ice-cold saline after treatment with effectors and removal of medium. Ice-cold acid is then added immediately and dishes subjected to a freeze–thaw cycle to ensure complete cell lysis. The extracts are removed from the dish and centrifuged (10 000 *g* for 10 min) and the supernatant taken for cAMP or cGMP assay (pellets can conveniently be retained for protein assay). Supernatants containing TCA must be extracted four times with water-saturated ether (4–5 times sample volume) to remove TCA. PCA extracts can be neutralized with 2 M $KHCO_3$ and the resulting precipitate removed by centrifugation. HCl extracts can be directly neutralized with an equal aliquot of 2 times assay buffer (sodium acetate) containing 0.1 M NaOH.

3.2 Assay procedures

Both the Gilman and Brown assays use the binding of cAMP to the regulatory subunit of the Type II cAMP-dependent protein kinase as the basis for the assay. The source of the kinase differs: in the Gilman method it is partially purified enzyme from rabbit skeletal muscle whereas Brown uses a crude supernatant from bovine adrenal medulla. In the latter case it is essential to add phosphodiesterase inhibitors to block cAMP

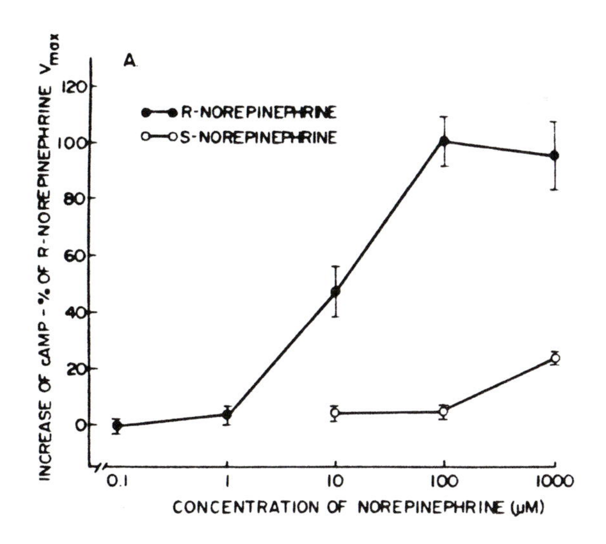

Figure 1. Dose–response curves for (R)- and (S)-norepinephrine-induced cAMP accumulation in slices of rat limbic forebrain. Slices prepared as described in the text were incubated for 5 min in the presence of the catecholamines, cAMP was extracted and measured using the Gilman assay. Results are expressed as the percentage of V_{max} of the (R)-isomer ± SEM. Basal levels of cAMP were 21.3 ± 0.6 pmol/mg protein, maximal response to (R)-norepinephrine was 172.2 ± 14.9 pmol/mg protein (from ref. 49, with permission.

breakdown by the tissue extract. We have found that a partially purified enzyme works as well in the Brown assay with less chance of cAMP breakdown. The Brown assay uses charcoal to adsorb unbound cAMP, the whole assay can thus be performed in a single tube provided that a suitable centrifuge with adapters for spinning that type of tube is available (e.g. glass 12 × 75 mm test tubes can be centrifuged in either an IEC or Sorvall RC5B equipped with an HS4 rotor and appropriate adapters). The Gilman assay uses adsorption of the bound cAMP to cellulose nitrate filters, thus a multiple filtration apparatus is invaluable (e.g. Brandel Instruments, Gaithersburg, MD, USA). For this reason the Brown assay is more convenient when large numbers of samples are to be processed. RIA methodology is well-covered in ref. 18.

An example of results using the Gilman assay to measure cAMP accumulation in rat limbic forebrain slices is shown in *Figure 1*.

4. MEASUREMENT OF PROTEIN PHOSPHORYLATION IN NERVOUS TISSUE

The two classical approaches to measurement of protein phosphorylation are those of

(i) pre-incubation of intact tissue with [^{32}P]orthophosphate to label intracellular ATP pools, followed by stimulation of the tissue, solubilization in a denaturing buffer and then analysis of phosphorylated proteins by gel electrophoresis;

(ii) incubation of subcellular fractions with [γ-^{32}P]ATP in the presence of effectors (e.g. cAMP or Ca^{2+}) of specific protein kinases, followed by solubilization of proteins and analysis, again usually by gel electrophoresis.

Ideally, these two methods should be used in combination, as questions raised by one technique may be resolved using the other. Another more specialized technique, termed 'back phosphorylation', has proven to be useful under certain circumstances. This procedure involves treatment of non-labelled tissue with the effector followed by termination of the reaction under conditions where further changes in phosphorylation do not take place and where extraction of the protein(s) of interest can be achieved. Then a specific, purified protein kinase is added to the extract together with [γ-^{32}P]ATP. This allows estimation of sites on the substrate that were not previously 'cold' phosphorylated in the intact cell (see ref. 3 for an analysis of the requirements of this technique). Finally, immunoprecipitation using specific antibodies to particular phosphoproteins can be used in conjunction with *in vivo* or *in vitro* methodologies, and with back phosphorylation techniques, to facilitate quantification of changes in protein phosphorylation state.

The type of information that can be obtained from method (i) concerns the nature of phosphorylation processes occurring in a physiological environment. The addition of an effector may activate one or more of these processes and the magnitude of the response can frequently be compared with a known physiological event (e.g. secretion) suspected to be regulated by phosphorylation. The drawback of this method is the difficulty of measuring changes in phosphorylation of a particular protein against a large background of other 'non-specific' phosphorylation events. Application of two-dimensional isoelectric focusing (IEF)/SDS–PAGE or immunoprecipitation may overcome these difficulties.

In method (ii) it is possible to select for the activation of specific protein kinases or phosphatases by adding specific effectors. Thus it may be possible to magnify the phosphorylation (or dephosphorylation) of particular proteins by optimizing the *in vitro* phosphorylation conditions. Of course, the natural relationship between substrates, kinases and phosphatases is disrupted on homogenization, so data should be interpreted with caution (see Section 4.4). However, the *in vitro* approach can yield useful pointers to specific events that can then be focused on in intact cell experiments. For example, a substrate for a particular enzyme can be identified in a subcellular fraction, then some of its salient properties identified (possibly leading to purification and/or antibody production), rendering more amenable the study of its phosphorylation in intact cells or tissues (a classic example is Synapsin I; see 3). Ultimately, intact cell experiments are essential if the physiological relevance of a particular phosphorylation/dephosphorylation event is to be determined.

4.1 **Standard techniques**

4.1.1 *Gel electrophoresis*

Identification of phosphorylated proteins is almost always by one- or two-dimensional IEF/SDS–PAGE and autoradiography. Many of the experiments described below utilize SDS–PAGE as an analytical tool. The most frequently used method for one-dimensional SDS–PAGE is that of Laemmli (21) and for two-dimensional IEF/SDS–PAGE that of O'Farrell (22). Space does not allow us to detail these techniques, especially as they are well-documented and in common use in many laboratories. The reader is referred to other works for practical details, particularly another volume in this series (23, see also 24). Two-dimensional IEF/SDS–PAGE can be used for analysis of minor phosphoproteins that may be masked by background phosphorylation events in one-dimensional gels. A convenient method of 'stopping' reactions is to add an ionic detergent such as SDS (e.g. as in Laemmli sample buffer, 21). However, the presence of free SDS in a first dimension IEF gel is deleterious to the generation of a pH gradient and so must either be eliminated from the 'stop' solution altogether or swamped with excess non-ionic detergent [e.g. a 5-fold excess of Nonidet P-40 (NP-40) or Triton X-100] prior to loading on the IEF tube gel. The non-ionic detergent incorporates most of the free SDS into mixed micelles and these migrate to the anodic end of the IEF gel. We have had success with a 'stop' solution containing 0.5% SDS/9.5 M urea/50 mM 2-mercaptoethanol. Following incubation, samples are solubilized with this mixture and then aliquots (normalized for protein content) are mixed with an equal volume of a buffer containing 5% NP-40/9.5 M urea/4% ampholytes (usually a 60–40% mixture of pH 3–10 and 5–8 ampholytes)/50 mM 2-mercaptoethanol. Samples prepared in this manner are also suitable for one-dimensional SDS–PAGE (after addition of the appropriate sample buffer) and immunoprecipitation experiments (again after swamping out the free SDS with non-ionic detergent; see Section 4.6).

The advantage of terminating reactions with SDS is that there is excellent preservation of phosphorylation state with little chance for kinases or phosphatases to act during sample work-up. However, it is possible to stop reactions by other means, provided that adequate precautions to prevent dephosphorylation are followed. A specific example is provided by the work of Garrison (25). In order to analyse cytosolic proteins in phosphorylated hepatocyte suspensions by two-dimensional methods, cells were rapidly mixed with a digitonin-containing buffer layered on top of a hydrocarbon medium in a microcentrifuge tube. (Digitonin creates large pores in membranes from which soluble proteins escape rapidly.) Centrifugation allowed separation of the soluble proteins remaining at the top of the tube from pelleted particulate material. The supernatant could then be treated directly with a stop solution suitable for two-dimensional IEF/SDS–PAGE. This method is so rapid that addition of phosphatase inhibitors was found to be unnecessary. Gel analysis of cytosol proteins is simplified by the fact that many nuclear and membrane phosphoproteins as well as non-protein phosphorylated species, such as nucleic acids and lipids, are eliminated from the sample prior to analysis. This considerably reduces the background on both one- and two-dimensional gels and allows the identification of many hormone-sensitive phosphorylation events (25).

4.1.2 *Autoradiography*

The visualization of ^{32}P incorporation into gel-separated proteins or peptides is achieved by autoradiography. For this purpose it is convenient to dry one- or two-dimensional gels onto a filter paper backing (e.g. Whatman 3MM) prior to exposure. Gel driers are commercially available (e.g. Hoefer Scientific, San Francisco, CA, USA) but may be constructed in the laboratory quite easily (23,24). Standard X-ray film, specifically for autoradiography, is available (Kodak X-AR); this is a double-emulsion film, but single-emulsion films (e.g. Kodak SB5) can be employed to obtain a finer image (26). Frequently, intensifying screens (e.g. DuPont Lightning Plus) are used to speed up the exposure process (see Chapter 7). As film must be exposed at −70°C for screens to be effective, enclosing the dried gel/film/screen sandwich in a metal X-ray cassette is the most convenient method.

Development of film is best accomplished in an automatic X-ray film machine (Kodak X-Omat) found in hospital Radiology departments or commercially available for laboratory applications (Kodak model M35). Alternatively, manual development of film is relatively straightforward (23; see also Chapter 7, Table 5). An excellent guide to practical aspects of autoradiography has recently been published (26).

4.1.3 *Quantification of incorporated radiolabel*

This is achieved either by direct counting of pieces cut from gels (absolute values) or by densitometric scanning of autoradiograms (relative values). It should be noted that gel pieces containing ^{32}P do not have to be 'solubilized' prior to counting as quenching is negligible for this isotope when using scintillation fluid. Alternatively, ^{32}P can be counted using 'Cerenkov' radiation [light emission caused by emitted particles from the isotope striking molecules of the surrounding solvent or the walls of the scintillation vial; for a discussion see (52)]. The efficiency of this method ranges from 25–50% and the ^{3}H-channel of the scintillation counter is used.

Preparation of autoradiographs for quantitative densitometry is covered in (26). Commercially available densitometers range from the simple and inexpensive (e.g. Model GS300, Hoefer Scientific) to the sophisticated and expensive (e.g. LKB Ultroscan XL or the digital analysis systems from companies like Joyce-Loebl). The value of one of these machines must be assessed in the light of individual project requirements; for example, if many two-dimensional gels are to be compared and quantified with respect to radioactive incorporation into a large number of proteins, clearly some kind of automated system with peak integration capacity and data storage ability is going to be needed.

The *stoichiometry* of phosphate incorporation into protein is often an important parameter to measure. While this is straightforward when dealing with purified substrates and enzymes *in vitro*, using ATP of known specific radioactivity, problems may arise in more complex circumstances. Thus, when dealing with *in vitro* phosphorylation of complex protein preparations, some method of estimating accurately the amount of the specific substrate of interest in the mixture is needed. This may be difficult to achieve if the substrate is a minor component. Sometimes a rough estimate can be obtained from a determination of the relative amount of the substrate obtained by scanning of stained gels, coupled with a conventional assay for total protein. However, a more accurate estimate, for example by some sort of quantitative immunoassay, is preferable. The situation is much more difficult in intact cells since the specific radioactivity of the intracellular ATP pool must be determined and, ideally, analysed further for percentage of radiolabel in the γ-position (25). The amount and degree of labelling of the substrate of interest should also be assessed. Again, quantification of substrate concentration may be extremely difficult without immunoassay.

4.1.4 *Protein estimation*

Assays are carried out using standard methodologies (27,28). For samples solubilized in either Laemmli or IEF sample buffer (components of which strongly interfere with most protein assays), prior elimination of interfering substances can be effected using deoxycholate/TCA precipitation before the conventional Lowry assay (27).

4.1.5 *Precautions*

When working with ^{32}P it is advisable to shield samples at all times. This is particularly true whilst pre-labelling intact preparations, when millicurie quantities of the isotope are used. ^{32}P is a strong β-emitter and lead shielding is effective, but can result in the production of 'Bremstrahlung' (secondary radiation akin to X-rays). The best shielding is Plexiglass (Lucite) of 1 cm thickness or greater, which has the advantage of transparency. Such material can easily be fashioned into boxes of various sizes, or simply mounted as a shield; these can be obtained from commercial suppliers but are much cheaper to construct oneself. It is also advisable to wear plastic goggles when dealing with unshielded samples. It should also be noted that glass and water are effective radiation blocks, but that styrofoam is not! A geiger counter equipped with a β-probe should be on hand at all times to monitor working surfaces and effectiveness of shielding. Needless to say, plastic disposable gloves should be worn at all times when dealing with radioactive materials.

4.2 **Phosphorylation in brain slices**

Two types of experimental approach have been used with brain slices: conventional pre-labelling followed by one- or two-dimensional gel electrophoresis, and 'back phosphorylation'.

Table 3. Pre-labelling and stimulation of brain slices.

1. Prepare 0.3 mm slices of neocortex and pre-incubate for 20 min as described in *Table 2*.
2. Remove KRB and wash the slices with 10 ml of phosphate-free KRB.
3. Add phosphate-free KRB containing 0.5–1.0 mCi/ml [^{32}P]orthophosphate (carrier-free) to each vial, then purge the vials with 95% O_2/5% CO_2, cap and incubate for 40 min with gentle shaking (enough to keep the slices moving back and forth on the bottom of the vial).
4. For studies with ACh, pre-treat some samples with physostigmine for 10 min and leave others as controls. Then add ACh for a further 2 min. After each addition, re-purge the vials with the gas mixture.
5. Terminate the reaction by aspiration of the solution and freeze slices in the vials by the addition of liquid N_2.
6. Warm vials to −20°C, then plunge them into a boiling water bath with simultaneous addition of 0.5 ml hot 1% SDS and follow by capping each vial (loosely to allow for gaseous expansion). After 4 min, remove vials, allow to cool and then transfer the mixtures to homogenizing vessels.
7. After homogenization, use an aliquot for protein determination (28) and estimate ^{32}P incorporation into the acid/organic solvent-insoluble fraction (pipette 10 μl aliquots onto 2 × 2 cm Whatman 3MM paper squares, wash in TCA and organic solvents; see ref. 29).
8. Adjust sample aliquots to contain equal amounts of acid/organic solvent-insoluble radioactivity and prepare for two-dimensional gel electrophoresis by adding an equal aliquot of a mixture containing 6.3% NP-40/16.8% 2-mercaptoethanol/6.7% ampholines (pH range used: one part 3–10, four parts 5–7) followed by solid urea (90 mg/100 μl total aliquot volume).
9. Load samples (10 μl, containing ~30 μg of protein; *Figure 2*) onto 120 × 1.8 mm IEF tube gels and subject to equilibrium IEF followed by SDS–PAGE (22). Stain and de-stain the gels, then dry and autoradiograph them using conventional methodology.

4.2.1 *Pre-labelling and two-dimensional IEF/SDS–PAGE*

An example of this procedure is an experiment designed to identify proteins phosphorylated in response to acetylcholine (ACh) in slices of neocortex (*Table 3*). Typical autoradiograms obtained from such an experiment are shown in *Figure 2*.

4.2.2 *Back phosphorylation*

To illustrate this rather specialized technique we describe experiments to estimate the effects of norepinephrine (noradrenaline) on Synapsin I phosphorylation in slices from rat neocortex (30, and for other examples, see 3). In this method, Zn^{2+} salts are used to precipitate proteins from the homogenate, a procedure that has been found to inactivate both kinases and phosphatases (3). Synapsin I has the advantage of remaining soluble in acidic solutions because of its extremely basic pI (10.3). Thus, acid treatment of the Zn^{2+} pellet extracts Synapsin I and only a few other proteins; after neutralization this mixture is suitable as a substrate for exogenous catalytic subunit of cAMP-dependent protein kinase ('C'). It is conceivable that a similar approach could be designed for other proteins, taking advantage of some peculiarity in their behaviour but extensive preliminary work would be necessary to establish the correct conditions. If antibodies are available, it may be possible to immunoprecipitate the protein from a crude sample and add back the appropriate kinase to the immunoprecipitate.

The methodology of such an experiment is described in *Table 4* and an example of the results is shown in *Figure 3*. Addition of increasing concentrations of norepinephrine to the brain slices leads to a *decrease* in the amount of Synapsin I that can be phosphorylated subsequently in the *in vitro* reaction. The implication therefore is that sites on Synapsin I have been phosphorylated in the intact tissue by endogenous ATP

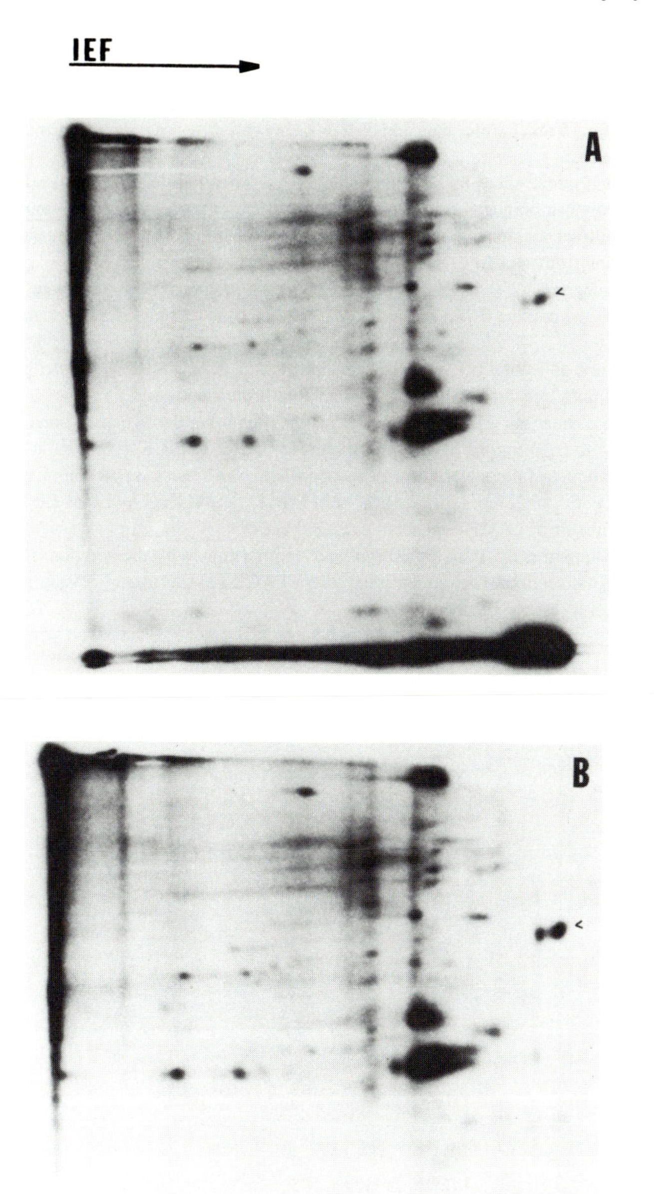

Figure 2. Autoradiogram of phosphorylated proteins in pre-labelled rat cortical slices exposed to acetylcholine (see text for details of labelling and incubation conditions). Shown are two-dimensional IEF/SDS gels of (**A**) control and (**B**) ACh (1 mM)-treated slices. The arrows point to an acidic (pI ~4.5) protein of molecular weight 85 000 (probably equivalent to 87 kd, see *Table 1*) whose phosphorylation is stimulated by ACh. (P.Mobley, unpublished results.)

Table 4. Back phosphorylation of Synapsin I from brain slices.

1. Prepare slices and pre-incubate them exactly as described in *Table 2*. Add isobutylmethylxanthine (5 μM) to some samples for 15 min before, then add different concentrations of norepinephrine (1–1000 μM) for a further 5 min.
2. Aspirate buffer and freeze slices by addition of liquid N_2. Then homogenize in 10 ml of 5 mM zinc acetate using a Polytron homogenizer. (For this step it is important to remove as much of the KRB as possible before freezing since the presence of bicarbonate and phosphate can interfere with the extraction of proteins from the Zn^{2+} precipitate.)
3. Centrifuge the homogenate (2000 *g* for 15 min) and extract the resultant pellet twice with 1 ml of 11 mM citric acid (pH 2.7) containing 0.1% Triton X-100. Extraction is accomplished by vigorous vortexing of the samples.
4. Pool the extracts and adjust the pH to 6.0 by the addition of 0.5 M Na_2HPO_4. Centrifuge the mixture (12 000 *g* for 15 min) and remove the supernatant for further analysis.
5. Assay for protein content[a] an aliquot of the supernatant containing the Synapsin I and adjust the protein concentration to 0.33 mg/ml by addition of further buffer at pH 6.0.
6. Analyse another aliquot of the material for dephospho-Synapsin I using a reaction mixture containing 10 μM [γ-^{32}P]ATP, 50 mM Hepes (pH 7.4), 10 mM $MgCl_2$, 1 mM EGTA, 1 mM EDTA and 7.5 nM of the catalytic subunit of cAMP-dependent protein kinase[b].
7. Incubate reaction mixtures at 30°C for 30 min and then terminate by the addition of Laemmli stop solution. Perform conventional one-dimensional SDS–PAGE as usual; stain, de-stain, dry and finally autoradiograph gel to assess ^{32}P incorporation.

[a]See ref. 28.
[b]Prepared as described in ref. 31; can also be purchased from Sigma.

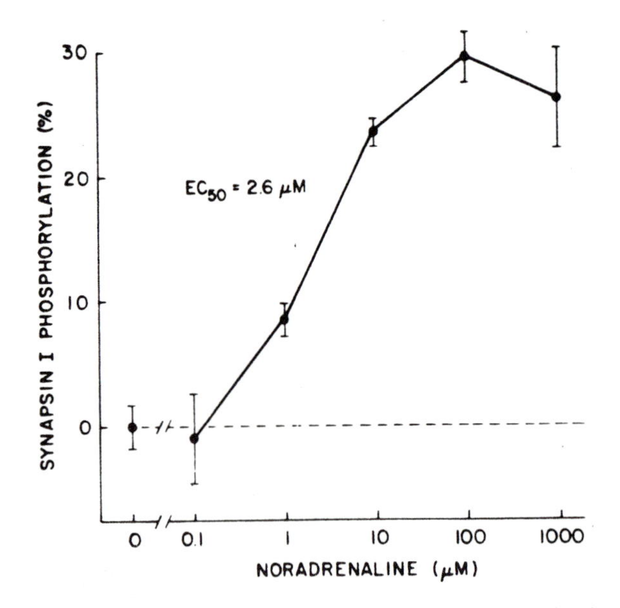

Figure 3. Back phosphorylation of Synapsin I. Slices from rat frontal cortex were incubated in the presence of (R)-norepinephrine, Synapsin I extracted and phosphorylated *in vitro* as described in the text. The figure shows that the amount of dephospho-Synapsin I decreases as the catecholamine concentration is raised (from ref. 30, with permission).

and cAMP-dependent protein kinase, and are thus unavailable for subsequent phosphorylation in the *in vitro* reaction. A similar approach has been used to show that serotonin regulates the state of phosphorylation of Synapsin I in slices from rat

Table 5. Depolarization-dependent phosphorylation in pre-labelled synaptosomes.

1. Prepare a P_2 pellet from three rat brains (33; see also Chapter 1) then resuspend to 8–10 mg/ml in 0.32 M sucrose/5 mM Hepes (pH 7.4) and fractionate on a discontinuous sucrose density gradient (2 ml of 1.5 M, 10 ml of 1.2 M, 6 ml of 1 M and 7 ml of 0.8 M; all made in 5 mM Hepes, pH 7.4). Centrifuge at 90 000 *g* for 2 h in a Beckman SW25.1 rotor which serves to separate myelin, synaptosomes and mitochondria.
2. Dilute the synaptosomal fraction to isotonicity with ice-cold 5 mM Hepes and centrifuge at 12 000 *g* for 30 min. (Sorvall SS34 rotor). Resuspend the pelleted synaptosomes to a protein concentration of 3–5 mg/ml in an oxygenated buffered salt solution (132 mM NaCl, 5 mM KCl, 2.5 mM $MgSO_4$, 0.1 mM EGTA, 10 mM glucose, 20 mM Hepes, adjusted to pH 7.4).
3. Transfer the synaptosome suspension to a screw-cap tube, purge with O_2, and incubate for 5 min at 37°C. Then add 0.25 mCi/ml carrier-free [^{32}P]orthophosphate, re-purge the tube with O_2, seal and continue incubation for 30 min at 37°C.
4. Remove aliquots (50–200 μl) from the pre-labelled synaptosomes and transfer to fresh tubes containing equal volumes of the effectors to be tested (e.g. veratridine, Ca^{2+}, high K^+ concentration etc.) pre-equilibrated at 37°C. After appropriate incubation times, terminate the reactions by direct addition of Laemmli sample buffer and boil the samples for 1 min. Alternatively, stop samples with an IEF sample buffer (see Section 4.1.1) and subject to two-dimensional IEF/SDS–PAGE.
5. Electrophorese samples of equal volume and appropriate protein concentration (40–100 μg) by one-dimensional SDS–PAGE, then stain, de-stain and autoradiograph gels by conventional procedures. Care must be taken with such gels as they contain large amounts of free ^{32}P. This migrates ahead of the dye front (if bromophenol blue is used) or slightly behind it (if pyronin Y is used). Ideally, most of the free ^{32}P should be left in the gel itself (i.e. dye-front >3 cm from the end of the gel) the end of which can be discarded to solid waste; if the ^{32}P is allowed to electrophorese out of the gel it ends up in the lower tank buffer which must be discarded to liquid radioactive waste.

facial motor nucleus and that a number of conditions alter the state of phosphorylation of Synapsin I in slices from bovine superior cervical ganglion (see ref. 3 for review).

4.3 Phosphorylation in synaptosomes

Viable synaptosomes possess actively metabolizing mitochondria and pre-incubation with [^{32}P]orthophosphate readily labels the internal ATP pool; protein phosphorylation in response to various stimuli can then be studied (*Table 5*). It is important to remember that during the pre-labelling period the [^{32}P]phosphoprotein content should achieve constant specific radioactivity because changes in phosphorylation on addition of effectors will then be easier to interpret (see 32 for a review of this point). In synaptosomes, such an equilibrium is attained within 30 min at 37°C, after which aliquots of the pre-incubated material can be subjected to different stimulation conditions. Probably the most elegant example of this approach is the demonstration of depolarization-dependent phosphorylation of Synapsin I first demonstrated by Krueger *et al.* (33, cf. 34). An example of the results of such an experiment on synaptosomes is shown in *Figure 4* (34).

4.4 Phosphorylation in subcellular fractions: *in vitro* techniques

Direct phosphorylation of subcellular fractions can be achieved by the addition of [γ-^{32}P]ATP under appropriate conditions. These conditions should optimize the activity of kinases endogenous to the preparation and will usually include sufficient Mg^{2+} (most kinases require this) and appropriate activators, for example cyclic nucleotides, calcium, calmodulin, phosphatidylserine. In this respect, 'endogenous' phosphoryla-

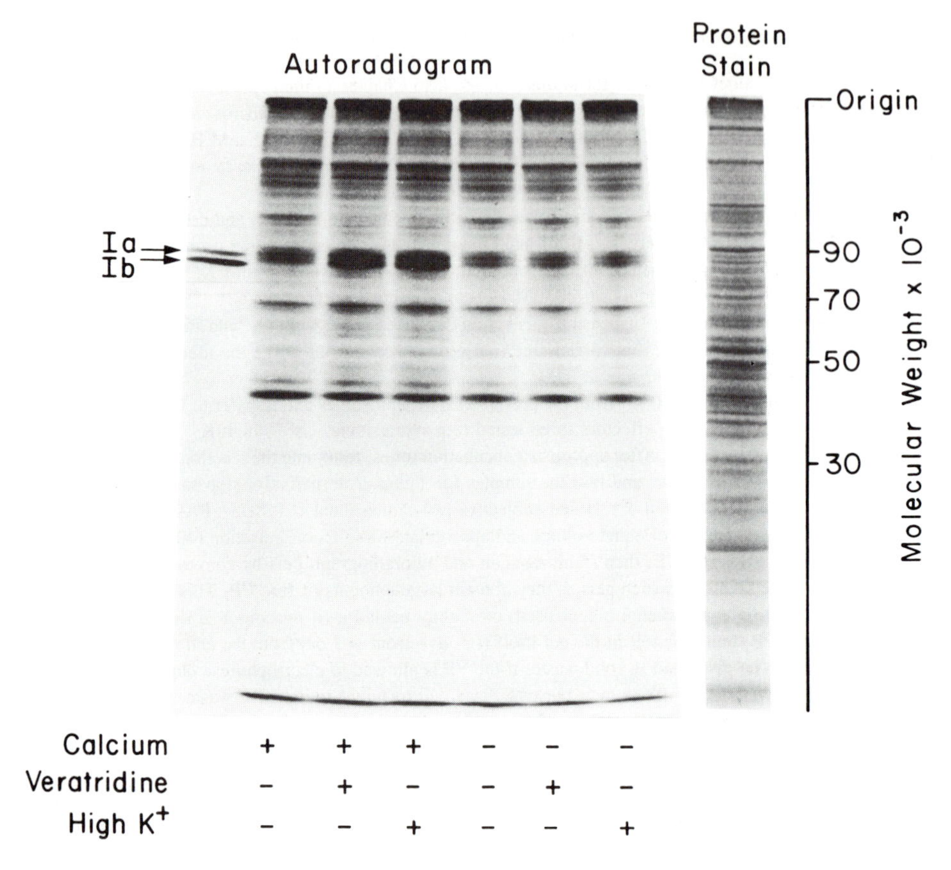

Figure 4. Phosphorylation in intact synaptosomes. Rat brain synaptosomes were prepared and phosphorylated as described in the text. Proteins were separated on SDS−8% polyacrylamide gels and phosphoproteins localized by autoradiography (**A**). Note the increased phosphorylation of Synapsin I (Ia and Ib) on depolarization of the synaptosomes with veratridine (100 μM) or high K^+ (60 mM) and the dependence of this on extracellular Ca^{2+}. (**B**) Coomassie blue stain of synaptosomal proteins. (From ref. 34, with permission.)

tion, as such experiments are often called, closely resembles 'exogenous' phosphorylation (i.e. addition of a substrate, as, for example, in the assay of the kinases themselves, see Section 6) without the exogenous substrate.

It should be remembered that one must be cautious in making direct inferences about phosphorylation events occurring in an intact cell or tissue from an *in vitro* analysis of subcellular fractions. An example of this problem is the controversy that arose over the physiological significance of synaptic membrane protein phosphorylation (see ref. 35 for relevant literature and one view of the problem). Early work showed that cAMP rapidly stimulated the phosphorylation of certain proteins in isolated synaptic membrane preparations and that maximal phosphorylation was attained in a matter of seconds when the ATP concentration in the *in vitro* reaction was low (10 μM), suggesting that such events could be important in the regulation of short-term synaptic activity. Later investigators found that phosphorylation in synaptic membranes was considerably slowed when the ATP concentration was raised closer to physiological levels (1 mM). These

Table 6. Analysis of substrates for individual protein kinases by *in vitro* phosphorylation.

1. Stun and decapitate 150–200 g male rats and remove the brain and spinal cord, placing them in ice-cold buffer A [10 mM Tris-HCl, pH 7.4, 2 mM EDTA, 1 mM dithiothreitol, 50 units/ml Trasylol, 2 μg/ml pepstatin A, 0.1 mM phenylmethylsulphonylfluoride (PMSF)].
2. Dissect brain regions[a] and pool samples from 4–6 animals. Weigh samples and then homogenize them in 10 vols of buffer A using a Teflon–glass homogenizer.
3. Centrifuge the homogenate in an ultracentrifuge at 150 000 *g* for 30 min to separate crude particulate and soluble fractions. Resuspend the particulate fraction using the Teflon–glass homogenizer in twice the original homogenization volume of buffer A. (Save soluble fractions for *in vitro* analysis if required.)
4. Subject aliquots of particulate (or soluble) fractions (20–150 μg protein) to *in vitro* phosphorylation. Conduct the reactions in either 12 × 75 mm disposable glass test tubes or microcentrifuge tubes.
   Ingredients (final concentrations in 100 μl volume):
   25 mM Tris-HCl, pH 7.4
   6 mM $MgSO_4$
   1 mM EDTA
   1 mM EGTA
   1 mM dithiothreitol
   In addition, for the detection of substrates for endogenous cAMP-dependent protein kinase add:
   2 μM cAMP + 1 mM isobutylmethylxanthine (to alternate tubes).
   For the detection of substrates for calmodulin-dependent protein kinases add:
   1.5 mM $CaCl_2$ (final free [Ca^{2+}] 0.5 mM) + 10 μg/ml CaM (to alternate tubes)[b]
   For the detection of endogenous substrates for protein kinase C, add:
   1.5 mM $CaCl_2$ + 50 μg/ml phosphatidylserine (to alternate tubes).
5. Prepare reaction mixtures on ice and pre-incubate for 60–90 sec at 30°C. Initiate the reaction by the addition of [γ-^{32}P]ATP (final concentration 2 μM, ~0.5–1 μCi or 5000–10 000 c.p.m./pmol; this high specific activity is used because of the low levels of endogenous substrate protein in the assay). Allow the reaction to proceed for 10 sec at 30°C, then terminate the reaction by adding Laemmli sample buffer (e.g. 50 μl of a 3 times stock) and boil for 1 min.
6. Electrophorese samples of 50 μg protein on SDS–7% or –12% polyacrylamide gels and then process gels for autoradiography as usual.

[a]As described in ref. 36.
[b]N.B. because soluble fractions contain large amounts of endogenous CaM, they must be depleted of this by fractionation prior to *in vitro* assay designed to show exogenous CaM dependence; see ref. 36.

data suggested that such events were too slow to play any role in rapid synaptic events. In fact, it is impossible to conclude anything about the duration of phosphorylation events at the synapse, or indeed in any intact preparation, from *in vitro* studies alone. This is because the disruption of the normal relationship between substrates, kinases and phosphatases may distort the rates, extent and reversibility of phosphorylative processes. For example, if cytosolic phosphatases were responsible for dephosphorylating synaptic membrane proteins, their absence in the *in vitro* reaction would lead to artificially 'long duration' labelling of endogenous substrates (such an interpretation could explain the incomplete dephosphorylation seen, e.g. in ref. 35). Caution is thus required in the interpretation of such experiments.

Several methods exist for the sub-fractionation of nervous tissue: a complete procedure designed to separate and purify a number of subcellular components is described in ref. 16. One fraction that has been extensively studied is the 'synaptic plasma membrane' usually derived from lysed synaptosomal preparations (see Chapter 1). Unfortunately, even the best preparations of this material are impure; moreover, improving 'purity' by more rigorous sub-fractionation techniques extends the time involved

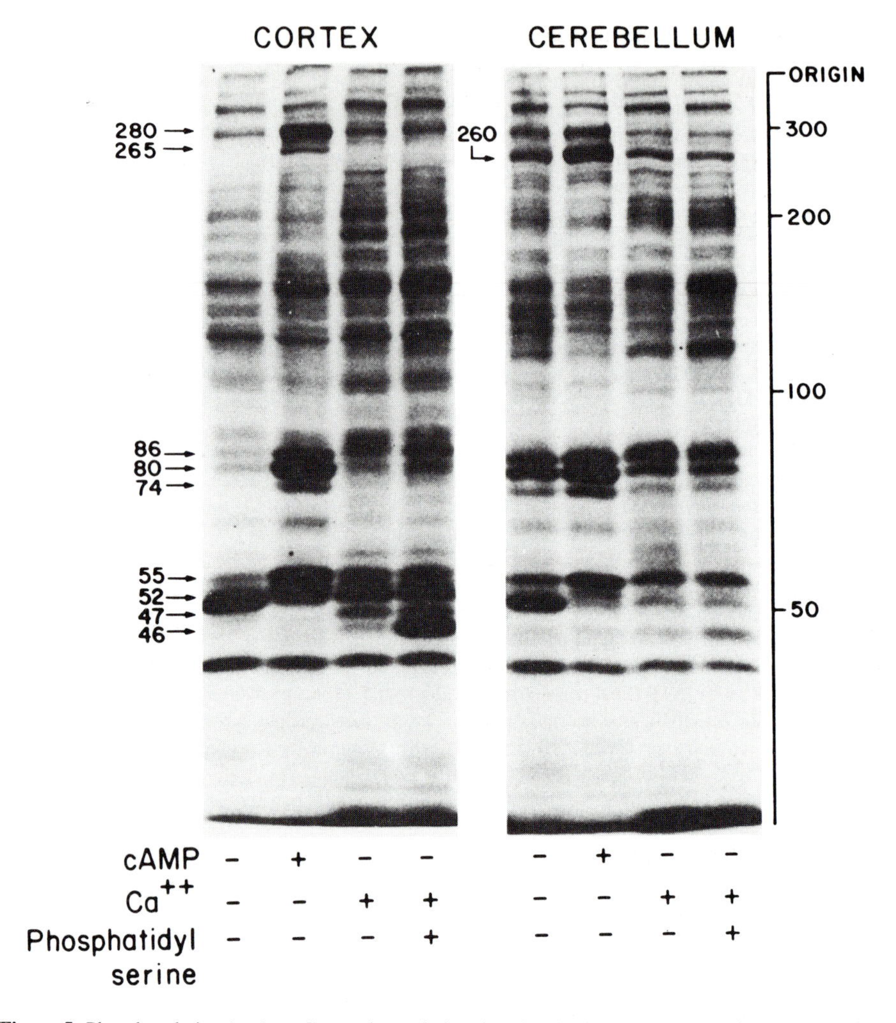

Figure 5. Phosphorylation *in vitro*. Comparison of phosphorylated substrates in particulate fractions from rat cortex and cerebellum. Fractions were prepared as described in the text and phosphorylated in the presence of [γ-^{32}P]ATP with the additions indicated. Aliquots containing 50 μg of protein were separated by SDS−7% PAGE and autoradiographed. Molecular weights $\times$ 10^{-3} are indicated. Note the presence of several cAMP-stimulated phosphoproteins in both brain regions, including those of molecular weight 280 000 (MAP2), 86 000 and 80 000 (Synapsin Ia and b), 74 000 and 55 000 (Protein IIIa and b, though the lower molecular weight band also contains some autophosphorylated regulatory subunit of cAMP-dependent protein kinase). Note the presence of a phosphoprotein of molecular weight 260 000 in cerebellum that is not present in cortex. This protein has been shown to be specific for cerebellar Purkinje cells (50). (From ref. 36, with permission.)

prior to assay and increases the probability of inactivating or removing important components. Typical investigations of synaptic membrane and post-synaptic density phosphorylation with relevant practical details can be found in (4) (see also 16,35). In *Table 6* we present, as an example of *in vitro* methodology, an experiment designed to determine substrates for different protein kinases in particulate fractions from cortex and cerebellum (36).

An example of such an experiment is shown in *Figure 5*. Note that very low 'unphysiological' levels of ATP are used in this experiment. However, the purpose is not to determine rates of phosphorylation, but to maximize the incorporation of isotope into substrates for specific kinases, such that they can be distinguished from the general background phosphorylation. A wealth of information has been generated by this approach (36) but the physiological relevance of individual substrates can only be determined by more laborious experiments on intact tissue.

4.5 Phosphorylation in tissue culture: use of immunoprecipitation

Tissue cultured cells of nervous system origin are a convenient tool in the analysis of phosphorylation systems. Here, we consider two examples: primary cultures of oligodendroglia and cultures of the transformed rat pituitary cell line, GH_3. Pre-labelling of intact cells with [^{32}P]orthophosphate is coupled with a most powerful tool in the study of phosphorylated proteins, immunoprecipitation. A necessary prerequisite for this technique is a suitable monoclonal or polyclonal antibody to the protein(s) of interest. A discussion of antibody generation is beyond the scope of this chapter (see for example, ref. 24), but it should be noted at the outset that antibodies that show positive reactions in some immunological tests (e.g. the 'Western' blot) are not necessarily functionally useful in immunoprecipitation assays. Serum antibodies need to be of high titre, or to be affinity purified from whole serum, before they can be useful in immunoprecipitation work. The ability of the particular antibody to precipitate the antigen quantitatively from reaction mixtures, such as those obtained in whole cell pre-labelling experiments, must also be ascertained. If these criteria can be fulfilled then this technique undoubtedly offers the most convenient and accurate method of assessing phosphorylation of a particular protein in intact cells. Such methodology has been applied to several phosphoproteins in many cell types. Another approach that may ultimately yield extremely sensitive methods of analysing phosphoproteins in intact tissues is the possibility of obtaining antibodies specific for the phosphorylated form of a protein (see ref. 55 for an example).

Immunoprecipitation can also be coupled to *in vitro* phosphorylation or back phosphorylation techniques. Coupling to back phosphorylation has rarely been exploited, perhaps because immunoprecipitation can probably be used effectively with pre-labelled samples in such systems, but also because of the fear that phosphorylation sites on the protein of interest may be masked in the immune precipitate. Nevertheless, this method may be useful when working with a protein containing multiple phosphorylation sites.

The pre-labelling methodology used in immunoprecipitation experiments is similar to that described above for synaptosomes; cells are pre-incubated in [^{32}P]orthophosphate to label intracellular ATP pools to constant specific activity; effectors are then added for a short period and the reaction is stopped. Next, extracts are made from the cells, care being taken to avoid dephosphorylation or proteolysis and the extracts are then incubated with antibody at a suitable dilution and the antigen–antibody complex precipitated either with immobilized Protein A or second antibody. Here we shall consider two specific examples: those of precipitation of myelin basic protein (*Table 7*) from cultured oligodendrocytes and of the 100 000-dalton calmodulin (CaM) kinase III substrate from cultured GH_3 cells (*Table 8*).

Table 7. Immunoprecipitation, example 1; myelin basic protein.

1. Label both unattached and attached oligodendrocyte cultures (in 6 cm plastic culture dishes) with phosphate-free Dulbecco's modified Eagle medium (Irvine Scientific) containing 0.1 mCi/ml [^{32}P]orthophosphate for 30 min at 37°C.
2. Immediately after the pre-labelling period, take cells for analysis or incubate them for various periods with different effectors (e.g. phorbol esters, cyclic nucleotides).
3. Terminate the reactions by the addition of 1 ml of 10% TCA directly to the dish.
4. Remove the material from the dish with a Pasteur pipette and transfer it to microcentrifuge tubes. Centrifuge to obtain the TCA pellet (note that the supernatant at this stage is highly radioactive and should be discarded into an appropriate container).
5. Wash the pellet twice with ether:ethanol (1:1) to remove TCA, then re-dissolve the pellet in 100 μl of 2% SDS, boil for 1 min and then add 4 vol of immunoprecipitation buffer (50 mM Tris-HCl, pH 7.4, 190 mM NaCl, 2.5% Triton X-100 and 1 mM PMSF) to each sample.
6. Add an appropriate amount of anti-MBP antiserum[a] and continue incubation overnight at 4°C with gentle rotation. Precipitate antigen–antibody complexes with 40 μl of a 100 mg/ml suspension of protein A–Sepharose (Pharmacia) prepared in immunoprecipitation buffer (3 h incubation). Centrifuge in a microcentrifuge to obtain pellet.
7. Wash the pellet six times with immunoprecipitation buffer by repeated centrifugation and aspiration and finally take it up in 100 μl of Laemmli sample buffer. Boil (1 min) and re-centrifuge to obtain a supernatant and load onto SDS–12% polyacrylamide gels for analysis.

[a]See ref. 37.

4.5.1 *Immunoprecipitation of myelin basic protein (MBP) from oligodendrocyte cultures (37)*

These experiments were performed in the context of investigations of the onset of myelination in cultured oligodendrocytes. These cells can be isolated from ovine white matter as described (37) and cultured either in suspension or attached to a polylysine-coated sub-stratum. In the latter case, myelinogenic functions are initiated and MBP becomes phosphorylated. The methodology is described in *Table 7* and typical results are shown in *Figure 6*.

4.5.2 *Immunoprecipitation of the 100 K protein from GH_3 cells*

The GH_3 cell line of rat pituitary origin possesses voltage-sensitive Ca^{2+}-channels that can be activated by depolarizing stimuli such as elevated extracellular K^+ concentrations. We have described a calmodulin-dependent protein kinase (CaM-PKIII) existing in many mammalian cells that phosphorylates a protein of molecular weight 100 000 daltons (100 K; refs 38,39). Using antiserum to this substrate we have demonstrated that the phosphorylation of the 100 K protein is markedly increased upon depolarization. The method for immunoprecipitation of 100 K is given in *Table 8*.

An autoradiogram from a typical experiment is shown in *Figure 7*.

The two methods described above differ somewhat in detail, but achieve the same end, the immunoprecipitation of a specific protein. When available, this is clearly the method of choice for quantification of phosphorylation of proteins in intact cells and can easily be adapted to *in vitro* phosphorylation when straightforward gel electrophoresis is unsatisfactory.

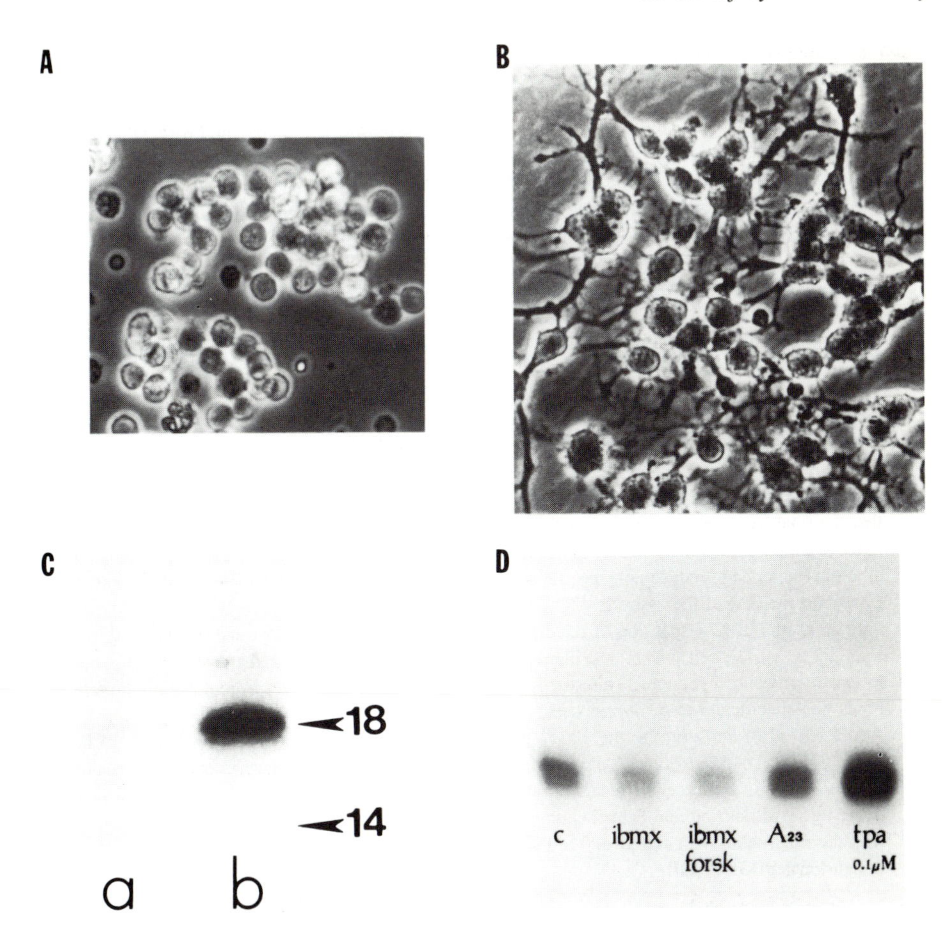

Figure 6. Immunoprecipitation of phosphorylated myelin basic protein (MBP) from cultured oligodendroglia. Cells were cultured in either an unattached (**A**) or attached (**B**) state. (**C**) Auto-radiogram of immunoprecipitated [^{32}P]MBP from floating (**a**) and attached (**b**) cultures: attachment promotes phosphorylation of this protein (using [^{3}H]amino acid labelling, the absolute amount of protein is comparable in the two types of culture). (**D**) Autoradiogram showing modulation of MBP phosphorylation in attached cultures. Pre-labelled cells were incubated in the presence of isobutylmethylxanthine (ibmx, 50 μM), ibmx + forskolin (1 μM), A23187 (A23, 0.1 μM), or 12-O-tetradecanoyl phorbol acetate (tpa, 0.1 μM) for 20 min. Protein kinase C seems to be the sole positive modulator of MBP phosphorylation in these cultures and its activation may be responsible for the attachment-dependent phosphorylation shown in (**c**). (Modified from ref. 37, with permission.)

5. PRELIMINARY ANALYSIS AND SEPARATION OF PHOSPHORYLATION SITES IN PROTEINS

It is frequently of great value to subject labelled phosphoproteins to further analysis, especially when multiple phosphorylation sites are suspected. This is readily achieved by proteolytic cleavage and peptide mapping and by analysis of phosphoamino acids. Such analyses can be performed on material cut from polyacrylamide gels. Dried, primary one- or two-dimensional gels are autoradiographed to localize the proteins of

Table 8. Immunoprecipitation, example 2: 100 K CaM-PKIII substrate.

1. Grow GH_3 cells on 6 cm culture dishes in Ham's F12 medium + 15% horse serum. After 2 days, pre-label near-confluent cultures with ^{32}P by replacing the normal growth medium with phosphate-free Hepes buffered Ham's F12 + 15% dialysed horse serum + 0.25 mCi/ml [^{32}P]orthophosphate for 1 h in an air incubator. After this period cells have achieved a constant specific radioactivity in protein as determined by TCA precipitation and scintillation counting[a].
2. Wash the plates twice with 3 ml of phosphate-free medium without isotope and then incubate them for various times in the same medium (5 mM KCl) or in medium containing 50 mM KCl (replacing NaCl), or with other effectors.
3. After incubation, wash the cultures twice with 3 ml of ice-cold Tris-buffered saline (25 mM Tris-HCl, pH 7.4/150 mM NaCl) and then solubilize with IEF stop solution (1 ml/plate; for composition, see Section 4.1.1). Transfer solubilized samples to 12 × 75 mm plastic tubes (samples may initially be very viscous but can be transferred using an automatic pipettor or Pasteur pipette; viscosity is subsequently reduced by freezing and thawing and trituration).
4. Assay aliquots of each sample for protein (27) and subject equal protein aliquots to the immunoprecipitation procedure.
5. Mix samples of less than 60 μl vol with an appropriate amount of NET buffer[b] containing NP-40 (such that the final SDS:NP-40 ratio is less than 0.2) to a final volume of 200 μl. This may be performed at room temperature in microcentrifuge tubes. Add 20 μl of NET/0.1% NP-40/25 mg/ml bovine serum albumin (included to reduce non-specific adsorption of proteins from the sample in subsequent steps). Carry out the rest of the assay at 4°C.
6. Add 25 μl of a 10% suspension of fixed *Staphylococcus aureus* cells[c], mix and incubate for 10 min. Spin out *S. aureus* cell pellets in a microcentrifuge (10 000 *g*, 2 min) and transfer the supernatants to new microcentrifuge tubes. This step is termed pre-clearing and serves to remove any protein that binds non-specifically to *S. aureus* cells[d].
7. Add 5–10 μl of antiserum to the 100 K protein (38), mix and incubate for a further 1–24 h.
8. Again add 25 μl of *S. aureus* cells, incubate for 30 min and precipitate the conjugates of antigen–antibody–*S. aureus* cells in a microcentrifuge. Discard the supernatant (to radioactive waste) and wash the pellet twice by re-suspension and centrifugation using NET/0.1% NP-40 (0.7 ml/wash).
9. Mix the final pellet with 120 μl of 1× Laemmli sample buffer, boil for 1 min, and process for one-dimensional SDS–PAGE.

[a]Attainment of constant protein specific radioactivity can be assessed by saturation of TCA-precipitable counts as a function of time and should be determined for each cell type studied.
[b]150 mM NaCl/50 mM Tris-HCl, pH 7.4/1 mM EDTA.
[c]A typical preparation is Pansorbin from Calbiochem-Behring; the bacteria are washed three times in NET/0.1% NP-40 prior to use, by repeated centrifugation in a microcentrifuge and re-suspension.
[d]Practically, we have found that this step can sometimes be omitted when extracts are solubilized in high SDS/urea/NP-40 combinations.

interest. The autoradiograph and stained gel are then exactly aligned to enable an area from the gel to be marked (a mounted needle is useful to punch through the autoradiogram onto the gel surface) and cut out using a razor blade. Excised gel pieces can then be subjected to limited or complete proteolytic analysis and phosphoamino acid composition determined as follows.

5.1 Limited proteolysis: one-dimensional maps

This procedure is most frequently used to compare proteins from different sources (23, 38, 40), but in many cases can also be used to separate phosphorylation sites of interest. It may be particularly useful in circumstances where one is attempting to assess

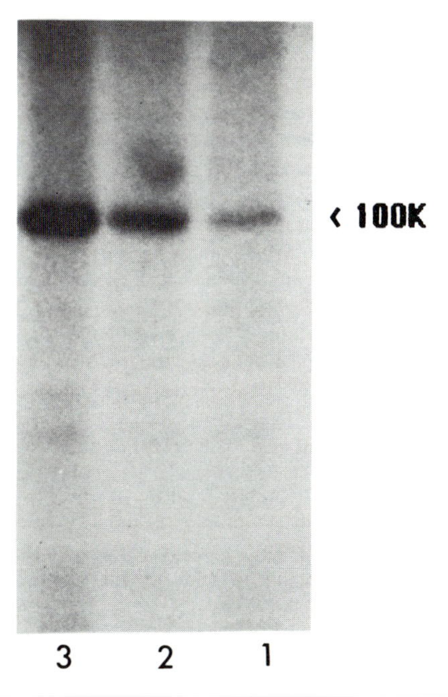

Figure 7. Immunoprecipitation of phosphorylated '100 K' protein from cultured GH_3 cells (see text for details). Pre-labelled cells were incubated in the presence of 5 mM KCl (**lane 1**) or 50 mM KCl for 5 min (**lane 2**) or 10 min (**lane 3**). These cells possess voltage-dependent Ca^{2+} channels, allowing Ca^{2+} to enter the cytoplasm and activate CaM-kinase III leading to phosphorylation of the 100 K protein. (H.C. Palfrey and J. Kuchibhotla, unpublished results.)

phosphorylation at a particular site(s) in a protein phosphorylated at multiple sites (e.g., see 53). The technique is descibed in *Table 9*. An example of this method using Synapsin I as substrate is given in *Figure 8*.

5.2 Limit proteolysis: two-dimensional maps (tryptic fingerprints)

This method is useful when multiple phosphorylation sites are suspected (e.g. 41) and can also be used for finer comparisons between phosphoproteins with supposed homology. To obtain autoradiographs in a reasonable length of time it is best to start with at least 1000 c.p.m./gel piece, but lower amounts may be used if, for example, only one or two phosphorylation sites are present, or if only qualitative data are sought. For quantification of ^{32}P content of individual peptides it is better to start with larger numbers of counts (3000–4000 c.p.m./gel piece; it may be necessary to combine gel pieces). It should also be noted that fingerprint and phosphoamino acid analyses are both conveniently obtained from the same hydrolysed gel piece(s) provided sufficient radiolabelled material is present. The methodology is outlined in *Table 10*. Tanks for running such analyses are commercially available (Savant Instruments, NY, USA), but can be easily constructed from plastic (Lucite) and platinum wire. An example of typical results of this procedure using immunoprecipitated tyrosine hydroxylase from superior cervical ganglia is shown in *Figure 9*.

Table 9. One-dimensional peptide mapping of phosphorylated proteins.

1. Re-swell pieces from the dried primary gel containing the protein of interest in 1× Laemmli sample buffer (this takes about 15–30 min at room temperature). Boil the whole sample (1 min), remove sample buffer immediately and replace with 1 ml of 125 mM Tris-HCl, pH 6.8/0.1% SDS.
2. Insert the gel pieces in the wells of a second SDS gel (usually 15% polyacrylamide to separate small peptide fragments) using a spatula (the lanes of the secondary gel should be wider than the lanes of the primary gel; the gel is best laid flat for this task) making sure that as few air bubbles as possible are trapped between the gel piece and the bottom of the well.
3. Mount the gel vertically in a gel tank as usual, add running buffer and overlay each gel piece with a solution of 60% upper gel buffer/40% glycerol/0.005% bromophenol blue. Apply over the first layer an appropriate protease solution[a], made in 90% upper gel buffer/10% glycerol/0.01% pyronin Y.
4. Electrophorese the gels as usual. Let the dye migrate through the stacking gel at reduced voltage (e.g. 60 V) to allow time for the protease to cleave the protein in the gel piece; voltage can then be increased to normal.
5. Wash the gel briefly in water before drying. Staining and de-staining are usually not necessary if an autoradiograph is desired. ^{14}C or pre-stained molecular weight markers can be included in one lane of the gel if peptide molecular weights are required. Carry out autoradiography as usual.

[a]The most useful concentration and type of protease for each protein must be determined empirically — we have found *S. aureus* V8 protease (various suppliers, including Miles Laboratories and Sigma) at concentrations of between 0.1 and 2.5 μg/lane to be the enzyme of choice in this regard.

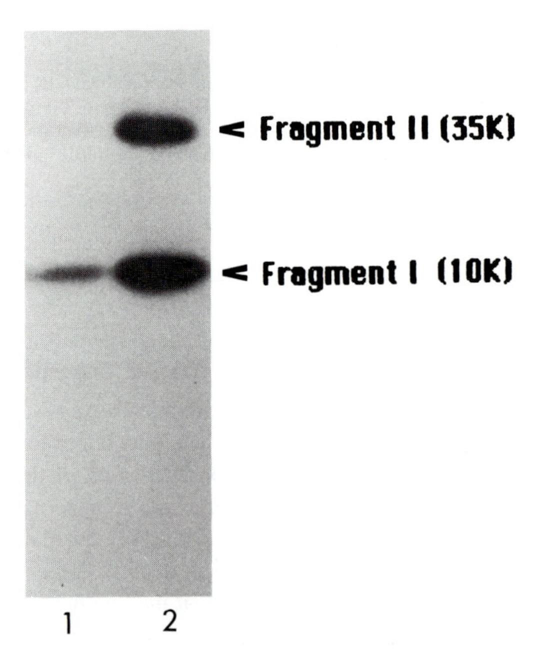

Figure 8. One-dimensional peptide mapping of [^{32}P]Synapsin I. Rabbit heart cytosol was fractionated on a DE52 (DEAE–cellulose) column (0–0.3 M NaCl gradient) and a fraction eluting at 75 mM NaCl (10 μg protein used) incubated *in vitro* with purified Synapsin I (10 μg) and [γ-^{32}P]ATP (50 μM) in the presence or absence of Ca^{2+} (0.2 mM) + CaM (20 μg/ml). Phosphorylation was terminated after 2 min at 30°C and the Synapsin I separated by one-dimensional SDS–PAGE. The Synapsin I bands were cut out from the dried gel and proteolysed as described in the text using 5 μg/lane of *S. aureus* V8 protease. This autoradiogram of the resultant peptides shows ^{32}P-labelled species at molecular weights of 10 000 (Fragment I) and 35 000 (Fragment II) whose phosphorylation increases with the addition of Ca^{2+} and CaM (**lane 2**). Different enzymes (CaM-kinases I and II) are responsible for the phosphorylation of the two fragments. (H.C.Palfrey, unpublished results.)

Table 10. Fingerprinting of phosphopeptides from proteins.

1. Re-swell the original gel piece in 25% methanol (~1 h) and wash twice in the same solution (1–2 h each, or overnight) to remove acid and SDS totally. This can conveniently be achieved in 20 ml scintillation vials: the radioactivity in the gel pieces can then be directly measured by Cerenkov radiation in the same container.
2. Blot each gel piece dry on tissue paper, then freeze-dry in separate plastic tubes.
3. Add a solution of an appropriate protease[a], made in 50 mM ammonium bicarbonate (pH 8.3), directly to each tube (~0.5–1 ml/gel piece used) and allow proteolysis to continue for several (often 24) h, or until proteolysis is complete.
4. Remove the supernatant (the small phosphopeptides having escaped from the gel during proteolysis) and freeze-dry. Re-suspend the residue in an appropriate volume (usually 20–100 μl) of the thin-layer electrophoresis buffer[b] (see below).
5. Subject the samples to thin-layer electrophoresis in the first dimension and chromatography in the second dimension, preferably on flexible plastic-backed 20 cm × 20 cm, 100 μm thick, cellulose plates (Eastman-Kodak or Merck).

   For electrophoresis.
   (i) Apply samples (ideally <30 μl) with an automatic pipettor to a point 3 cm from the bottom of the plate at the mid-line (if pH 3.5 buffer is to be used[b]) or in one corner (if buffers of pH 1.9 or 8.9 are used). Direct a steam of hot air at the plate to speed up drying. Take care to minimize the area of the applied spot so that the resultant peptides occupy as small an area as possible. It is often useful to add a small amount of either a basic or an acidic dye (e.g. phenol red or basic fuchsin) to the sample to monitor the progress of the electrophoresis. It is possible to standardize conditions from one run to another by allowing the dye to migrate a set distance.
   (ii) Wet the plate with electrophoresis buffer (use buffer-soaked Whatman 3MM filter paper sheets with a hole cut out so that the filter does not make direct contact with the origin, or spray buffer with an aerosol gun), taking care to avoid overwetting the plate.
   (iii) Electrophorese the plate in the tank at 400–1000 V for appropriate lengths of time (400–600 V.h. is a useful starting point).
   (iv) Dry the plate thoroughly and subject it to conventional ascending t.l.c. in the second dimension.

   For chromatography.
   (i) Place the dried cellulose plate at a right angle to the direction of electrophoresis in a glass chromatography tank containing about 150 ml of buffer[c].
   (ii) Remove the plate when the buffer front reaches 3 cm from the top, air-dry and autoradiograph as usual. Intensifying screens are important at this stage as the ^{32}P content of each individual spot may be low and conventional autoradiography may not reveal incorporation into minor peptides.
6. If desired, measure radioactive incorporation into individual fingerprint spots by scraping the appropriate part of the plate and counting directly in scintillation fluid.

[a]Trypsin at 50 μg/ml is a useful starting point, but chymotrypsin, thermolysin and proteases of different specificity may also be employed.
[b]Buffers of various pH values have been used to separate phosphopeptides (see ref. 24, Chapter 21); here we consider buffers of pH 1.9, 3.5 and 8.9. Of course, the migration of the phosphopeptides in each system will depend on their pI: most will be positively charged (and migrate towards the cathode) at pH 1.9 and will be negatively charged (and migrate towards the anode) at pH 8.3. pH 1.9: 2.5% formic acid/8.7% acetic acid/88.5% water; pH 3.5: 1% pyridine/10% acetic acid/89% water; pH 8.9: 1% ammonium carbonate/99% water.
[c]A typical buffer used in this dimension would be: 37.5% *n*-butanol/25% pyridine/7.5% acetic acid/30% water.

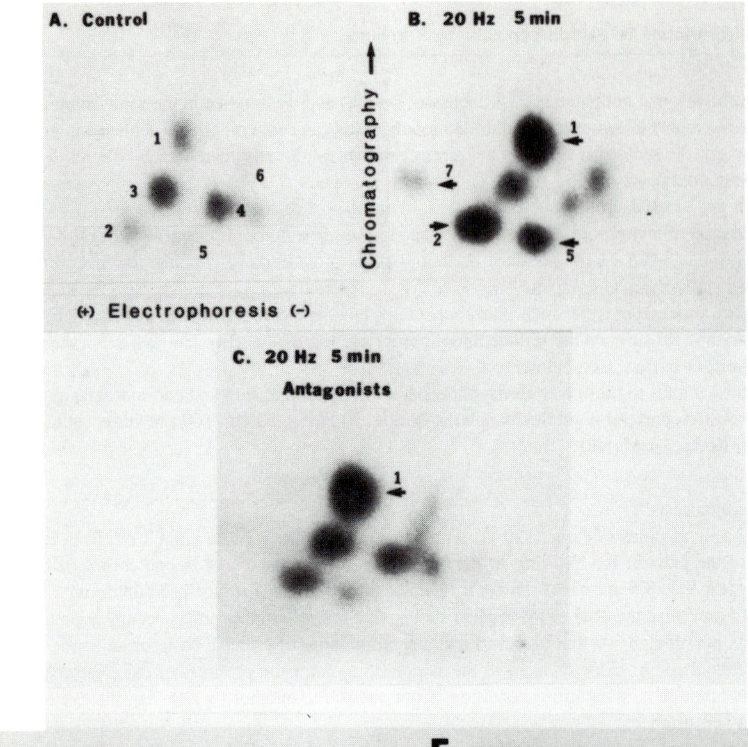

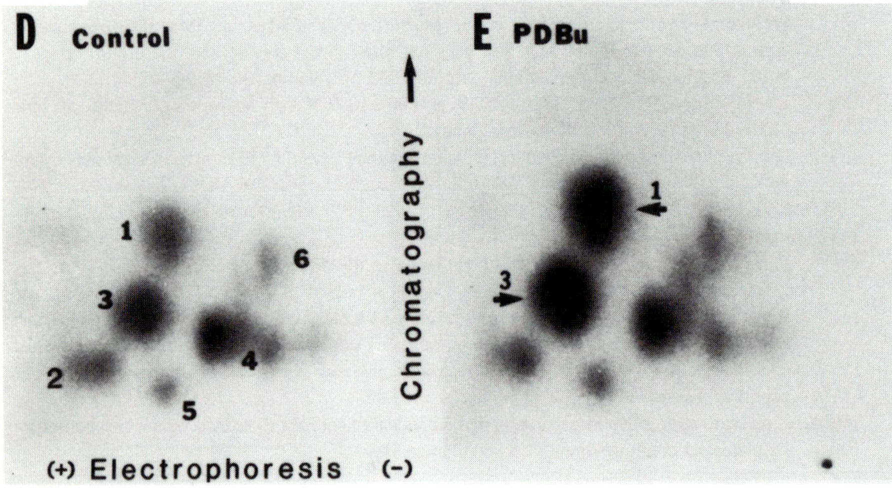

Figure 9. Two-dimensional tryptic mapping of ^{32}P-labelled tyrosine hydoxylase (TH) from rat superior cervical ganglia. Ganglia were incubated for 3 h in oxygenated physiological saline containing [^{32}P]orthophosphate and then stimulated pre-ganglionically for 5 min at 37°C (see ref. 51 for details) or incubated in phorbol dibutyrate (PDBu, 1 μM) for a similar period. TH was isolated from the ganglia by immunoprecipitation followed by SDS−PAGE as previously described (51). The TH band was cut from the gel and trypsinized as described in the text (50 μg TPCK-trypsin/ml). Peptides were separated by two-dimensional thin-layer electrophoresis/chromatography: electrophoresis was carried out with 1% $(NH_4)_2CO_3$, pH 8.9 at 500 V for 30 min; chromatography was run in *n*-butanol:pyridine:acetic acid:water (75:50:15:60), following which the plates were autoradiographed. The origin of each chromatograph is in the bottom right hand corner. Prominent phosphopeptides have been labelled 1−6; arrows indicate peptides whose phosphorylation has been increased by the specific treatments. (From A.Cahill and R.Perlman, in press.)

Table 11. Phosphoamino acid analysis.

1. After dissolution in thin-layer electrophoresis buffer (pH 1.9, see *Table 10*), transfer a sample (<20 μl) of the protease hydrolysate containing an appropriate number of counts to a flame-sealable borosilicate ampoule (e.g. Wheaton 176772). Add 200 μl of freshly-made 6M-HCl and seal the ampoule under vacuum (a freeze-drier with an appropriate alkali trap can be used), or under N_2.
2. Transfer the ampoule to a heater block or oven maintained at 110°C. Allow hydrolysis to proceed for 1–4 h[a].
3. Open the vial and freeze-dry the contents, using an alkali-trap again. Take up the residue in a minimum volume of pH 1.9 buffer and apply it as a spot at the origin of a cellulose t.l.c. plate (see below) together with marker (phenol-red dye) and 2 μg/ml of phosphoserine, phosphothreonine and phosphotyrosine (Sigma).
4. Single-dimension separations: several samples may be loaded side-by-side on the same plate. Load samples with an automatic pipettor 3 cm from one end of plate. Wet the plate evenly with pH 1.9 buffer using an aerosol gun. Place the plate in an electrophoresis apparatus similar to that used for fingerprinting. Electrophorese at 400–500 V (migration towards the anode) for 1–2 h or until the phenol-red marker is about two-thirds of the way across the plate. Phosphothreonine and phosphoserine migrate behind the dye and are well-separated.
5. Two-dimensional separations: prepare samples as in step 3 and apply each sample in one corner of a 20 cm × 20 cm cellulose t.l.c. plate. Wet the plate and electrophorese in pH 1.9 buffer as in step 4. Remove the plate and dry thoroughly. Spray the plate with pH 3.5 buffer (see *Table 10*) and again electrophorese at right angles to the pH 1.9 dimension (migration of dye and sample towards the anode again). Note that phenol-red migrates *behind* the phosphamino acids at pH 3.5, thus electrophoresis should be continued only until the dye has reached about half-way across the plate.
6. After electrophoresis with either method remove the plate, dry thoroughly and spray with 0.25% ninhydrin in acetone using an aerosol gun (use fume-hood!). Dry the plate in an oven to reveal the location of the marker phosphoamino acids (blue spots). Localize radioactivity by autoradiography and quantify radioactivity contained in individual spots by scraping and scintillation counting.

[a]It should be noted that this method is only semi-quantitative in that a time-dependent and variable hydrolysis of the protease hydrolysate to individual amino acids and then to free phosphate occurs (42,43). Samples to be compared should be hydrolysed at the same time.

5.3 Phosphoamino acid analysis

Part of the tryptic (or other protease) hydrolysate, or material eluted from individual fingerprint spots and dried down, can be used for this procedure (*Table 11*). As is well-known, phosphothreonine and phosphotyrosine are not separable by one-dimensional electrophoresis at pH 1.9, but can be separated by two-dimensional electrophoresis at pH 1.9 and then pH 3.5 (for a complete description of the analysis of phosphotyrosine in proteins, see ref. 43). It is also possible to separate all three phosphoamino acids in a single dimension by sequential electrophoresis at pH 1.9, followed by drying and re-electrophoresis at pH 3.5. If phosphotyrosine is known to be absent, the pH 1.9 procedure alone is sufficient, and is much more convenient for comparative purposes. An example of the separations achieved using both one- and two-dimensional methods (45) is shown in *Figure 10*.

5.4 Other procedures

Space does not allow us to consider more advanced techniques of phosphopeptide analysis, particularly h.p.l.c. The interested reader should consult appropriate reviews where such technology has been used (e.g. ref. 24, Chapter 5; ref. 46, Chapter 5). In addition, protein sequencing techniques can be used to identify phosphorylated amino acids in pure proteins (see, for example, ref. 44).

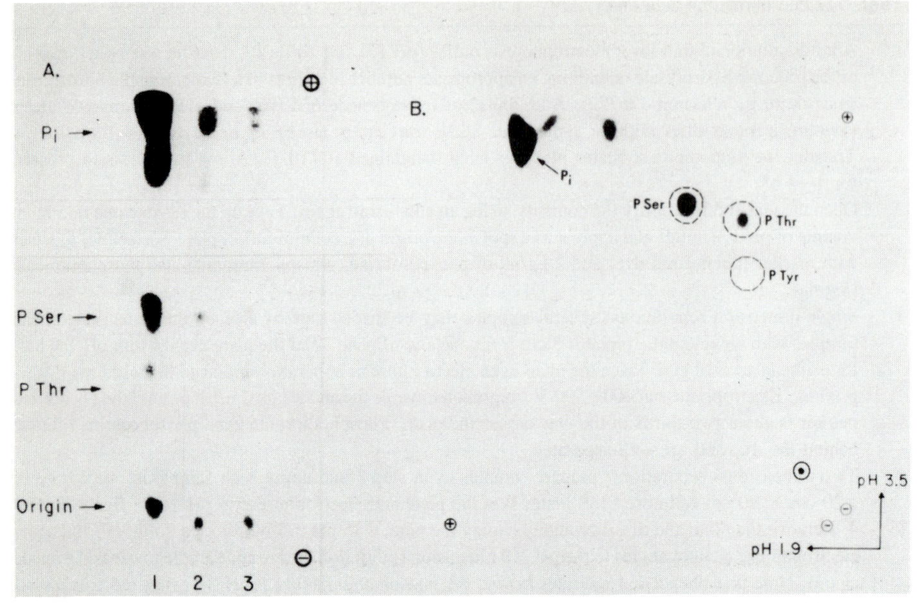

Figure 10. Phosphoamino acid analysis: examples of one-dimensional (**A**) and two-dimensional (**B**) analyses. Samples were ^{32}P-labelled CaM-PKII substrates phosphorylated in cytoplasmic extracts from *Torpedo* electric organ and digested as described in the text. (**A**) Autoradiogram of one-dimensional separation at pH 1.9: **lane 1**, 39-kd; **lane 2**, 54-kd and **lane 3**, 62-kd substrates phosphorylated *in vitro* in the presence of Ca^{2+} + CaM. (**B**) Autoradiogram of two-dimensional separation with electrophoresis in the first dimension at pH 1.9, then in the second dimension at pH 3.5. Sample was ^{32}P-labelled substrate (mol wt 39 000) from *Torpedo* cytosol phosphorylated in the presence of Ca^{2+} + CaM. The arrows in (**A**) and the dotted outlines in (**B**) indicate the position of ninhydrin-stained phosphoamino acid standards run simultaneously. (From ref. 45, with permission.)

6. ASSAY OF PROTEIN KINASES IN NERVOUS SYSTEM TISSUE

The complexity of protein phosphorylation reactions in the nervous system is well-exemplified by the plethora of identified protein kinases in such tissue (2,3). Of these, the best characterized are the cyclic nucleotide and Ca^{2+}-dependent enzymes, although important roles for tyrosine kinases and specific receptor kinases are beginning to emerge (2). Space does not permit us to cover protein phosphatases here (see ref. 54).

The activity of protein kinases is measured by the transfer of phosphate from [γ-^{32}P]ATP to (an) exogenous substrate(s) suitable for that particular enzyme. Different protein kinases have distinct substrate specificities and it is important to use the appropriate substrate; for example cAMP/cGMP-kinases: histone f2b; protein kinase C: histone H1; CaM-kinase II: Synapsin I, histone f3. As it is frequently useful to assess directly the level of activity of such enzymes in different preparations, we provide a brief outline here (see ref. 46 for further information).

Assay mixtures contain buffer, Mg^{2+} (usually 5–10 mM final concentration), substrate and effector (e.g. cAMP, Ca^{2+}) as well as suitable dilutions of extract or fraction to be assayed. The reaction mixtures are pre-incubated for a short period at the temperature of assay and reactions initiated by the addition of [γ-^{32}P]ATP (final concentration should be saturating for the particular enzyme being assayed). Methods of terminating the reaction depend on the method chosen for detection of radioactivity in substrate. Incorporation of radioactivity into substrate can be detected in various ways: for example precipitation with TCA and collection and washing of the precipitate either by centrifugation or filtration, binding to ion-exchange papers (e.g. histones and Synapsin I bind to phosphocellulose, 45) or gel electrophoresis. In general, SDS−PAGE offers the least background of the above methods but is the most time-consuming to perform.

As with any enzymic assay, it is important to ensure one is assaying initial reaction rates. This can be more difficult in crude samples because of the presence of phosphatases, thus attempts should be made to work with small amounts of extract, with sufficiently high ATP to ensure that substrate depletion does not occur during the assay period, and with short assay times. The inclusion of phosphatase inhibitors may also be useful, although non-specific inhibitors such as fluoride can also have effects on kinases.

6.1 Cyclic nucleotide-dependent protein kinases

These enzymes are most conveniently assayed using histone f2b. Assay methods are given in *Table 12*.

cGMP-dependent protein kinase can be assayed in a manner similar to that given in *Table 12* using cGMP as the stimulus rather than cAMP. It should be noted however that high concentrations of cGMP can activate the cAMP-dependent protein kinase, so it is wise to include the Walsh inhibitor (46), in these assays.

Table 12. Assay of cAMP-dependent protein kinase.

1. Pipette the following components into test tubes such that reaction mixtures (100 μl) contain (final concentrations): 50 mM Tris-HCl (pH 7.4), 5−10 mM $MgSO_4$, 1 mM dithiothreitol, 1 mM EGTA (the catalytic subunit is inhibited by Ca^{2+}), 0.4 mg/ml histone f2b, 50 μM isobutylmethylxanthine. In alternate tubes add 20 μM cAMP (Na^+ salt)[a].
2. Cool the tubes on ice and add tissue extracts. Pre-incubate tubes at 30°C for 1 min and then initiate the reaction by addition of [γ-^{32}P]ATP (50 μM final concentration; between 200 and 1000 d.p.m./pmol).
3. Incubate the reactions for a suitable time (depending on the concentration of kinase in the extract this may be 30 sec − 10 min or more) then stop the reaction by the addition of 10 μl of a solution containing 100 mM EDTA and 10 mM ATP (EDTA chelates the Mg^{2+} and cold ATP dilutes the isotope such that phosphorylation is effectively terminated). Transfer the tubes back to the ice bath.
4. Spot 50 μl of the total reaction mixture onto 2.2 × 1.2 cm phosphocellulose paper rectangles (cut from Whatman P81 phosphocellulose sheets) spread on a piece of aluminium foil and appropriately numbered in pencil. After application of sample it is not necessary to let the paper pieces dry.
5. Place the phosphocellulose rectangles in a plastic tri-pour beaker and swirl with about 50−100 ml of tap water for 5 min. About 95% of the applied [^{32}P]ATP is released from the pieces in this period and should be decanted into a liquid radioactive waste container. Wash the phosphocellulose pieces in running tap water for a further 15 min (use a shaped piece of plastic mesh clamped to the top of the beaker to retain the pieces).
6. Sort and dry phosphocellulose pieces on paper towelling then count by direct insertion into mini-vials with scintillant (alternatively any other liquid may be used and the samples counted using Cerenkov radiation). Background in this assay is less than 0.2% of the input counts.

[a]The activity due to the cAMP-dependent kinase is considered as the value of the samples in the presence of cAMP (+cAMP) minus those without the nucleotide. However, in some preparations there may be free catalytic subunit that may cause an elevated −cAMP value. To verify that this is indeed the catalytic subunit, addition of Walsh inhibitor (see ref. 46, Chapter 11 for preparation) can be used.

6.2 Calmodulin (CaM)-kinases I and II

These two enzymes are most conveniently assayed using the exogenous substrate Synapsin I, although alternative substrates such as MAP-2, histone f3, glycogen synthase and (less active) casein are also available for CaM-kinase II (2). An advantage with Synapsin I is that its phosphorylation can readily be quantified using the phosphocellulose paper technique described in the previous section. (Synapsin I is relatively easy to prepare from bovine brain in large quantities suitable for use as a substrate).

Assays are conducted in a similar manner to the method described in *Table 12* except that the reaction mixtures contain (final concentrations in 100 μl); 50 mM Tris-HCl (pH 7.4), 5−10 mM $MgSO_4$, 50 μM leupeptin, either 1 mM EGTA or 0.2 mM (free) Ca^{2+}, 20 μg/ml calmodulin, and substrate (e.g. Synapsin I, 100−200 μg/ml). Steps 2−5 from *Table 12* are then followed. The activity of the CaM-dependent enzymes is considered as the difference between the Ca^{2+}- and EGTA-containing tubes. In the case of Synapsin I, CaM-kinases I and II phosphorylate different sites on the protein that can be separated by one-dimensional partial proteolysis (see Section 5.1) and quantified independently. For this procedure, the reactions are stopped by the addition of Laemmli sample buffer and the mixture subjected to one-dimensional SDS−PAGE to separate Synapsin I prior to proteolysis. The other substrates mentioned above largely quantify CaM-kinase II, as the substrate specificity of CaM-kinase I is narrower than that of CaM-kinase II (2). Another CaM-requiring kinase, myosin light chain kinase, can be assayed in a similar manner, except that the substrate is myosin light chain and that incorporation is quantified by spotting reaction mixtures onto Whatman 3 MM filter paper followed by TCA washing.

6.3 Protein kinase C

The assay mixture contains 50 mM Tris-HCl (pH 7.4), 10 mM $MgSO_4$, 50 μM leupeptin, either 1 mM EGTA or 0.5 mM free Ca^{2+}, 25−100 μmol/ml phosphatidylserine, 300 μg/ml histone H1 (Type IIIS from Sigma) and optional diolein (48; increases Ca^{2+} sensitivity of enzyme). The performance of the reaction and analysis of incorporation are carried out exactly as described in *Table 12* (steps 2−5) using the phosphocellulose paper assay or as detailed in ref. 48.

7. ACKNOWLEDGEMENTS

We would like to thank Drs Ivar Walaas, Angus Nairn, Paul Greengard, Tim Vartanian, Anne Cahill and Robert Perlman for allowing us to use figures from their work.

8. REFERENCES

1. Nestler,E. and Greengard,P. (1983) *Nature*, **305**, 583.
2. Nairn,A.C., Hemmings,H.,Jr. and Greengard,P. (1985) *Annu. Rev. Biochem.*, **54**, 931.
3. Nestler,E. and Greengard,P. (1984) *Protein Phosphorylation in the Nervous System*. Wiley, New York.
4. Gispen,W.H. and Routtenberg,A. (eds) (1981) *Brain Phosphoproteins. Progress in Brain Research*. Elsevier, Amsterdam, Vol. 56.
5. Berridge,M. and Irvine,R. (1984) *Nature*, **312**, 315.
6. Nishizuka,Y. (1986) *Science*, **233**, 305.
7. Mitrius,J.C., Morgan,D.G. and Routtenberg,A. (1981) *Brain Res.*, **212**, 67.
8. Agrawal,H.C., Martenson,R.E. and Agrawal,D. (1982) *J. Neurochem.*, **39**, 1755.
9. Rodnight,R., Trotta,E. and Perrett,C. (1985) *J. Neurosci. Methods*, **13**, 87.
10. Daly,J.W. (1977) *Cyclic Nucleotides in the Nervous System*. Plenum Press, New York.
11. Medina,M.A., Dean,A.P. and Stavinoha,W.B. (1980) In *Cerebral Metabolism and Neuronal Func-*

tion. Passonneau,J.V., Hawkins,R.W., Lust,W.D. and Welsh,F.A. (eds), Williams and Wilkins, Baltimore, p. 56.
12. Lemos,J.R., Novak-Hofer,I. and Levitan,I.B. (1985) *J. Biol. Chem.*, **260**, 3207.
13. Kakiuchi,S. and Rall,T.W. (1968) *Mol. Pharmacol.*, **4**, 367.
14. Garthwaite,J., Woodhams,P.L., Collins,M.J. and Balazs,R. (1979) *Brain Res.*, **173**, 373.
15. Hollingsworth,E.B., McNeal,E.T., Burton,J.C., Williams,R.J., Daly,J.W. and Creveling,C.R. (1985) *J. Neurosci.*, **8**, 2240.
16. Ueda,T., Greengard,P., Berzins,K., Cohen,R.S., Blomberg,F., Grab,D. and Siekevitz,P. (1979) *J. Cell. Biol.*, **83**, 308.
17. Nelson,P.G. and Lieberman,M. (eds) (1981) *Excitable Cells in Tissue Culture*. Plenum, New York.
18. Brooker,G., Greengard,P. and Robison,G.A. (eds) (1979) *Advances in Cyclic Nucleotide Research*. Raven Press, New York, Vol. 10.
19. Brown,B.L., Ekins,R.P. and Albano,J.D.M. (1972) *Adv. Cyclic Nucleotide Res.*, **2**, 25.
20. Gilman,A.G. (1970) *Proc. Natl. Acad. Sci. USA*, **67**, 305.
21. Laemmli,U.K. (1970) *Nature*, **227**, 680.
22. O'Farrell,P.H. (1975) *J. Biol. Chem.*, **250**, 4407.
23. Hames,B.D. and Rickwood,D. (eds) (1983) *Gel Electrophoresis of Proteins – A Practical Approach*. IRL Press, Oxford and Washington, D.C.
24. Walker,J.M. (ed.) (1984) *Methods in Molecular Biology. Vol. 1, Proteins*. Humana Press, Clifton, New Jersey.
25. Garrison,J.C. (1983) In *Methods in Enzymology*, Corbin,J.D. and Hardman,J.G. (eds), Academic Press, New York, Vol. 99, p. 20.
26. Laskey,R.A. (1984) *Radioisotope Detection by Fluorography and Intensifying Screens*. Amersham, Review Guide No. 23.
27. Peterson,G.L. (1977) *Anal. Biochem.*, **83**, 346.
28. Bradford,M. (1976) *Anal. Biochem.*, **72**, 248.
29. Corbin,J.O. and Reimann,E. (1974) In *Methods in Enzymology*. Hardman,J.G. and O'Malley,B.W. (eds), Academic Press, New York, Vol. 38, p. 287.
30. Mobley,P.M. and Greengard,P. (1985) *Proc. Natl. Acad. Sci. USA*, **82**, 945.
31. Castelluci,V.F., Kandel,E.R., Schwartz,J.H., Wilson,F., Nairn,A.C. and Greengard,P. (1980) *Proc. Natl. Acad. Sci. USA*, **77**, 7492.
32. Rudolph,S.A., Beam,K.G. and Greengard,P. (1978) *Membrane Transport Processes*. Hoffman,J.F. (eds), Raven Press, New York, Vol. 1, p. 107.
33. Krueger,B.K., Forn,J. and Greengard,P. (1977) *J. Biol. Chem.*, **252**, 2764.
34. Wu,W.C.-S., Walaas,S.I., Nairn,A.C. and Greengard,P. (1982) *Proc. Natl. Acad. Sci. USA*, **79**, 5249.
35. Ng,M. and Matus,A. (1979) *Neuroscience*, **4**, 1265.
36. Walaas,S.I., Nairn,A.C. and Greengard,P. (1983) *J. Neurosci.*, **3**, 291.
37. Vartanian,T., Szuchet,S., Dawson,G. and Campagnoni,A. (1986) *Science*, **234**, 1395.
38. Palfrey,H.C. (1983) *FEBS Lett.*, **157**, 183.
39. Nairn,A.C., Bhagat,B. and Palfrey,H.C. (1985) *Proc. Natl. Acad. Sci. USA*, **82**, 7939.
40. Cleveland,D.W., Fischer,S.G., Kirschner,M.W. and Laemmli,U.K. (1977) *J. Biol. Chem.*, **252**, 1102.
41. Huttner,W.B. and Greengard,P. (1979) *Proc. Natl. Acad. Sci. USA*, **76**, 5402.
42. Bitte,L. and Kabat,D. (1974) In *Methods in Enzymology*. Moldave,R. and Grossmann,L. (eds), Academic Press, New York, Vol. 30, p. 563.
43. Cooper,J.A., Sefton,B.M. and Hunter,T. (1983) In *Methods in Enzymology*, Corbin,J.D. and Hardman,J.G. (eds), Academic Press, New York, Vol. 99, p. 387.
44. Williams,K.R., Hemmings,H.C.,Jr., LoPresti,M.B., Konigsberg,W.H. and Greengard,P. (1986) *J. Biol. Chem.*, **261**, 1890.
45. Palfrey,H.C., Rothlein,J. and Greengard,P. (1983) *J. Biol. Chem.*, **258**, 9496.
46. Corbin,J.D. and Hardman,J.G. (eds) (1983) *Methods in Enzymology*. Academic Press, New York, Vol. 99.
47. Blumberg,J.B., Vetulani,J., Stawarz,R.J. and Sulser,F. (1976) *Eur. J. Pharmacol.*, **37**, 357.
48. Kitano,T., Go,M., Kikkawa,U. and Nishizuka,Y. (1986) In *Methods in Enzymology*. P.M.Conn (ed.), Academic Press, New York, Vol. 124, p. 349.
49. Robinson,S.E., Mobley,P.L., Smith,H. and Sulser,F. (1978) *Naunyn-Schmiedbergs Arch. Pharmakol*, **303**, 175.
50. Walaas,S.I., Nairn,A.C. and Greengard,P. (1986) *J. Neurosci.*, **6**, 954.
51. Cahill,A.L. and Perlman,R.L. (1984) *Proc. Natl. Acad. Sci. USA*, **81**, 7243.
52. Berger,S.L. and Krug,M.S. (1985) *Biotechniques*, **3**, 38.
53. Alper,S.L., Palfrey,H.C., DeRiemer,S.A. and Greengard,P. (1980) *J. Biol. Chem.*, **256**, 11029.
54. Ingebritsen,T.S. and Cohen,P. (1983) *Science*, **221**, 331.
55. Nairn,A.C., Detre,J., Casnellie,J. and Greengard,P. (1982) *Nature*, **299**, 734.

CHAPTER 7

Physiological responses to receptor activation: phosphoinositide turnover

CATRIONA M.F.SIMPSON, IAN H.BATTY and JOHN N.HAWTHORNE

1. INTRODUCTION

Activation of many cell surface receptors is linked to the metabolism of phosphoinositides and there has been an almost explosive growth of interest in the field recently. Many, though not all, phosphoinositide-linked receptors raise intracellular free Ca^{2+} when activated. Berridge (1) suggests that inositol 1,4,5-trisphosphate (Ins-1,4,5-P_3) produced by receptor-induced hydrolysis of phosphatidylinositol 4,5-bisphosphate (PtdIns-4,5-P_2) is the chemical messenger releasing this Ca^{2+}. It is also suggested (2) that diacylglycerol, the other product of the phospholipase C-catalysed hydrolysis, provides a further intracellular signal by activating protein kinase C (3). In addition to the reviews quoted (2,3) there is a recent review by Michell (4). Most of the evidence for these hypotheses to date comes from experimental work using non-neural tissues. Though the polyphosphoinositides, phosphatidylinositol 4,5-bisphosphate and phosphatidylinositol 4-monophosphate (PtdIns-4-P), are most plentiful in nerve and brain it is hard at present to find experimental support for the concept that their neuronal function is simply concerned with Ca^{2+} mobilization (5).

2. THE BASIC TECHNIQUES

Radioactive isotopes such as [^{3}H]inositol and [^{32}P]orthophosphate are valuable tools in phosphoinositide research. Much of the early work in this field involved the measurement of increased ^{32}P incorporation into tissue phosphoinositides in response to agonist stimulation. However, the initial agonist-induced event is probably hydrolysis of PtdIns-4,5-P_2 and PtdIns-4-P and so measurement of the release of [^{3}H]inositol phosphates is a more direct method of study than measurement of ^{32}P labelling of the lipids which is really an indication of their subsequent resynthesis.

The pathways are shown in *Figure 1*. Though usually ignored at present, there is an active phosphatase converting PtdIns-4,5-P_2 to PtdIns-4-P and then to phosphatidylinositol (PtdIns). There are also phosphatases converting the water-soluble inositol trisphosphate ($InsP_3$) to inositol bisphosphate ($InsP_2$), inositol monophosphate (InsP) and free inositol. Lithium chloride inhibits the InsP phosphatase and is therefore used (6) to amplify phosphoinositide responses measured in terms of inositol phosphate release. Because of phosphatase action it is best to measure this release over a short time period, providing methods are sensitive enough, and also to measure all three inositol phosphates. Chemical measurements would be ideal but are rarely made since

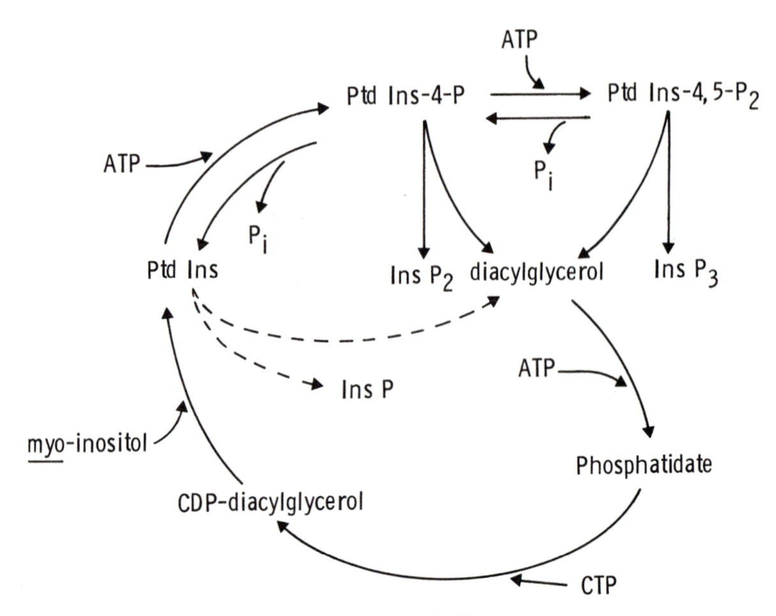

Figure 1. Some of the pathways of phosphoinositide metabolism.

labelling studies are easier. The tissue phosphoinositides are first labelled with either [^{32}P]orthophosphate or [^{3}H]inositol and some pitfalls can be avoided if changes in labelled phosphoinositides are measured as well as release of labelled inositol phosphates. As will be seen from *Figure 1*, the ratio of counts in the various inositol phosphates will differ between [^{32}P]orthophosphate and [^{3}H]inositol labelling. With [^{32}P]orthophosphate, labelling of PtdIns-4,5-P_2 and PtdIns-4-P will be rapid, while that of PtdIns will be relatively slow. With [^{3}H]inositol on the other hand, PtdIns will be more highly labelled than the polyphosphoinositides. Using cultured cells the pre-labelling period can be long enough for the three lipids to reach equilibrium with the precursor pool so that these differences do not arise, but with tissue slices such long periods would damage the tissue. Since the various inositol phosphates are usually measured only in terms of radioactivity these points are important to remember.

In what follows therefore, we suggest methods for separating and estimating the lipids themselves by thin-layer chromatography (t.l.c.) and for estimating the inositol phosphates by ion-exchange chromatography or high-pressure liquid chromatography (h.p.l.c.). The ion-exchange method was developed more than 20 years ago (7) and it is now in widespread use, more or less unchanged. A gas–liquid chromatography (g.l.c.) method for estimating inositol itself is also given.

Irvine and his colleagues have provided evidence that there is an inositol 1,3,4-trisphosphate as well as the inositol 1,4,5-trisphosphate released by the hydrolysis of PtdIns-4,5-P_2 (8). Batty *et al.* (9) have shown that stimulation of muscarinic receptors in rat cerebral cortex produces inositol 1,3,4,5-tetrakisphosphate. This is probably formed by phosphorylation of Ins-1,4,5-P_3 (10) rather than by hydrolysis of an inositol lipid with four phosphate groups, though the existence of such a lipid is difficult to rule out if it is very labile. These recent studies emphasize the need for good analytical methods for the extraction and separation of the phosphoinositides and the various inositol phosphates.

Berridge and Irvine (2) propose that both inositol trisphosphate and diacylglycerol have second messenger functions. We have not attempted to deal with diacylglycerol and its activation of protein kinase C (3) but several points are important to note.

(i) To our knowledge there is no *direct* evidence that kinase C is activated by the diacylglycerol released when receptor activation causes phosphoinositide hydrolysis. Moreover, the kinase is not activated better by the 1-stearoyl 2-arachidonoyl glycerol released from phosphoinositides than by diacylglycerols of other fatty acid composition, such as dioctanoyl glycerol.

(ii) Many workers have used phorbol esters such as 12-*O*-tetradecanoyl phorbol-13-acetate as activators of protein kinase C, but without checking to see that proteins are actually phosphorylated as a result. The tumour-promoting phorbol esters may have physiological effects which do not involve kinase C as an intermediate.

(iii) Use of [^{32}P]orthophosphate in tissues where receptor activation leads to phosphoinositide hydrolysis often shows rapid production of labelled phosphatidate. Since this is formed by phosphorylation of diacylglycerol it provides a measure of the formation of the latter.

Further information about the properties and assay of protein kinase C may be found in Chapter 6 and in reviews from Nishizuka's laboratory (11,12).

3. RADIOLABELLING OF THE PHOSPHOINOSITIDES

PtdIns-4-P and PtdIns-4,5-P_2 are found in higher concentrations in myelin-rich nervous tissue than in most other cell types, though even in brain and nerve they represent only a minor fraction of the total cellular phospholipid. The low concentrations of the polyphosphoinositides make chemical analysis and detection difficult and therefore most investigations of the role of inositol lipids in transmembrane signalling involve radiolabelling with either [^{32}P]orthophosphate or [^{3}H]inositol. Receptor-stimulated phosphoinositide metabolism has most frequently been observed in incubations of intact cells and therefore systems such as brain slices, ganglia and cultured neurones have been employed in many lipid labelling studies. However, the phosphoinositide effect is also demonstrable in synaptosomes.

The monoester phosphate groups of the polyphosphoinositides (i.e. the 4-phosphate of PtdIns-4-P and the 4- and 5-phosphates of PtdIns-4,5-P_2) turn over more rapidly than the 1-phosphate and inositol moieties. As a consequence, PtdIns-4-P and PtdIns-4,5-P_2 are more rapidly labelled with [^{32}P]orthophosphate than is PtdIns. In contrast, PtdIns is more rapidly labelled with [^{3}H]inositol than are the polyphosphoinositides.

Phosphoinositide labelling studies may be used in a number of ways.

(i) To compare the time courses of incorporation of [^{32}P]orthophosphate and [^{3}H]inositol into the inositol phospholipids in the presence and absence of agonist. With this approach it is important to establish that any observed increase in agonist-induced [^{32}P]labelling is truly a consequence of stimulated phosphoinositide turnover and is not merely a reflection of an increase in the specific activity of the ^{32}P-labelled γ-phosphate group of ATP (13).

(ii) To label the phosphoinositides for long enough to allow their phosphate or inositol groups to come close to isotopic equilibrium with their metabolic precursors (i.e. the [$\gamma^{32}P$]phosphate group of ATP or free, cytosolic [^3H]inositol). This is more readily achieved in cultured cells than in tissue slices since the former can be incubated for longer periods of time without incurring risk of cell damage. Upon addition of an agonist, a direct measurement of the rate of phosphoinositide hydrolysis may be obtained by determining the rate of disappearance of labelled lipids and/or the rate of release of labelled inositol phosphates. Inclusion of 5−10 mM LiCl in the incubations will inhibit the inositol-1-phosphatase and thereby amplify agonist-dependent Ins-1-P accumulation (6). Agonist-induced phosphoinositide breakdown occurs very rapidly and experiments of this nature should be conducted over a short time course (10−15 min) with initial samples taken at perhaps 10 or 20 sec intervals. It should be noted that in experiments with brain slices the clarity with which the effect of an agonist upon phosphoinositide hydrolysis is seen will depend upon the diameter of the tissue section. The agonist will take time to penetrate the slice and the lag between stimulation of cells at the external surface of the tissue and stimulation of those at the centre will increase as the cross-section of the slice increases.

(iii) To estimate the rate of metabolic turnover of the individual phosphate groups of PtdIns-4-P and PtdIns-4,5-P_2. Hawkins *et al.* (14) have described methods for the assay of: (a) the 5-phosphate from PtdIns-4,5-P_2 using a 5-phosphate-specific phosphomonoesterase from human erythrocyte membranes, and (b) the 4-phosphate from PtdIns-4-P and the total monoester phosphate content of PtdIns-4,5-P_2 using alkaline phosphatase from bovine intestine.

Procedures for the extraction, separation and quantification of the radiolabelled phosphoinositides and inositol phosphates are described in Sections 4 and 5.

4. MEASUREMENT OF CHANGES IN PHOSPHOINOSITIDE METABOLISM

4.1 Extraction of phosphoinositides

Analysis of the phosphoinositides must be preceded by their quantitative isolation intact and free of non-lipid contaminants. The polyphosphoinositides were first discovered as protein−lipid complexes in brain (15) and they are now believed to be ubiquitous components of eukaryotic cells. They comprise a small proportion of the total cell phospholipid and have unusual solubility characteristics. Most membrane-associated phospholipids may be extracted using a mixture of chloroform and methanol (16). The latter is a polar solvent which disrupts the hydrogen bonding and electrostatic forces between membrane lipids and proteins. However, the polyphosphoinositides, being anionic at physiological pH, are more soluble in aqueous than organic solvents. The three and five negative charges of PtdIns-4-P and PtdIns-4,5-P_2 respectively enable these molecules to disperse in water in the form of spherical micelles with their polar headgroups oriented outwards. The quantitative extraction of the polyphosphoinositides is achieved only with acidified chloroform/methanol or with chloroform/methanol in the presence of a high concentration of KCl.

Table 1 outlines the method of Griffin and Hawthorne (17) for the extraction of the

Table 1. Extraction of phosphoinositides from brain synaptosomes using the method of Griffin and Hawthorne (17).

1.	Add 1 ml of 20% (w/v) TCA to a 1 ml suspension of synaptosomes. Centrifuge at 1000 *g* for 5 min and wash the pellet once with 1.5 ml of 5% (w/v) TCA containing 1 mM EDTA, and once with 1.5 ml of water.
2.	Resuspend the washed pellet in 1.5 ml of chloroform/methanol/concentrated HCl (100:100:1, by vol). Allow to stand for 30 min. Centrifuge at 1000 *g* for 5 min. Retain the supernatant and repeat this procedure.
3.	Resuspend the pellet in 1.5 ml of chloroform/methanol/concentrated HCl (200:100:1, by vol). Centrifuge and retain the supernatant.
4.	Add 1.5 ml of chloroform to the combined supernatants. The total volume is now 6 ml and the ratio of chloroform to methanol is 2:1 (v/v).
5.	Add 1.5 ml of 0.1 M HCl to give a mixture of chloroform/methanol/HCl in the ratio 8:4:3 (by vol). Mix well and then centrifuge. The solvents partition into a lower phase of composition chloroform/methanol/HCl in the ratio 86:14:1 (by vol) and an upper phase in which the proportions are 3:48:47 (by vol), respectively. The lower phase, which comprises approximately 60% of the total volume contains the purified lipid and the upper phase contains the non-lipid contaminants. Remove the upper phase and the protein interface.
6.	Wash the lower phase by adding 3 ml of acidified synthetic upper phase (*Table 2*). Mix well and centrifuge. Remove the upper phase and wash the lower phase again with 3 ml of synthetic upper phase containing 2 M KCl (*Table 2*).
7.	Dry the lipid extract under a stream of nitrogen and re-dissolve in chloroform/methanol (2:1, v/v).

Table 2. Preparation of acidified and neutral synthetic upper phases.

Either	
1.	Mix chloroform/methanol/0.1 M HCl or chloroform/methanol/2 M KCl (8:4:3, by vol) in a separating funnel and allow to stand for 1 h before taking the upper phase.
Or	
2.	Mix chloroform/methanol/0.1 M HCl or chloroform/methanol/2 M KCl in the proportions 3:48:47 (by vol).

phosphoinositides and other phospholipids from trichloroacetic acid (TCA) precipitates of brain synaptosomes. [This method may also be used to extract the phosphoinositides from intact neural tissue or cell suspensions by homogenizing in chloroform/methanol/concentrated HCl (100:100:1, by vol) and then proceeding as for TCA precipitates.] The final lipid extract is washed with acidified synthetic upper phase (*Table 2*), the composition of which is designed to ensure both the removal of any non-lipid, water-soluble contaminants and the retention of the polyphosphoinositides in the chloroform phase. After washing, the lipid extract is dried under a stream of nitrogen and re-dissolved in a suitable volume of chloroform/methanol (2:1, v/v). Concentration of a phospholipid extract washed with acidified synthetic upper phase may result in acid hydrolysis of the lipid. To avoid this possibility the extract may be washed finally with synthetic upper phase containing KCl at a concentration high enough to prevent loss of the polyphosphoinositides from the chloroform layer (*Table 2*). In radiolabelling studies, contamination of the lipid extract with [^{32}P]orthophosphate or [^{3}H]inositol may be avoided by a final wash with neutral upper phase containing either 0.5 M phosphate buffer or 10 mM inositol.

Table 3. Extraction of polyphosphoinositides from brain using the method of Michell *et al.* (18).

1.	Homogenize the tissue in 10 vols of chloroform/methanol (1:1, v/v). Centrifuge at 1000 *g* for 5 min and discard the supernatant which contains the bulk phospholipid. Wash the tissue residue three times with 10 vols of chloroform/methanol (2:1, v/v).
2.	Suspend the washed residue in aqueous 2 M KCl (4–6 ml/g tissue). Add chloroform/methanol (1:2, v/v) (3.75 ml/ml 2 M KCl). Mix well and allow to stand for 30 min.
3.	Add chloroform and then water (1.25 ml of each/ml 2 M KCl) to give a mixture of chloroform/methanol/KCl in the ratio 10:10:9 (by vol). Mix well and separate the phases by centrifugation.
4.	Retain the lower phase and wash the upper phase with chloroform equal in volume to the original lower layer. Combine the lower phases.
5.	Dry the lipid extract under a stream of nitrogen and re-dissolve in chloroform/methanol (2:1, v/v).

Table 3 shows the procedure of Michell *et al.* (18), based upon that of Bligh and Dyer (19), for the extraction of the polyphosphoinositides from brain using chloroform/methanol in the presence of 2 M KCl. The method describes an optional preliminary extraction with neutral chloroform/methanol which will remove the bulk of the phospholipids leaving a residue enriched with respect to the polyphosphoinositides. It is important to note that PtdIns will be extracted by the neutral solvent. An initial neutral extraction may be valuable in procedures for the purification of the polyphosphoinositides but for quantitative analyses it is advisable to check that some PtdIns-4-P and PtdIns-4,5-P_2 are not lost with the neutral solvent. The extraction procedure utilizing acidified chloroform/methanol has been successfully applied to many cell types. However, inexplicably, the KCl extraction appears to be effective only for brain polyphosphoinositides.

During extraction and storage, precautions should be taken to minimize the degradation of the very labile polyphosphoinositides. Extraction procedures should be conducted at or below room temperature to retard enzyme-catalysed hydrolysis. Polyunsaturated fatty acids will auto-oxidize very rapidly if left unprotected in air. Anti-oxidants such as butylated hydroxytoluene may be included in the solvent systems. Solvent should always be removed from lipid extracts under vacuum in a rotary evaporator or under a stream of nitrogen. Phospholipid extracts should never be left dry for lengthy periods of time but should always be re-dissolved in solvent as soon as possible and stored under nitrogen at $-20°C$.

4.2 Separation of phospholipids by thin layer chromatography

In t.l.c., the separation of the components of a phospholipid mixture is achieved by utilizing the different degrees to which the lipids are adsorbed on to a solid support coating a glass plate and then eluted by the passage of a suitable solvent system.

Silica gel, a highly porous amorphous silicic acid, is the most commonly used adsorbant for the separation of phospholipids by t.l.c. Ready-made t.l.c. plates may be obtained from a number of suppliers (e.g. Merck, Desaga, Camlab). The silica gel on these pre-coated plates generally contains a binder, $CaSO_4$, in order to facilitate adhesion of the adsorbent to the glass plate. Plates of dimensions 20×20 cm coated with a 0.25 mm layer of silica gel 60 (i.e. 60 Å pore diameter) with a mean particle size 2–25 μm are used routinely for analytical studies. Plates with layers of thickness 0.5

Table 4. Preparation of thin layer plates for the separation of phosphoinositides.

1.	Wash glass plates (20 × 20 cm, 3 mm in thickness) with chloroform/methanol (1:1, v/v). It is important that the surface of the glass is completely free of dust and grease.
2.	Place the plates in a plate-making apparatus (available commercially from Shandon or Desaga) and clamp into position.
3.	Measure the clearance between the plates and the leading edge of the spreader using a feeler gauge and adjust this if necessary using the screws on the spreader.
4.	Mix 40 g of silica gel 60 with 95 ml of a 1% (w/v) solution of potassium oxalate in water. The consistency should be such that the mixture drips off a spatula easily. This recipe should provide enough adsorbent to spread six plates with a layer 0.4 mm thick.
5.	Pour the slurry into the spreader and coat the plates.
6.	Allow the plates to harden before removing them from the apparatus.
7.	Activate the plates by heating at 120°C for approximately 1 h to ensure the removal of moisture.
8.	Store plates in an oven at 120°C or in a desiccator.

or 1.0 mm may be required for preparative work. For small quantities of sample, high performance thin layer plates of dimensions 10 × 10 cm coated with a 0.2 mm layer of silica gel 60 with a mean particle size 5–10 μm give separations quickly and with high resolution.

Ready-made plates have some disadvantages in that the binder may interfere with certain investigations and they are also expensive. T.l.c. plates may be prepared easily to required specifications and at a low price in the laboratory using the method outlined in *Table 4*. Some practice with the plate-spreading apparatus may be required before uniform and unflawed coating of the plates is consistently achieved. Plates which lack binder are rather fragile and extra care is required to ensure that the layer of adsorbent is not damaged.

Chromatographic separation of phospholipids on thin layer plates may be achieved by development in either one or two dimensions using selected solvent systems. Development in a second direction may resolve components of complex lipid mixtures which are inseparable by chromatography in one direction.

The best separations of the phosphoinositides by t.l.c. are attained on oxalate-impregnated plates (20). *Table 4* describes the preparation of oxalate plates using a plate-spreading apparatus. Ready-made plates may be impregnated with potassium oxalate using the following method.

(i) Develop plates in 150 ml methanol/water (2:3, v/v) containing 1% (w/v) potassium oxalate and allow the solvent front to run to the top of the plates. This will take a long time (~2 h for 10 × 10 cm plates) because of the high proportion of water in the solvent system.

(ii) Activate the plates by heating at 110°C for 15 min. Allow to cool and store in a desiccator.

Phospholipid samples are applied to thin layer plates and the plates are developed using the following procedure.

(i) Dissolve the phospholipid in as small a volume of chloroform/methanol (2:1, v/v) as possible. Guidelines for the quantity of phospholipid to use are approximately 0.5 mg for one-dimensional separation on 20 × 20 cm plates, approximately 1.0 mg for two-dimensional separation on 20 × 20 cm plates and

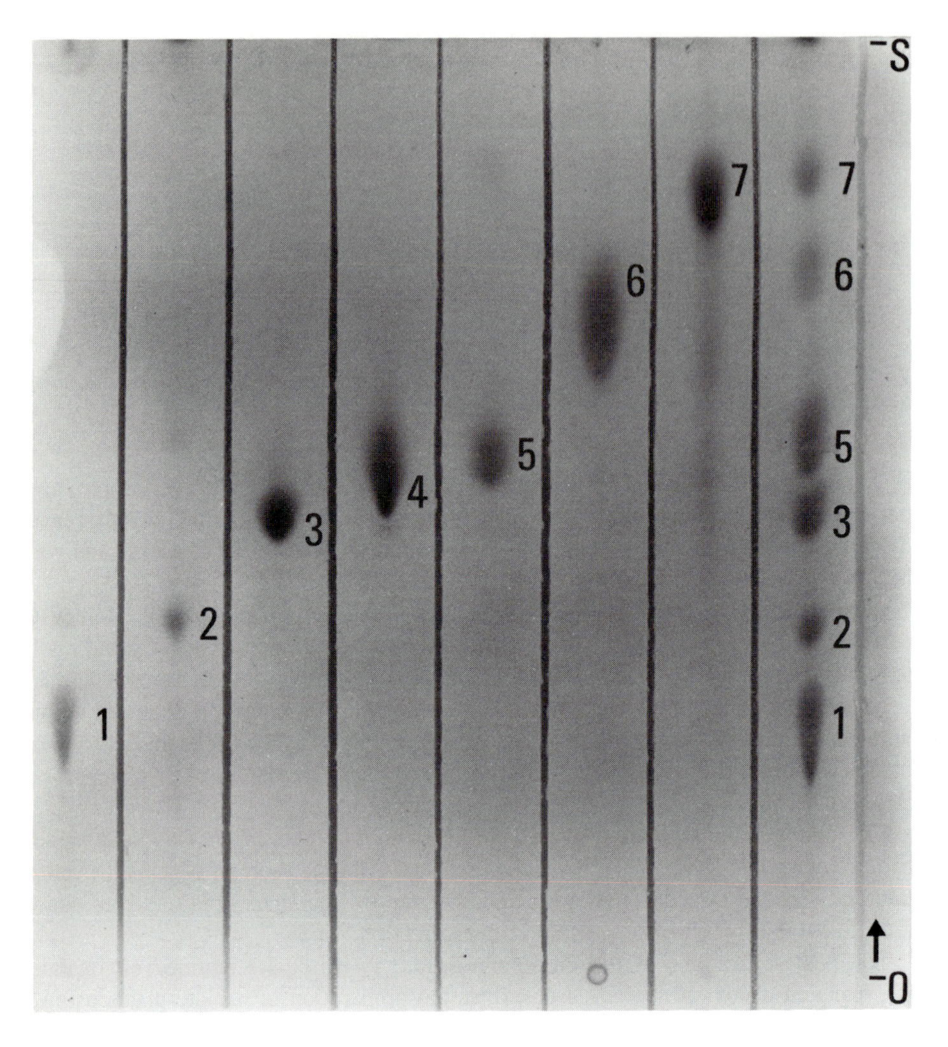

Figure 2. One-dimensional separation of phospholipid standards. **O**, origin; **S**, solvent front; **1**, PtdIns-4,5-P_2; **2**, PtdIns-4-P; **3**, PtdIns; **4**, PtdCho; **5**, PtdSer; **6**, PtdEtn; **7**, PtdA. The arrow represents direction of development.

approximately 50 μg for two-dimensional separation on 10 × 10 cm high performance t.l.c. plates.

(ii) Using a microsyringe apply the sample as a discrete spot or narrow band approximately 2 cm from the bottom of the plate. During application, the spot or band may be dried using a gentle stream of nitrogen. For one-dimensional separation, several samples may be run on the same plate. For two dimensional separation, apply the lipid to the left hand corner of the plate (2 cm away from the bottom and left hand edges), develop the chromatogram in the first solvent, dry thoroughly (i.e. for at least 1 h) at room temperature, turn through 90° in an anti-clockwise direction and then develop in the second solvent. The plates should be chromatographed immediately after application of the sample.

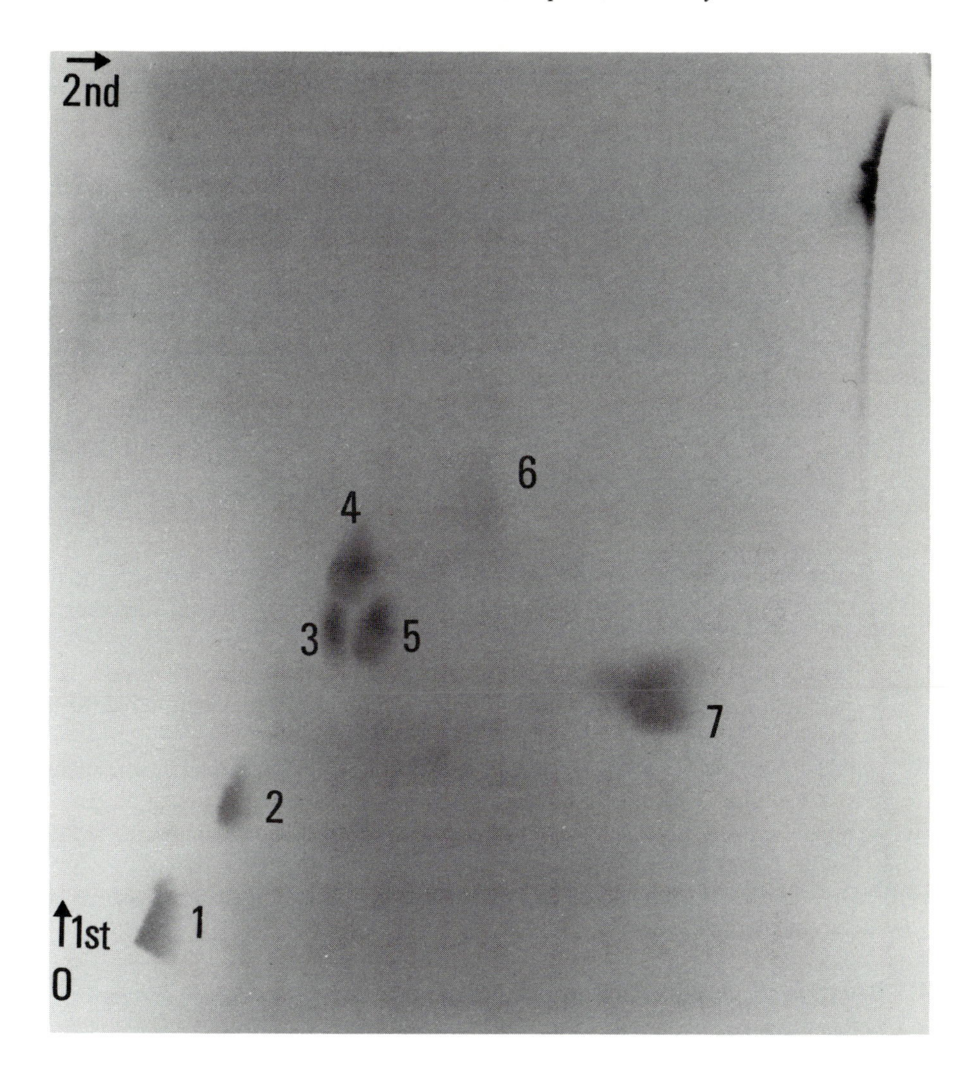

Figure 3. Two-dimensional separation of phospholipid standards. **O**, origin; **1**, PtdIns-4,5-P_2; **2**, PtdIns-4-P; **3**, PtdIns; **4**, PtdCho; **5**, PtdSer; **6**, PtdEtn; **7**, PtdA. The arrows represent directions of development.

Clear separation of the phosphoinositides is achieved by developing in chloroform/acetone/methanol/glacial acetic acid/water (40:15:13:12:7, by vol) in one dimension (*Figure 2*). However, using this method the resolution of phosphatidylserine and phosphatidylcholine is often poor. This problem may be overcome in a two-dimensional separation using chloroform/methanol/water/ammonia, specific gravity 0.88 (90:70:16:4, by vol) in the first dimension and the above acidic solvent system in the second (*Figure 3*). The method for the development of the t.l.c. plates is as described below.

(i) Prepare the solvent systems, pour them into tanks and allow to equilibrate for at least 1 h before use. Line the tanks with filter paper so that the atmosphere inside the tank is saturated with solvent vapour. This will hasten development and may improve resolution. Butylated hydroxytoluene (0.01%, w/v) may be added to the solvent. It migrates with the solvent front and thus protects the lipid from oxidation without interfering with the analysis.

(ii) Place the plate in a tank containing approximately 150 ml of eluting solvent. Take care that the solvent level does not reach the lipid sample at the origin. Tanks are available which will hold up to five 20 × 20 cm plates or eleven 10 × 10 cm plates. Cover the tank with a ground glass lid.

(iii) Remove the plate from the tank when the solvent front is approximately 1 cm away from the top of the plate. Allow to dry at room temperature.

Phospholipids separated by t.l.c. may be detected and identified using the methods described in Section 4.3 with reference to known standard compounds. Methods for quantitative analysis of phospholipid on thin layer plates are outlined in Section 4.4.

Phosphoinositides separated by t.l.c. may be recovered from the plates using the following procedure, adapted from that of Griffin and Hawthorne (17), for elution from silica gel.

(i) Scrape the adsorbent from a lipid-containing zone and suspend it in 1.5 ml of chloroform/methanol/concentrated HCl (100:100:1, by vol.). Allow to stand for 15–30 min. Centrifuge at 1000 *g* for 5 min and retain the filtered supernatant. Repeat.

(ii) Resuspend the silica in 1.5 ml of chloroform/methanol/concentrated HCl (200:100:1, by vol). Centrifuge at 1000 *g* for 5 min and retain the filtered supernatant.

(iii) Add 1.5 ml each of chloroform and 0.1 M HCl to the combined supernatants. Mix well and centrifuge to separate the phases. Discard the upper phase.

(iv) Wash the lower phase with 3 ml of synthetic upper phase containing 2 M KCl (*Table* 2).

(v) Dry the lipid extract under a stream of nitrogen and re-dissolve in chloroform/methanol (2:1, v/v).

4.3 **Detection of phospholipid on thin layer plates**

4.3.1 *Staining methods*

The following are general methods for visualizing phospholipids.

(i) *Charring.* Spray the plate with 3% (w/v) cupric acetate in 8% (v/v) phosphoric acid and heat at 160°C for 20 min. Phospholipids (and any other organic compounds) become visible as black deposits of carbon. This method is destructive, but very sensitive, with a lower detection limit of approximately 1 μg of phospholipid.

(ii) *Iodine vapour.* Place the plate in a tank with iodine crystals. Iodine dissolves in phospholipids and the latter appear as yellow/brown spots. With light staining the iodine will sublime from the plates when kept overnight under vacuum. However, heavier staining is destructive as the iodine reacts to some extent with polyunsaturated fatty acids.

(iii) *Spray plates with 0.1% (w/v) 2,7-dichlorofluorescein in 95% (v/v) methanol.* Phospholipids are visualized under u.v. light. This is a non-destructive technique as 2,7-dichlorofluorescein may be removed from eluted lipid samples by shaking with 2% (w/v) $KHCO_3$.

There are also a number of specific staining reagents.

(i) *Phosphate stain.* Method 1 (21). Dissolve 10 g of sodium molybdate in 100 ml of 4 M HCl and 1 g of hydrazine hydrochloride in 100 ml of water. Mix the solutions and heat in a boiling water bath for 5 min. Cool and make up to 1 litre. Spray the plates with reagent to reveal phosphorus-containing lipids as blue spots.

Method 2 (22). Dissolve 1.6 g of ammonium molybdate in 12 ml of water to give solution (a). Add 8 ml of solution (a) to 4 ml of concentrated HCl and 1 ml of mercury. Shake vigorously for 30 min and filter to give solution (b). Add 20 ml of concentrated H_2SO_4 to the remainder of solution (a) and all of solution (b). Cool and dilute to 100 ml with water. Spray the plates with the reagent to visualize phosphorus-containing lipids as blue spots.

(ii) *Periodate–Schiff stain (23).* This is specific for compounds containing vicinal diol groups. Prepare Schiff's reagent, that is a 1% (w/v) solution of *p*-rosaniline hydrocholoride in water decolourized by saturation with SO_2. If a yellow colour persists, the solution can be treated with charcoal and filtered. Spray the plate until it is saturated with a solution of 1% (w/v) sodium periodate. Allow oxidation to proceed for 5–10 min, then place the plate in a tank of SO_2 until the brown colour of the liberated iodine is completely removed. Spray the plate lightly with Schiff's reagent. Phosphatidylglycerol, which gives formaldehyde upon periodate oxidation, is revealed immediately as a blue spot. Glycolipids give aldehydes on periodate oxidation and also appear blue, but this colour is slower to develop. Phosphatidylinositol gives a yellow colour characteristic of those compounds forming malondialdehyde on periodate oxidation.

4.3.2 *Autoradiography*

Radioactive isotopes were originally discovered by their ability to blacken photographic

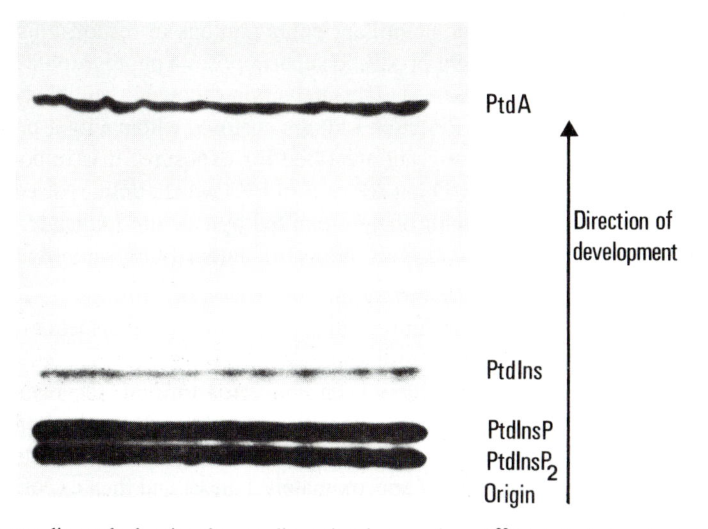

Figure 4. Autoradiograph showing the one-dimensional separation of ^{32}P-labelled phospholipids from the rectal gland of the spotted dogfish, *Scyliorhinus canicula.*

Table 5. Detection of radiolabelled phospholipids on thin layer plates by direct autoradiography.

1.	Use a dark room equipped with a red light.
2.	Place a sheet of film (e.g. Kodak DEF-2 direct exposure film) over the chromatogram and clip it securely into place. It is inadvisable to place ^{32}P-labelled thin layer plates on top of each other without two intervening glass plates because highly energetic β-particles will penetrate a thin layer of glass and give false spots on other autoradiograms.
3.	Wrap the plate and film in a black bag and place in a black-lined box.
4.	After the required exposure time, remove the film from the plate and immerse it in developer (e.g. Kodak LX24 X-ray developer diluted 1 + 5 with water) for 4 min at 20°C. Agitate gently during this time. Rinse in water for 30 sec.
5.	Fix the film (e.g. in Kodak FX-40 X-ray liquid fixer diluted 1 + 4 with water) for 5 min.
6.	Switch on the light and rinse the autoradiograph in running water for 20 min.
7.	Dip it into wetting agent (e.g. Kodak 'Photo-flo' solution diluted 1 + 200 with water) and dry.
8.	Align the autoradiograph with the chromatogram to locate radioactive spots.

emulsions. Today autoradiography using photographic film remains a convenient method for isotope detection (*Figure 4*).

The procedure for the location of radiolabelled phospholipids on thin layer plates by direct autoradiography is described in *Table 5*. The exposure time required will depend upon the amount of radioactivity in the chromatogram and upon the energy of the radiation. Provided that the film makes close contact with the plate, direct autoradiography will clearly reveal ^{14}C-labelled spots of activity 3000 c.p.m. within 48 h and ^{32}P-labelled spots of similar activity overnight. Emissions of ^{3}H are of such low energy that they are internally absorbed within the sample and fail to reach the film. The opposite problem may limit sensitivity for highly energetic β-particles such as those from ^{32}P. These emissions pass through and beyond the film so that only a small proportion of the total energy is recorded by an autoradiograph. The sensitivity of detection of low amounts of ^{32}P may be greatly increased by converting the emitted energy to light using an intensifying screen. A calcium tungstate screen is placed behind the film and this dense inorganic scintillant emits photons in response to β-particles which penetrate the film. The light produced superimposes a photographic image upon the autoradiographic image (*Figure 5*). The intensifying screens available commercially (e.g. from Dupont or Genetic Research Ltd) are enclosed within a light-proof cassette into which the film and chromatogram are fitted for exposure. It is important to use 'screen-type' film (e.g. Kodak X-Omat AR or Fuji RX) which, unlike 'direct exposure' film, is sensitive to the wavelength of light emitted by calcium tungstate. There are, however, disadvantages associated with the use of intensifying screens.

(i) The gain in sensitivity achieved by the conversion of emissions to light may be partly offset by a loss of resolution which occurs due to the dispersion of primary emissions and secondary scintillations.

(ii) Once radioactive emissions have been converted to light, the response of the film is no longer linear and low intensities of light produce disproportionately faint images. This problem may be overcome by pre-exposing the film to a hypersensitizing light flash of approximately 1 msec and then exposing the film at −70°C (24). The cassette should be allowed to warm to room temperature before removing and developing the film. The procedure for the use of intensifying screens is thus rather more complex than that for direct autoradiography.

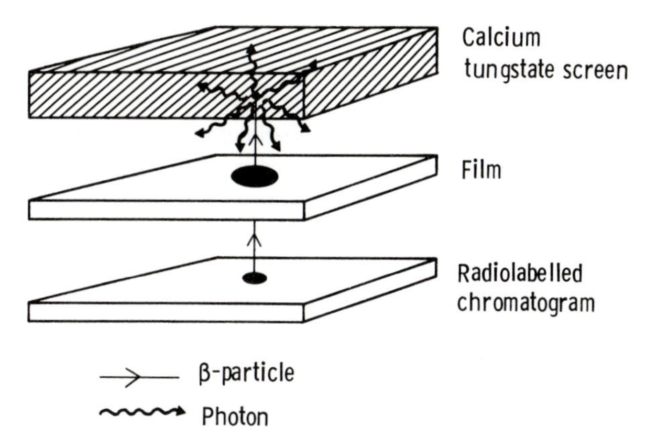

Figure 5. Diagram to illustrate the use of an intensifying screen to convert β-particles to light in order to overcome the problem of the penetration of high energy emissions through and beyond the film.

4.3.3 *Radioisotope scanning*

Radiolabelled phospholipids separated by one-dimensional t.l.c. may be located using a radiochromatogram scanner (e.g. from Panax or Desaga). The plate is scanned by a gas-flow counter and signals are transmitted to a recorder which gives a continuous trace illustrating peaks of radioactivity. Radiolabelled compounds can be identified using the detection methods described previously. For quantitative results, radiolabelled zones should be scraped and counted in a scintillation counter (see Section 4.4.4).

4.4 **Quantitative analysis of phospholipid on thin layer plates**

4.4.1 *Photodensitometry*

Components of a phospholipid mixture separated by t.l.c. in one dimension may be quantified by a photodensitometric scan of a charred plate. A light beam from a tungsten lamp (370–700 nm) scans a lane across the plate and the absorbance of each of the charred spots is recorded as a peak on a trace. Peak area is proportional to the amount of carbon in a spot and therefore to the amount of lipid originally present. Modern densitometers (e.g. the Desaga CD50) are linked to microcomputers which will integrate peaks, express the data as weight percentage of total lipid and, after calibration with internal or external standards, calculate the concentrations of individual components.

This method of quantification is quick and easy to perform. However, it does have drawbacks in that:

(i) The accuracy is dependent upon complete separation of components by one-dimensional t.l.c., which is not always possible.
(ii) The yield of carbon is influenced by the amount of charring reagent applied to the plate. Practice is required to obtain uniformly sprayed plates and reproducible results. It may be possible to achieve a more even application by dipping the plate in charring reagent.
(iii) The lipid samples are destroyed by charring.

4.4.2 *Fluorometry*

Dyes such as 2,7-dichlorofluorescein and Rhodamine 6G fluoresce in the presence of lipid and may therefore be used as stains for the detection of phospholipid as described in Section 4.3.1. Phospholipids separated by one-dimensional t.l.c. on plates impregnated with a fluorescent indicator may be quantified by measuring the fluorescence of each lipid spot with a scanning fluorometer. The fluorescence is proportional to the amount of lipid present but depends upon the structure of the lipid (e.g. double bonds in alkyl chains cause quenching) and so calibration curves should be prepared using known amounts of a variety of standards. Plates with a fluorescent indicator incorporated into the adsorbent may be obtained commercially (e.g. from Merck, Desaga, Camlab) or prepared in the laboratory. Alternatively, plates may be sprayed with a solution of the indicator. The most commonly used spray reagents for fluorometry are:

(i) 0.01% (w/v) Rhodamine 6G in water
(ii) 0.01% (w/v) 1-anilino-8-naphthalene sulphonate in water

Densitometers such as the Desaga CD50 are fitted with a deuterium lamp as well as a tungsten lamp and so serve also as fluorometers which will excite samples with light in the range 200–370 nm.

Fluorometry has disadvantages similar to those for photodensitometry. Again the accuracy is dependent upon resolution of components of a mixture by one-dimensional t.l.c. and the fluorescence is dependent upon the concentration of dye applied to the plate. It is probably easier to obtain reproducible results using plates impregnated with indicator than using plates which are sprayed with the dye. Nevertheless the method is quick, easy and also has the advantage of being non-destructive so that the lipids can be recovered for further analysis.

4.4.3 *Analysis of lipid phosphorus*

(i) *Method 1.* This assay, modified from that of Bartlett (25), involves the oxidation of phospholipid by perchloric acid with the release of inorganic phosphate which reacts with ammonium molybdate to form phosphomolybdic acid. Reduction of the latter yields a blue complex which may be measured spectrophotometrically. The reducing agent (ANSA) is prepared in the following manner:

(1) Add 2 g of hydrated sodium sulphite and then 0.5 g of 1-amino-2-naphtho-4-sulphonic acid to 200 ml of freshly-made 7.5% (w/v) sodium metabisulphite in water.
(2) Filter and store in the dark at 4°C. The reagent is stable for approximately 1 month.

For the assay it is advisable to use tubes which have been washed in a strong, phosphate-free detergent. The assay procedure is as follows.

(1) Dry the phospholipid samples, which are in solvent, under a stream of nitrogen.
(2) Add 0.35 ml of 70% (v/v) perchloric acid to the dried samples and heat at 180°C in a heating block until they are clear. This reaction usually takes 1–2 h.
(3) Cool. Add 2 ml of water, 0.1 ml of 5% (w/v) ammonium molybdate and 0.1 ml of the ANSA reagent. Heat at 90°C for 20 min.
(4) Cool. Measure the absorbance at 830 nm. The colour remains stable for 24 h.

A blank and a series of standard samples should be assayed simultaneously. A standard solution of 21.93 mg of KH_2PO_4 in 100 ml of water contains 5 μg of phosphorus per 100 μl. The amount of phosphorus in the experimental samples is read from a calibration curve. With phospholipid samples which have been separated on thin layer plates there is no necessity to elute the lipid from the silica before conducting the assay. The perchloric acid oxidation may be carried out in the presence of silica provided that the suspension is centrifuged at 800 *g* and the supernatant decanted prior to absorbance measurement. This assay is very sensitive and has a lower limit of approximately 0.4 μg of phosphorus.

(ii) *Method 2 (26)*. This method for the quantitative determination of phospholipid does not involve initial acid digestion of the lipid to release inorganic phosphate. Phospholipid samples are heated with a chromogenic solution which is prepared by the addition of 45 ml of methanol, 5 ml of chloroform and 20 ml of water to 25 ml of the undiluted spray reagent of Vaskovsky and Kostetsky described in Section 4.3.1. The procedure for the assay is as follows.

(1) Dry the phospholipid samples, which are in solvent, under a stream of nitrogen.
(2) Add 0.4 ml of chloroform and 0.1 ml of the chromogenic solution to each sample.
(3) Place the tubes in a boiling water bath for 90 sec.
(4) Cool to room temperature and add 5 ml of chloroform. Mix well.
(5) Allow to stand for 30 min. A short spin in the bench centrifuge may be required to effect separation of aqueous and organic phases.
(6) Remove the lower organic layers and read the absorbance in quartz cuvettes at 710 nm. The blue complex is stable for at least 3 h.

A blank and a series of phospholipid standard samples should be assayed simultaneously. This assay is quick because it obviates the perchloric acid oxidation step. However, it is less sensitive than the Bartlett method described above, having a lower limit of approximately 1 μg of lipid phosphorus, and it also has the disadvantage of requiring that samples are first eluted from silica gel.

In calculating molar amounts of phospholipid from phosphorus analyses it should be remembered that PtdIns-4-P and PtdIns-4,5-P_2 have 2 and 3 mol of phosphorus/mol of total lipid, respectively.

4.4.4 *Counting of radioactivity*

The procedure for the quantification of radiolabelled phospholipids separated by t.l.c. is as follows.

(i) Locate phospholipids by staining the plate lightly with iodine vapour. Mark the lipid zones and allow the iodine to sublime in order to minimize the risk of quenching.
(ii) Scrape the adsorbent from the phospholipid-containing zones into vials and add 5 ml of scintillation cocktail. (While scraping the plates, great care should be taken not to inhale any radioactive silica dust nor to contaminate the laboratory). Mix thoroughly before counting to facilitate the elution of the radiolabelled lipids from the silica gel.
(iii) Count in a scintillation counter.

5. MEASUREMENT OF RELEASE OF INOSITOL PHOSPHATES

5.1 Sample preparation

Agonist-induced phosphoinositide breakdown may be studied directly by following the release of radiolabelled inositol phosphates from lipids labelled to (or close to) isotopic equilibrium. Separation of the inositol phosphates may be achieved either by anion-exchange chromatography (Section 5.2) or by h.p.l.c. (Section 5.4). The latter is a more powerful analytical tool. [^{3}H]Inositol is most commonly used to label tissue samples as the use of [^{32}P]orthophosphate presents difficulties in the separation of ^{32}P-labelled inositol phosphates from other water-soluble ^{32}P-phosphate containing compounds. The protocol below describes a method for the preparation of [^{3}H]inositol-labelled brain slices.

(i) Dissect out rat cerebral cortex on ice and cross-cut at 90° on a McIlwain tissue chopper at 350 μm.

(ii) Transfer to a small volume ($\sim$25 ml) of modified Krebs–Henseleit bicarbonate Ringer solution (containing 1.3 mM $CaCl_2$) and shake to distribute the slices.

(iii) Wash the slices twice in 100 ml of Ringer and pre-incubate under O_2/CO_2 (95:5, v/v) for 60 min at 37°C. Replace the medium at 15 min intervals.

(iv) Aspirate the medium, wash the tissue again with 100 ml of Ringer and then allow the slices to settle under gravity.

(v) Transfer 50 μl aliquots of the tissue suspension to 5 ml capacity flat-bottomed vials containing 0.5–5.0 μCi [^{3}H]inositol in 240 μl of medium. Incubate under O_2/CO_2 (95:5, v/v) for 60 min as in (iii). After labelling, the slices may be washed with fresh medium containing unlabelled inositol and then incubated for a further 60 min in this solution to ensure removal of all [^{3}H]inositol from the tissue. This treatment will prevent any agonist-stimulated phosphoinositide labelling but it may result in a reduced yield of labelled inositol phosphates and the longer the incubation period the greater the risk of causing cell damage.

(vi) Add the appropriate agonist in 10 μl of medium and incubate for the required time. 5–10 mM LiCl may be used to amplify the agonist-induced accumulation of the inositol phosphates. If it is used, it should be added to the tissue suspension 5 min before adding the agonist. It may be necessary, by trial and error, to determine labelling conditions which permit detection of potentially low levels of higher inositol polyphosphates but still allow clear resolution of lower phosphates accumulating to much greater concentrations.

(vii) Terminate the reaction with 300 μl of 1 M TCA. Leave samples on ice for 20 min then sonicate or vortex vigorously and sediment the tissue by centrifugation.

(viii) Remove an aliquot of supernatant and wash five times with 2 vol of water-saturated diethyl ether. If the stimulus results in very low levels of the inositol polyphosphates it may be necessary to pool several samples for analysis.

(ix) The samples may now be analysed by anion-exchange chromatography (Section 5.2) or by h.p.l.c. (Section 5.4).

5.2 Separation of inositol phosphates by anion-exchange chromatography

The water-soluble metabolites of the phosphoinositides may be separated, according to their charge, on small columns containing the formate form of a strongly basic anion-

Table 6. Solvent mixtures required for the separation of water-soluble metabolites of the phosphoinositides by anion-exchange chromatography.

Eluant solvent mixture	*Eluant volume (ml)*	*Eluate*
(i) Distilled water	15–20	Inositol
(ii) 5 mM Sodium tetraborate/ 60 mM ammonium formate	10–12	Glycerophospho-inositol
(iii) 0.1 M Formic acid/ 0.2 M ammonium formate	15–20	Ins-1-P
(iv) 0.1 M Formic acid/ 0.5 M ammonium formate	15–20	Ins-1,4-P_2
(v) 0.1 M Formic acid/ 0.8 M ammonium formate	15–20	Ins-1,4,5-P_3
(vi) 0.1 M Formic acid/ 1.0 M ammonium formate	15–20	Ins-1,3,4,5-P_4

exchange resin (7, 27). Dowex-1 chloride (8% cross-linked; 200–400 mesh; from Sigma) is a suitable resin which may be converted to the formate form in the following manner.

(i) Pack the column to a volume of 1 ml with the Dowex.
(ii) Wash the Dowex with 20 vol of 1 M NaOH (the resin will turn brown).
(iii) Wash the column with water until the eluate is neutral.
(iv) Wash the column with 5 ml of 1 M formic acid (the resin will turn yellow) and wash again with water until the eluate is neutral. The column is now ready for use.

Alternatively the resin may be bought in the formate form (e.g. AG 1; 8% cross-linked; 200–400 mesh; from Bio-Rad). In each case the resin is usually used once and then discarded.

The separation procedure is as follows.

(i) Neutralize the diethyl ether-washed TCA extracts containing the inositol phosphates (see Section 5.1) with 5 mM $NaHCO_3$.
(ii) Elute the inositol phosphates by the step-wise addition of solutions containing increasing concentrations of formate. Details of the solutions are given in *Table 6*. The eluant volumes are quoted to serve as a guideline.
(iii) Take an aliquot of each fraction for liquid scintillation counting or for total phosphorus determination (Section 4.4.3).

A typical profile of the separation of inositol phosphates on an anion-exchange column is given in *Figure 6*.

[^{3}H]Inositol tends to break down upon storage to release degradation products which elute from the ion-exchange column at high formate concentrations. To avoid the possibility of these products carrying over from the tissue incubation and confusing the results, it is advisable to store the [^{3}H]inositol with approximately 2 mg of the formate form of the resin present in the stock bottle.

5.3 Deacylation procedures

In addition to the methods for the measurement of the polyphosphoinositides described in Section 4, quantification may also be achieved by deacylation of the lipids and separation of the resulting water-soluble glycerophosphoinositol (GroPIns), glycerophos-

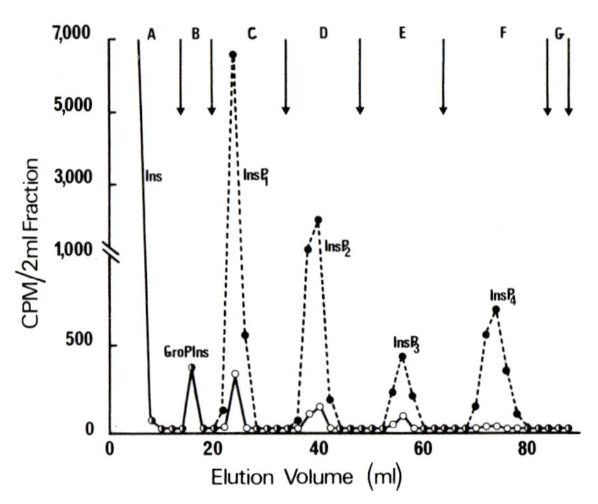

Figure 6. The separation of [^{3}H]inositol phosphates by anion-exchange chromatography on a 3 × 0.6 cm column, AG 1 × 8, 200–400 mesh, formate form (BioRad). Sample preparation was as described in Section 5.1 except labelling with [^{3}H]inositol was for only 30 min prior to agonist addition. (●) and (○) represent samples subsequently incubated (15 min) in the presence and absence of 1 mM carbachol, respectively. **A** = sample followed by water to elute free [^{3}H]inositol; **B** = 0.06 M $HCOONH_4$/0.005 M $Na_2B_4O_7$; **C** = 0.20 M $HCOONH_4$; **D** = 0.50 M $HCOONH_4$/0.1 M HCOOH; **E** = 0.8 M $HCOONH_4$/0.1 M HCOOH; **F** = 1.0 M $HCOONH_4$/0.1 M HCOOH; **G** = 2.0 M $HCOONH_4$/0.1 M HCOOH.

phoinositol-4-phosphate (GroPInsP) and glycerophosphoinositol-4,5-bisphosphate (GroPInsP$_2$) on Dowex-1 anion-exchange columns. Two deacylation procedures are described below.

(i) *Method 1.* Deacylation by mild alkaline hydrolysis (7, 27)

(1) Re-dissolve the dried phospholipid extract in 1 ml of chloroform.
(2) Add 0.2 ml of methanol and 0.4 ml of 0.5 M NaOH in methanol/water (19:1, v/v).
(3) Allow to stand at room temperature for 20 min and then add 1 ml of chloroform, 0.6 ml of methanol and 0.6 ml of water. Mix vigorously.
(4) Separate the phases by centrifugation and remove 1 ml of the upper phase which contains the deacylated phospholipids. Dilute this sample to 5 ml with sodium tetraborate so that the latter has a final concentration of 5 mM. Sodium tetraborate complexes with vicinal hydroxyl groups in a manner dependent upon their configuration and hence it aids the separation of the inositol phosphates. Load samples on to anion-exchange columns.

(ii) *Method 2.* Deacylation of phospholipids by the O→N-transacylation of fatty acids to monomethylamine (28).

(1) Prepare the methylamine reagent by bubbling monomethylamine gas into 40 ml of methanol/water/*n*-butanol (4:3:1, by vol) until the volume of the mixture increases to 65 ml. The reagent is stable to storage at −20°C.
(2) Wash the phospholipid extract with synthetic upper phase containing 0.1 M HCl and either 0.1 M EDTA or 0.1 M sodium cyclohexane-1,2-diaminetetra-acetate in order to remove Ca^{2+} and Mg^{2+} ions which may interfere with the deacylation reaction.

Table 7. Solvent mixtures required for the separation by anion-exchange chromatography of water-soluble products obtained by the deacylation of the phosphoinositides.

Eluant solvent mixture	*Eluant volume (ml)*	*Eluate*
(i) Distilled water	15–20	Inositol
(ii) 5 mM sodium tetraborate/ 0.18 M ammonium formate	10–12	GroPIns
(5 mM sodium tetraborate/ 0.025–0.06 M ammonium formate)	(15–20)	(GroPIns separate from $InsP_1$)
(0.15–0.2 M ammonium formate alone)	(15–20)	($InsP_1$ separate from GroPIns)
(iii) 0.1 M Formic acid/ 0.3 M ammonium formate	15–20	GroPInsP
(0.05 M formic acid/ 0.225–0.25 M ammonium formate)	(15–20)	(GroPInsP separate from $InsP_2$)
(0.1 M formic acid/ 0.4 M ammonium formate)	(15–20)	($InsP_2$ separate from GroPInsP)
(iv) 0.1 M Formic acid/ 0.75 M ammonium formate	15–20	$GroPInsP_2$
(0.1 M formic acid/ 0.5 M ammonium formate)	(20–30)	($GroPInsP_2$ separate from $InsP_3$)
(0.1 M formic acid/ 0.8 M ammonium formate)	(15–20)	($InsP_3$ separate from $GroPInsP_2$)
(0.1 M formic acid/ 1 M ammonium formate)	(15–20)	($InsP_4$ separate from $InsP_3$ and $GroPInsP_2$)

(3) Incubate the dried phospholipid extract in 3 ml of monomethylamine/methanol/water/butanol (5:4:3:1, by vol) at 53°C for 30 min.

(4) Cool the mixture on ice, add 1.5 ml of ice-cold *n*-propanol and evaporate in a rotary evaporator. The *n*-propanol prevents bumping.

(5) Add 1 ml of water and 1.2 ml of *n*-butanol/light petroleum (b.p. 40–60°C)/ethyl formate (20:4:1, by vol). Mix well.

(6) Remove the upper phase and wash the lower aqueous phase with 0.75 ml of the above solvent mixture. Apply the deacylation products, present in the lower phase, to anion-exchange columns.

Table 7 gives details of the solutions used to elute the deacylated products from the Dowex-1 columns. The eluant volumes are quoted to serve as a guideline. The information in parentheses may be useful to check that the deacylation reaction does not result in further hydrolysis of the glycerophosphoinositols. It may also be useful to determine whether any glycerophosphoinositols are produced upon agonist-stimulated phosphoinositide breakdown. Again aliquots of each fraction may be taken for liquid scintillation counting or for total phosphorus determination (Section 4.4.3).

5.4 Analysis of inositol phosphates by high pressure liquid chromatography

The development of and requirement for more powerful analytical systems for the resolution of inositol metabolites has paralleled expanding interest in receptor-mediated hydrolysis of phosphoinositides, which now appears to be more complex than previously thought (see 29 and 30). The demonstrations of isomeric forms of $InsP_3$ accumulating in stimulated cells (8), of the potential occurrence of cyclic and non-cyclic phosphoinositide hydrolysis products (31, 32), and of the occurrence of more polar inositol phosphates ($InsP_4$, $InsP_5$ and $InsP_6$; 9, 33) reveal the need for the more rigorous analysis which h.p.l.c. can provide.

The principle by which the h.p.l.c. systems so far described achieve resolution of the inositol phosphates is essentially the same as that relied upon in less sophisticated ion-exchange systems. The varying charge carried by different glycerophospho- and phosphoinositols provides an ideal basis for their crude fractionation, as by Dowex chromatography. Isomeric inositol phosphates should possess identical charge and separation of these must rely upon more subtle molecular features, for example slight variations in p*K*a and in charge distribution induced by substitution of the inositol ring at different positions. By virtue of the greater density of sites capable of interacting with sample molecules, h.p.l.c. systems should be able to exploit the finer details of molecular structure to provide increased resolution.

5.4.1 *Practical advantages/disadvantages of h.p.l.c. analysis*

The application of h.p.l.c. to the separation of water-soluble inositol phosphates is in its early stages and to date no single chromatographic system has been reported to resolve all those now identified. Given its present limited application, it may be useful when considering h.p.l.c. analysis to note some of the other major advantages and disadvantages over the separation methods detailed in Section 5.2.

(i) H.p.l.c. is the only method at present available for the resolution of Ins-1,4,5-P_3 and Ins-1,3,4-P_3 and thus the only means for accurate estimation of the putative Ca^{2+}-mobilizing second messenger. Additionally, several other inositol metabolites may be resolved more rapidly than can be achieved by other procedures.

(ii) H.p.l.c. analysis is limited to sequential handling of samples and requires 30 min−2 h per run involving fraction collection at frequent intervals (0.2−1 min). This is costly and time consuming compared with Dowex chromatography with which tens of samples may be analysed simultaneously.

(iii) The requirement for collection of large numbers of fractions from h.p.l.c. columns could be overcome by use of an on-line detection system. At present detection of inositol phosphates relies largely on measurement of accumulated radioactivity identified by its retention time in the system used. Radioactivity corresponding to the inositol polyphosphates may be extremely low. This situation is aggravated in the gradient elution h.p.l.c. systems, where the high salt concentrations used reduce counting efficiency. This may present problems when using an on-line radioactivity monitor equipped with a low volume flow-through cell.

(iv) The absence of detector/data recording facilities for constant monitoring of column eluate will prevent use of an autosampler system which might otherwise facilitate automation of the h.p.l.c. analysis and allow more rapid and convenient handling of samples.

(v) A problem related to (ii) and (iii) is the detection of unlabelled compounds. The large numbers of fractions and extremely low phosphate levels in some samples preclude analysis by phosphate assay. Additionally, some of the h.p.l.c. systems

Table 8. High pressure liquid chromatography systems available for resolution of inositol phosphates.

Elution: Gradient or Isocratic	*Buffer(s)*	*Column (quoted supplier)*	*Phosphoinositols resolved*	*Comments*	*Refs.*
1. Isocratic	0.075 M ammonium formate, pH 4.0	Micropak-NH_2 (Varian Instruments)	Ins-1-P, Ins-2-P		34
2. Isocratic	0.075 M ammonium formate, pH 4.0	μBondapak-NH_2 (Waters Associates)	Ins, Ins-1-P, Ins-2-P	A third, inositol-containing compound also partially resolved but not identified.	35
3. Combined	Ammonium acetate/ acetic acid, pH 4.0	LiChrosorb-NH_2 (Merck) or μBondapak (Waters Associates)	Ins, Ins-1-P, Ins-2-P, Ins-4-P, $InsP_2$ and $InsP_3$	Compare with 4.	36
4. Combined	Ammonium acetate/ acetic acid, pH 4.0	μBondapak (Waters Associates)	Ins, GroPIns, Ins-1,2-cyclic P, Ins-1-P, Ins-2-P or Ins-4-P, GroPIns-4-P, $InsP_2$, GroPIns-4,5-P_2, $InsP_3$	Less good resolution of $InsP_1$ isomers than reported for 3. No separation $InsP_3$ isomers. Retention times for $InsP_4$–$InsP_6$ not reported.	37
5. Combined	Ammonium formate, pH 6.25	Whatman Partisil, 10 SAX (Corbert Associates)	Separates cyclic and non-cyclic $InsP_1$, $InsP_2$ and $InsP_3$	Retention times for $InsP_3$ isomers and glycerophospho-inositols not reported.	32
6. Combined	Ammonium formate, pH 3.7, H_3PO_4	Partisil, 10 SAX (Technicol)	Separates $InsP_1$, $InsP_2$, Ins-1,3,4-P_3 and Ins-1,4,5-P_3 and glycerophospho-inositols	Only separation to resolve $InsP_3$ isomers. Gradient extension and modification allow resolution of $InsP_4$-$InsP_6$ also.	38, 9, 10, 33, 39

described require relatively high concentrations of orthophosphoric acid. Thus, characterization of column separations using cold inositol phosphate standards is not always possible.

In view of these limitations, unless resolution of inositol phosphates to the level of isomeric species is critical to the experimental design, analyses as described in Section 5.2 will be preferable to h.p.l.c.

5.4.2 *H.p.l.c. separations available for inositol phosphates*

Since the requirement for h.p.l.c. analysis of complex mixtures of phosphoinositols is only now being recognized, the availability of suitable systems is extremely limited. The technique has been applied with a variety of aims to different groups of inositol phosphates over the last 5–6 years. Choice of a particular method depends on experimental design. *Table 8* shows a summary of the systems developed and should provide a guide to which of these will be most appropriate. However, as current interest in phosphoinositide hydrolysis centres on quantification of the putative second messenger, Ins-1,4,5-P_3, the following sections will concentrate on the single method so far developed to separate this from Ins-1,3,4-P_3, a second $InsP_3$ isomer which also accumulates in stimulated cells. This system was first described by Irvine *et al.* (38) and has since appeared in various modified forms (9,10,33,39). By adaptation of the gradient elution used, rapid and economical quantification of Ins-1,3,4-P_3 and Ins-1,4,5-P_3 can be achieved. Further modification of buffer strength and gradient allows separation of $InsP_1$, $InsP_2$ and $InsP_3$ from glycerophosphoinositols and clear resolution of $InsP_4$, $InsP_5$ and $InsP_6$. In terms of the range of compounds separated and the resolution obtained, it would appear the most useful h.p.l.c. system so far described. Application of the method to $InsP_1$ and $InsP_2$ isomers has not been reported but in view of its viability for $InsP_3$ isomers, it may provide a convenient starting point. The major disadvantages of the method are the requirements for high salt concentrations and inclusion of phosphoric acid in the solvents used (see Section 5.4.1).

5.4.3 *Setting up h.p.l.c. for gradient elution*

(i) *Assembling components of the system*. Although the following sections refer to method 6 (*Table 8*), the equipment required is basically the same for the other gradient systems listed, with the obvious exception of the column used. Specific details for other systems are available in the quoted references.

Most available gradient elution systems employ a programmable computerized mechanism to control the output volume from two separate h.p.l.c. pumps feeding into a common mixing chamber and thence to the column. The inlets to the separate pumps are connected via Teflon tubing terminating in inlet filters to two reservoirs containing the solvents which comprise the components of the gradient. The angle of the gradient is dependent on the rate of change of volume output from each pump. The systems quoted below allow for pump A (water) and pump B (ammonium formate, pH 3.7, H_3PO_4). Thus, at 100% of gradient only pump B operates. The gradient is set according to pre-determined values for the % maximum to be achieved from pump B by a defined time (see *Table 9*). These are programmed, together with the flow-rate, using the controller software. For the illustrated examples equipment comprised an ACS

Table 9. Programmed pre-set values for generating gradients illustrated.

Figure	*7*		*8*		*9*	
	Time (min)	*% B*	*Time (min)*	*% B*	*Time (min)*	*% B*
	0	0	0	0	0	0
	12.0	0	10	0	10	0
	12.1	2	15	44	15	44
	16.0	2	17	44	17	44
	16.1	5	23	59	25	59
Pre-set	24.0	5	33	100	35	100
values	46.0	59	40	100	40	100
	56.0	100	41	0	41	0
	64.0	100	45	0	45	0
	66.0	0				
	70.0	0				

Gradients are for a two-solvent system comprising: A = water, B = 1.7 M $HCOONH_4$, pH 3.7, H_3PO_4. Flow-rate = 1.2 ml/min. Fraction collection at 1, 0.3 or 0.2 min intervals as illustrated in figures.

gradient pumping system controlled via an Apple computer, programmed using the Profiler gradient system (Drew Scientific). The selection of an appropriate gradient is considered in (ii) below.

Connect the outlet from the pump mixing chamber to a standard six port h.p.l.c. injection valve via the minimum length of stainless steel tubing and connect the valve to the column inlet in the same manner. The injection loop (2 ml capacity) is attached directly to the valve such that when the latter is set to load, sample may be inserted from a syringe directly into the loop while pumped solvent by-passes the loop to the column. When the valve is switched to the inject position, solvent flushes sample from the loop onto the column and elution begins. Separations described are achieved using a Partisil 10 SAX analytical column, 25 cm × 4.6 cm (available from HPLC Technology, Technicol, Whatman). If a guard or pre-column is used this should be matched to the separating column and connected between the valve column outlet and main column inlet. If samples are filtered prior to injection, a pre-column may not be essential but it is a useful precaution preventing accumulation of particulate matter on the main column and thus prolonging its life. Diminishing quality of separation with column usage may result from accumulated particulate material on the surface packing material and it may be advisable to examine this and remove contaminated packing (which will show as brown colouration, rather than white if the column is clean). Care should be taken not to remove more than the surface 1−2 mm of packing. The space created is re-packed with 10 μm silica or 10 μm SAX material, being sure to leave no void volume otherwise anomalous peak splitting may occur. Use of a guard column and its periodic replacement will eliminate these difficulties.

Next, connect the column outlet via Teflon tubing to the inlet of an appropriate/u.v. detector equipped with a flow-through cell and adjusted to 254 nm [for detection of u.v. markers, if these are to be used, see (ii)]. The detector outlet should then be connected to the fraction collector or on-line radiation monitor. It should be noted that, between the pump outlet and fraction collector/u.v. monitor, the intermediate components will have a volume of several millilitres. At a flow-rate of 1 ml/min an ap-

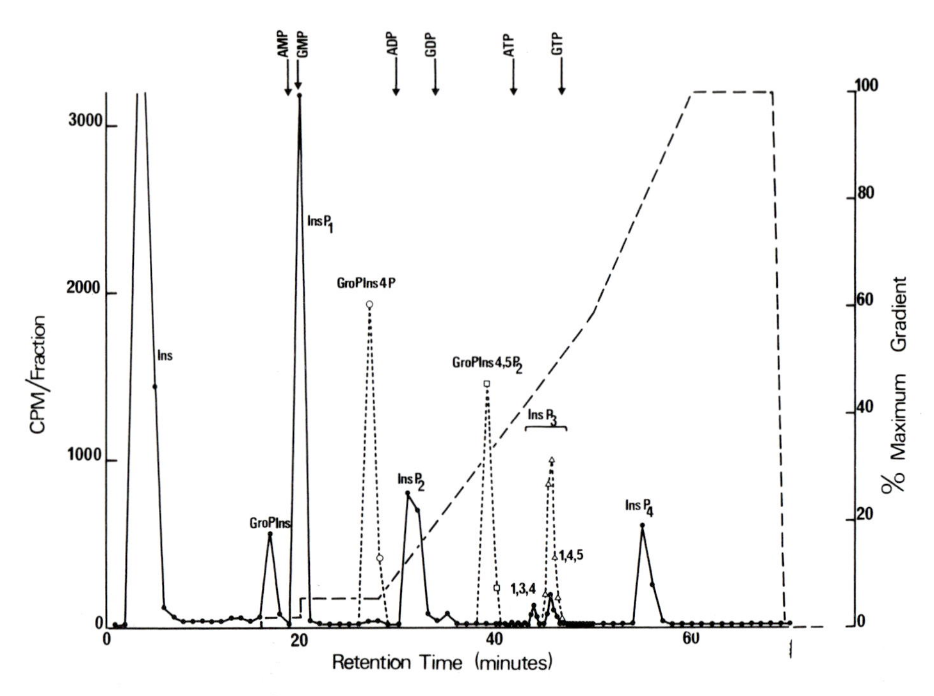

Figure 7. The separation of [^{3}H]inositol phosphates achieved by h.p.l.c. analysis using the gradient listed in *Table 9*. (•) represents [^{3}H]inositol-labelled compounds accumulating in rat cerebral cortical slices prepared as described in Section 5.1 and incubated 15 min in the presence of 1 mM carbachol. (△), (○), (□) demonstrate retention times observed in successive runs for ^{3}H standards of Ins-1,4,5-P_3, GroPIns-4-P and GroPIns-4,5-P_2, respectively. (---) Shows the gradient superimposed as a percentage of maximum solvent molarity.

propriate compensation of several minutes must be made when superimposing gradients on the elution profiles obtained.

Finally, prepare an appropriate volume of 1.7 M or 3.4 M ammonium formate, adjust pH to 3.7 with orthophosphoric acid, and filter through Millipore 0.2 μm cellulose acetate filters (or equivalent). Additionally, filter an equal volume of distilled de-ionized water (2.0 litres of each solvent will be sufficient for at least 20 separations). De-gas both solvents on preparation, and before each day's use, by bubbling helium gas through them and then transfer the buffers to the reservoirs that are connected to the h.p.l.c. pumps. Analysis can then begin after selection and programming of an appropriate gradient.

(ii) *Selecting an appropriate gradient/checking and modifying gradient.* Selection of the most suitable gradient depends on the range of compounds of experimental interest. For example, if quantification of only Ins-1,3,4-P_3 and Ins-1,4,5-P_3 is required, the total run time per sample may be considerably shortened by first displacing lower phosphates from the column using a steep gradient then reverting to a shallow gradient over the appropriate range of buffer molarity to achieve the desired resolution (see *Table 9* and *Figures 8* and *9*). If a wider range of compounds is of interest, a shallower gradient may be used and run time per sample will therefore increase. Secondly, it is important to ensure that, at the completion of each gradient, all the components of the original sample are removed from the column. If this is not achieved over successive

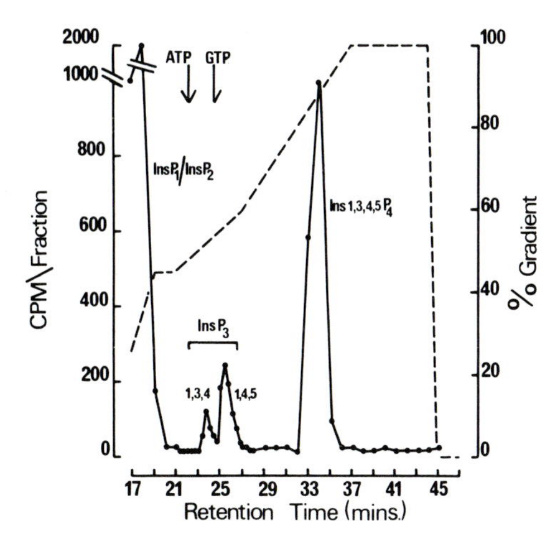

Figure 8. The use of an acute gradient initially to achieve early displacement of $InsP_1$ and $InsP_2$ from the column (see *Table 9*) and allow more rapid, subsequent estimation of $InsP_3$ isomers and $InsP_4$. Sample is as described for *Figure 7*.

runs, material accumulated on the column may begin to elute producing spurious peaks.

Heslop *et al.* (33) have recently reported a system which, over the range 0–100% of a gradient buffer of 3.4 M ammonium formate, pH 3.7, H_3PO_4 will elute inositol phosphates up to $InsP_6$ from a Partisil 10 SAX column. Initial examination of samples in this system will allow characterization of the most polar components likely to be present in samples and thus indicate the maximum buffer molarity required. From this starting point appropriate gradients can be developed.

The example gradients detailed in *Table 9* and *Figures 7–9* are suitable for separation of inositol phosphates up to $InsP_4$ and are simple modifications of gradients described in refs 38 and 9. For the gradients shown, attempts have been made to verify that the pre-set values quoted are an accurate reflection of those achieved by the pump system used, particularly at points low on the gradient where inaccuracies will be most significant. It may be useful to note when setting up new systems that, because of the widely differing ionic charge carried by GroPIns up to $InsP_6$, separation of all inositol phosphates using a single solvent may prove difficult unless the h.p.l.c. pumps used are capable of high accuracy at low output volumes. For example, using 3.4 M ammonium formate/H_3PO_4, pH 3.7, GroPIns may be expected to elute at 1% or less of the gradient maximum. At a flow-rate of 1.2 ml/min 1% corresponds to an output of only 12 μl/min from pump B. Unless equipment of such accuracy is available, resolution of compounds with short retention times may be impaired using high molarity buffers. If this proves a problem it may be necessary to elute weakly bound inositol phosphates from the column using a more dilute solvent and then switch to a higher molarity solvent for compounds retained longer. Many pumps allow two such buffers to be connected to a single pump, with switching between the two controlled by an appropriate valve. The accuracy of the pumping system can be easily estimated by pumping a shallow or stepwise gradient of a solution of a compound with suitable u.v. absorption. Provided that the relationship beween concentration and u.v. absorption is linear over the concentration range used, a visible estimate of pumping accuracy will be obtained. The validity of the il-

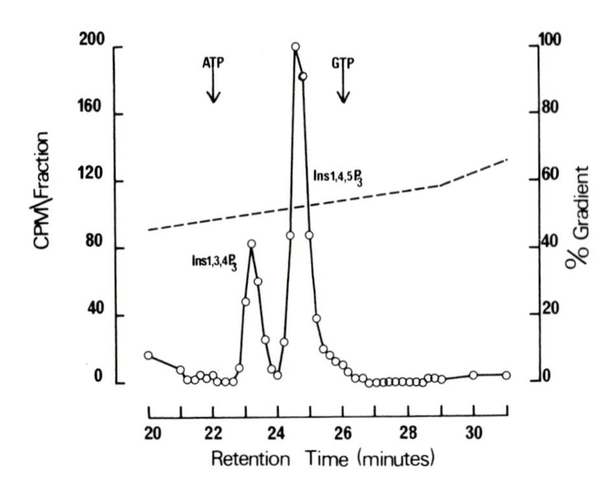

Figure 9. Improved resolution of $InsP_3$ isomers achieved by more frequent fraction collection and slight shallowing of the appropriate section of the gradient (cf. *Figures 7* and *8*) as described in *Table 9*. Sample preparation as for *Figure 7*.

lustrated gradients has been checked in this manner using a 25 × 4.6 cm Hypersil ODS column (HPLC Technology) with 0.5% acetone to give u.v. absorption. This has indicated some pumping error at low values and thus some variation from the reported retention times may be observed using different systems. However, the separations shown have been found to be very reproducible from run to run with both radioactive samples and standards, and with u.v. markers.

To achieve optimal resolution it may be necessary to tailor gradients to experimental requirements. As inositol phosphates are detected by measurement of the radioactivity of fractions, optimization of the system with samples or standards can be tedious and costly. In this respect, u.v. marker compounds may prove useful as these are readily detected by the on-line monitor. AMP, ADP and ATP have previously been used as approximate markers for $InsP_1$, $InsP_2$ and Ins-1,3,4-P_3, respectively (38,39). In modifying gradients to our own requirements we have used these, together with guanine nucleotides, and routinely spike samples with a mixture (see Section 5.4.4) to obtain a u.v. trace as a measure of the reproducibility of the separation achieved. The approximate retention times for these compounds are indicated in the figures. ATP and GTP elute, respectively, just before and after Ins-1,3,4-P_3 and Ins-1,4,5-P_3. Thus, by monitoring the ability of different gradients to improve resolution between ATP/GTP, an *approximate* measure of the corresponding separation of $InsP_3$ isomers should be obtained. Similarly, ADP/GDP and AMP/GMP may be useful indications for modified gradients aimed at resolution of bis- and mono-phosphate isomers. Caution should be taken in using these compounds as absolute markers for inositol phosphates particularly in other systems where molecular properties other than charge are exploited to achieve separation. Indeed, even using the present system, the nucleotides must *not* be regarded as internal standards for inositol phosphates. Irvine *et al.* (38) and a second recent report (39) demonstrate exact co-elution of Ins-1,3,4-P_3 and ATP whereas, beween columns of identical type, we have observed variable coincidence of these.

Finally, when establishing new gradients it is useful to recall the following general features.

(1) Shallower gradients, while increasing resolution, also increase peak width. Sharper gradients lead to sharper peaks but decreased resolution.
(2) Increased resolution may be achieved by more frequent fraction collection, that is as is employed to separate Ins-1,3,4-P_3 and Ins-1,4,5-P_3. Conversely, increased flow-rates with constant frequency of fraction collection result in sharper peaks but decreased resolution. Large sample volumes may present problems by lowering counting efficiency for radioactive samples.

5.4.4 *Preparation of samples for h.p.l.c. analysis*

Tissue samples are incubated and extracted as described in Section 5.1 then treated as below.

(i) Extract tissue as in Section 5.1 ensuring maximum precipitation of TCA-insoluble material during the centrifugation step.
(ii) Pool the samples if necessary, wash with ether and neutralize.
(iii) Remove a 1–2 ml aliquot of sample and spike with 400–500 c.p.m. of appropriate internal standards (if required, see Secion 5.4.5) and/or u.v. markers. When using nucleotide markers, ensure in preliminary separations that the concentrations used give a detectable recorder deflection above the baseline absorption. This will increase with the gradient molarity. At detector and recorder sensitivities of 2.0 absorbance units full scale and 10 mV respectively, 50 μl of a mixture which is 1 mM with respect to each marker should be sufficient.
(iv) Adjust the sample volume to 2.0 ml with water and filter through 0.2 μm cellulose acetate filters (or equivalent) ensuring minimal loss of volume.
(v) Load the sample via the injection valve into the loop (see Section 5.4.3). Transfer the sample on to the column by moving the valve to inject and begin elution.
(vi) Counting fractions: the high salt concentration of the h.p.l.c. solvent may be incompatible with some scintillants. Addition of 50% methanol to the sample, to give an approximate ratio of 1:4:15 (sample: 50% CH_3OH: scintillant), may help overcome this problem.

5.4.5 *Characterization of separation with standard materials*

The analysis of phosphoinositides by h.p.l.c. allows detailed separation but, if this is to be achieved within a run time suitable for handling multiple samples, resolution may be limited. The sequence of phosphate elution should remain constant within a specific system but identification of peaks can only be achieved by co-chromatography of sample and standard materials. For the systems described above, it is possible that retention times may vary slightly with continuing column usage (39). For this reason it may be advisable to spike samples with internal standards particularly for peaks where resolution is limited (i.e. between Ins-1,3,4-P_3 and Ins-1,4,5-P_3). Most recent studies of InsP_3 isomers (38,9,33,39) have employed this precaution, spiking ^{3}H-labelled samples with ^{32}P labelled Ins-1,4,5-P_3 (see below) prior to h.p.l.c. analysis, followed by quantifying the radioactivity of fractions by dual isotope counting.

For the illustrated separations, identification of each peak is based on co-chromatography of sample and labelled standards in successive runs. Exceptions to this are $InsP_2$ isomers identified by the retention times relative to standards of GroPIns-4-P and GroPIns-4,5-P_2 and Ins-1,3,4-P_3 identified by its elution position relative to standard [^{3}H]Ins-1,4,5-P_3 (Amersham International). Standard glycerophosphoinositols were obtained by deacylation of the corresponding authentic ^{3}H-labelled lipids (New England Nuclear or Amersham International) according to Section 5.3. The identity of $InsP_1$ is based on co-elution with [U^{14}C]L-myo-inositol-l-monophosphate (Amersham International). The identification of $InsP_4$ has recently been detailed (9). Tritiated D-myo-inositol-l-monophosphate and 1,4-bisphosphate are also available commercially (New England Nuclear or Amersham International).

In the absence of commercially available materials a number of inositol phosphates may be prepared either as pure compounds or as previously identified mixtures. The preparation of Ins-1-P, Ins-2-P and Ins-1,2-cyclic P from labelled PtdIns has been described (37). Appropriate ^{3}H- or ^{14}C-labelled phosphoinositides may be extracted from tissue samples and separated by t.l.c. according to Section 4.2. Ins-1,4-P_2 and Ins-1,4,5-P_3 labelled with ^{32}P may be prepared using erythrocyte ghosts according to (27).

Methods for comparatively large-scale preparation of Ins-1,3,4-P_3 and Ins-1,3,4,5-P_4 have yet to be described. However, small quantities of radiolabelled materials can be obtained from appropriately stimulated tissues. The occurrence of Ins-1,3,4-P_3 was first reported in rat parotid glands exposed to 1 mM carbachol (8) and time courses for its accumulation relative to Ins-1,4,5-P_3 have been characterized (38). After stimulation with this agonist for 15 min, the predominant $InsP_3$ isomer accumulating in this tissue is Ins-1,3,4-P_3. Therefore, preparation of appropriate samples according to (38) and subsequent h.p.l.c. analysis will allow identification of Ins-1,3,4-P_3 by virtue of its documented predominance over Ins-1,4,5-P_3 and/or by its retention relative to a standard of the latter.

[^{3}H]Ins-1,3,4,5-P_4 may be prepared in small quantities ($1-2 \times 10^5$ d.p.m.) from rat cerebral cortical slices incubated according to the protocol described in Section 5.1, if 15−20 tissue aliquots labelled with 5 μCi [^{3}H]inositol/50 μl slices are pooled. $InsP_4$ may be separated from $InsP_1$, P_2 and P_3 by Dowex chromatography and subsequently de-salted according to (9), although recoveries may be low unless the $InsP_4$ fraction is spiked with nanomolar quantities of cold $InsP_3$. Alternatively, $InsP_4$ may be very conveniently prepared from [^{3}H]Ins-1,4,5-P_3 using a crude kinase preparation from one of several tissue sources which have very recently been described (10). Provided $InsP_3$ is available as substrate, this method is preferable since a high percentage of conversion can be achieved by simple procedures.

6. DETERMINATION OF FREE AND LIPID-BOUND INOSITOL

The concentration of cytosolic free inositol and total lipid-bound inositol in nervous tissue can be determined by g.l.c. Using this technique, components of a mixture are separated in the vapour state by partitioning between a mobile gas phase and a stationary non-volatile liquid phase dispersed on an inert solid. Sugars and sugar alcohols are non-volatile and therefore must be converted to volatile trimethylsilyl ether derivatives for analysis by g.l.c. (40).

Gas chromatographs are available from many manufacturers (e.g. Perkin-Elmer, Hewlett Packard and Pye Unicam). The types of column used for analysis by g.l.c. are given below.

(i) *Packed columns.* These are stainless steel tubes approximately 2 m in length and with an internal diameter of approximately 3 mm which are packed with a diatomite material (e.g. Chromosorb or Gas-chrom) coated with the liquid phase. Nitrogen is used as the carrier gas. Packed columns have a high inherent resistance to carrier gas flow which puts practical limit upon the length of the column.

(ii) *Open tubular capillary columns.* These are capillary tubes made of glass or fused silica up to approximately 60 m in length and with an internal diameter of 0.25−0.75 mm. There are two types of capillary column. (a) Wall-coated, in which a thin film (0.1−0.5 μm) of stationary phase is coated directly onto the inside wall of the capillary tubing. (b) Support-coated, in which a thin layer of liquid-coated support lines the inside wall of the tubing. The recommended carrier gas for these columns is helium. An open tubular column, even of capillary dimensions, has very little resistance to gas flow and so it is possible to manufacture capillary columns of much greater lengths than packed columns. The resolving power of open tubular capillary columns is generally much greater than that of packed columns.

There are a wide range of liquid phases available for g.l.c. Some liquids suitable for the analysis of trimethylsilyl ether derivatives of sugars and sugar alcohols are the non-polar methyl silicone gums SE-30 and DB-1 and the non-polar methyl silicone fluid OV-101.

Table 10. Procedure for the release of inositol from phosphoinositides by acid hydrolysis.

1.	Wash the phospholipid-containing $ZnSO_4/Ba(OH)_2$ precipitates with 1.5 ml of water.
2.	Extract the phospholipid from the washed precipitates using the method described in *Table 1*.
3.	Transfer the lipid extracts to glass ampoules and dry under a stream of nitrogen. Add 1 ml of 6 M HCl to each sample and seal the ampoules with a gas/O_2 torch.
4.	Heat the samples at 120°C for 20 h.
5.	Allow the ampoules to cool, break off the tops and allow the acid to evaporate by heating in a block at 120°C for 1 h.
6.	Extract the free inositol three times with 1 ml of water.
7.	Add a known amount (e.g. 50 μg) of α-methyl-D-mannoside to each of the inositol extracts and lyophilize.

Table 11. Preparation of trimethylsilyl ether derivatives of sugars and sugar alcohols (40).

1.	Prepare the silylation reagent by mixing anhydrous pyridine (dried over KOH pellets or a molecular sieve), hexamethyldisilazane and trimethylchlorosilane in the proportions 5:1:0.5 (by vol). Upon the addition of trimethylchlorosilane the solution will become cloudy due to the formation of a fine precipitate of ammonium chloride. This precipitate does not interfere with the derivatization of the sugars and sugar alcohols.
2.	Add 200 μl of the silylation reagent to ≤2 mg of lyophilized sample.
3.	Shake gently at room temperature for 5−10 min.
4.	Add 1 ml of hexane followed by 2 ml of water to the reaction mixture and mix thoroughly. Centrifuge at 1000 g for 5 min to separate the phases. The pyridine is miscible with the water and the trimethylsilyl ether derivatives are extracted in the upper hexane layer. Retain the upper phase.
5.	Keep the trimethylsilyl ether derivatives free of moisture as they are susceptible to hydrolysis.

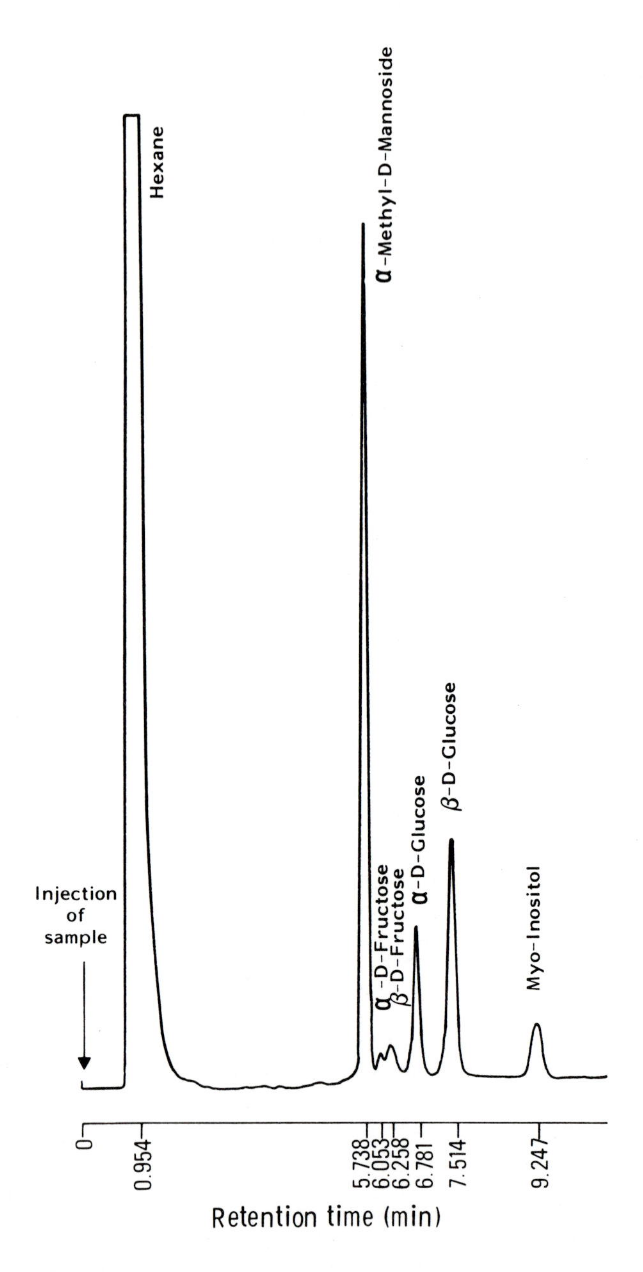

Figure 10. Use of g.l.c. to separate free sugars and sugar alcohols extracted from the dorsal root ganglia of a rat with streptozotocin-induced diabetes. The sample was analysed using a Perkin-Elmer 8310 gas chromatograph equipped with a stainless steel, glass-lined column 2 m in length with an internal diameter of 3 mm and packed with Chromosorb W HP (80–100 mesh) coated with 3% SE-30. The injection port was operated at 250°C and the flame ionization detector at 300°C. The column was held at 195°C for 2 min after injecting the sample and subsequently the oven temperature was programmed to increase at a rate of 15°C/min to a final temperature of 250°C which was retained for 5 min.

Sugars and sugar alcohols are extracted and derivatized for g.l.c. using the following procedure.

(i) Homogenize the tissue (up to 20 mg) in 1.5 ml of water/0.17 M $ZnSO_4$/0.15 M $Ba(OH)_2$ (13:1:1, by vol).

(ii) Centrifuge at 1000 *g* for 10 min to precipitate the protein and the phospholipid.

(iii) Remove the supernatant which contains the free sugars and sugar alcohols and add a known amount (e.g. 50 μg) of α-methyl-D-mannoside to serve as an internal standard. Lyophilize.

(iv) Extract the phosphoinositides from the precipitate and acid hydrolyse the lipid to release inositol using the method described in *Table 10*. This hydrolysis procedure may result in a low yield of inositol (~60%). A convenient way to measure the yield is by the addition of a known amount of *scyllo*-inositol as an internal standard. The loss of *scyllo*-inositol and *myo*-inositol is similar under these reaction conditions (W.R.Sherman, personal communication).

(v) Prepare trimethylsilyl ether derivatives of the lyophilized sugar and sugar alcohol samples using the method outlined in *Table 11*.

(vi) Dry the hexane extract of the trimethylsilyl ether derivatives under a stream of nitrogen and re-dissolve in a small volume (e.g. 50 μl) of hexane.

(vii) Apply a 0.5−2.0 μl sample to the column of the gas chromatograph.

The injection port is held at a temperature high enough (~250°C) to volatilize the trimethylsilyl ether derivatives as they are applied to the column. The column temperature can be programmed according to requirements. With capillary columns, samples are normally analysed over a lower temperature range than with packed columns. Recommended programme rates are also lower with wall-coated (0.5−5°C/min) and support-coated (5−10°C/min) open-tubular columns than with packed columns (up to 15°C/min). A flame ionization detector is used for most g.l.c. analyses. Eluted components are burned in a flame of hydrogen and air (~300°C) to form ions which are detected and measured by an electronic system.

Figure 10 illustrates the separation of α-D-fructose, β-D-fructose, α-D-glucose, β-D-glucose and *myo*-inositol by g.l.c. using a packed column coated with 3% SE-30. The peaks are identified with reference to known standards and quantification is effected by relating the peak area of sugar or sugar alcohol to that of internal standard. Modern gas chromatographs are equipped with data handling systems. α-Methyl-D-mannoside is chosen as an internal standard because its retention time is such that it does not overlap with the peaks of the other components in the sample.

Methods for the separation of inositol phosphates by g.l.c. are being investigated and trimethylsilyl ether derivatives of Ins-1-P and Ins-2-P have been resolved using this technique (31).

7. ACKNOWLEDGEMENTS

The authors wish to thank Jim Strupish for his assistance in setting up the h.p.l.c. equipment and Sylvia Millett for her help with the diagrams and photographs.

8. REFERENCES

1. Berridge,M.J. (1983) *Biochem. J.*, **212**, 849.
2. Berridge,M.J. and Irvine,R.F. (1984) *Nature*, **312**, 315.

3. Nishizuka,Y. (1984) *Nature*, **308**, 693.
4. Michell,R.H. (1985) In *Molecular Mechanisms of Transmembrane Signalling*. Cohen,P. and Houslay,M.D. (eds), Elsevier, Amsterdam, p. 3.
5. Hawthorne,J.N. (1986) *Int. Rev. Neurobiol.*, **28**, 241.
6. Berridge,M.J., Downes,C.P. and Hanley,M.R. (1982) *Biochem. J.*, **206**, 587.
7. Ellis,R.B., Galliard,T. and Hawthorne,J.N. (1963) *Biochem. J.*, **88**, 125.
8. Irvine,R.F., Letcher,A.J., Lander,D.J. and Downes,C.P. (1984) *Biochem. J.*, **223**, 237.
9. Batty,I.H., Nahorski,S.R. and Irvine,R.F. (1985) *Biochem. J.*, **232**, 211.
10. Irvine,R.F., Letcher,A.J., Heslop,J.P. and Berridge,M.J. (1986) *Nature*, **320**, 631.
11. Kikkawa,N. and Nishizuka,Y. (1986) In *The Enzymes*. Krebs,E.G. (ed.), Academic Press, New York, Vol. 17, p. 167.
12. Kitano,Y., Go,M., Kikkawa,N. and Nishizuka,Y. (1987) *Methods in Enzymology*. Academic Press, New York, in press.
13. Hawkins,P.T., Michell,R.H. and Kirk,C.J. (1983) *Biochem. J.*, **210**, 717.
14. Hawkins,P.T., Michell,R.H. and Kirk,C.J. (1984) *Biochem. J.*, **218**, 785.
15. Folch,J. (1949) *J. Biol. Chem.*, **177**, 497.
16. Folch,J., Lees,M. and Stanley,G.H.S. (1957) *J. Biol. Chem.*, **226**, 497.
17. Griffin,H.D. and Hawthorne,J.N. (1978) *Biochem. J.*, **176**, 541.
18. Michell,R.H., Hawthorne,J.N., Coleman,R. and Karnovsky,M.L. (1970) *Biochim. Biophys. Acta*, **210**, 86.
19. Bligh,E.G. and Dyer,W.J. (1959) *Can. J. Biochem. Physiol.*, **37**, 911.
20. Jolles,J., Zwiers,H., Dekker,A., Wirtz,K.W.A. and Gispen,W.H. (1981) *Biochem. J.*, **194**, 283.
21. Vaskovsky,V.E. and Svetashev,V.I. (1972) *J. Chromatogr.*, **65**, 451.
22. Vaskovsky,V.E. and Kostetsky,E.Y. (1968) *J. Lipid Res.*, **9**, 396.
23. Shaw,N. (1968) *Biochim. Biophys. Acta*, **164**, 435.
24. Laskey,R.A. (1984) *Amersham Review,* No. 23.
25. Bartlett,G.R. (1959) *J. Biol. Chem.*, **234**, 466.
26. Raheja,R.K., Kaur,C., Singh,A. and Bhatia,I.S. (1973) *J. Lipid Res.*, **14**, 695.
27. Downes,C.P. and Michell,R.H. (1981) *Biochem. J.*, **198**, 133.
28. Clarke,N.G. and Dawson,R.M.C. (1981) *Biochem. J.*, **195**, 301.
29. Michell,R.H. (1986) *Nature*, **319**, 176.
30. Nahorski,S.R. and Batty,I.H. (1986) *Trends Pharmacol. Sci.*, **7**, 83.
31. Wilson,D.B., Bross,T.E., Sherman,W.R., Berger,R.A. and Majerus,P.W. (1985) *Proc. Natl. Acad. Sci. USA*, **82**, 4013.
32. Wilson,D.B., Connolly,T.M., Bross,T.E., Majerus,P.W., Sherman,W.R., Tyler,A.N., Rubin,L.J. and Brown,J.E. (1985) *J. Biol. Chem.*, **260**, 13496.
33. Heslop,J.P., Irvine,R.F., Tashjian,A. and Berridge,M.J. (1985) *J. Exp. Biol.*, **119**, 395.
34. Hallcher,L.M. and Sherman,W.R. (1980) *J. Biol. Chem.*, **255**, 10896.
35. Hokin-Neaverson,M. and Sadeghian,K. (1984) *J. Biol. Chem.*, **259**, 4346.
36. Siess,W. (1985) *FEBS Lett.*, **185**, 151.
37. Binder,H., Weber,P.C. and Siess,W. (1985) *Anal, Biochem.*, **148**, 220.
38. Irvine,R.F., Anggard,E.E., Letcher,A.J. and Downes,C.P. (1985) *Biochem. J.*, **229**, 505.
39. Burgess,G.M., McKinney,J.S., Irvine,R.F. and Putney,J.W. (1985) *Biochem. J.*, **232**, 237.
40. Sweeley,C.C., Bentley,R., Makita,M. and Wells,W.W. (1963) *J. Am. Chem. Soc.*, **85**, 2495.

CHAPTER 8

Isolation procedures and *in vitro* applications of cell nuclei from the mammalian brain

R.J.THOMPSON

1. INTRODUCTION

The preparation of cell nuclei from the mammalian brain is complicated by the considerable diversity of cell types present and by the difficulties of unequivocal morphological identification of individual nuclei once isolated. As a generalization, neuronal nuclei (at least from the cerebral cortex) are large (up to 15 μm in diameter) with pale dispersed nucleoplasm and one (or possibly two) very prominent nucleoli; astrocyte nuclei are similar but with two or more nucleoli (or nucleolus-like inclusions), while oligodendroglial and microglial nuclei are small (up to ~6 μm in diameter) with dark condensed nucleoplasm and lacking prominent nucleoli (1–3). Several methods have been devised to isolate cell nuclei from the mammalian brain and to separate them into cell-specific populations on roughly the above criteria. Two general approaches are possible, firstly to isolate different cell types in bulk from the mammalian brain (4) and then to prepare individual nuclei from these, or to prepare a mixed population of purified cell nuclei directly from the brain and then attempt to separate these into cell-specific populations. The first approach — isolating specific cell types in bulk from the brain — will not be discussed further here.

Prior to 1973 several methods had been devised in the second category (1–3 and references therein). These usually involved separations in hypertonic sucrose and were often lengthy, involving several centrifugation steps in complicated 'step gradients'. While the inclusion of a divalent cation (usually magnesium) was universally agreed to be essential to preserve nuclear structure, these methods often varied considerably in other details such as the inclusion of buffers in the sucrose solutions (probably unnecessary) or of detergents (generally believed to remove the outer layer of the nuclear envelope). One pitfall which was overlooked by some workers was the use of whole brain (i.e. including the cerebellum) in the starting material rather than just the cerebral cortex. The cerebellum contains at least five times more DNA per gram wet weight than the cerebrum, because of the enormous numbers of small neurones in the granule cell layer (5). These have small dark nuclei which are difficult to distinguish from 'non-astrocytic glial' nuclei (5). Inclusion of the cerebellum in the initial starting material means that at least half of the isolated nuclei come from these cells and there is no doubt that a proportion of 'glial' nuclei in some of the early separation procedures in fact came from granule cell neurones (5).

The method described in this chapter was originally developed for the isolation of 'neuronal' and 'glial' cell nuclei from the adult rabbit cerebral cortex (2). Since its original description it has been taken up and used by several other workers on rat, guinea-pig, mouse, bovine (and even human) cerebral cortex with reasonable success (2,3,8–11). Also included in this chapter is a previously unpublished method for the isolation of nuclear envelopes from neuronal nuclei. Other workers have described the isolation of nucleoli from brain nuclei (6,7).

While there is no current evidence that the RNA polymerases and DNA polymerases in brain nuclei are fundamentally any different from those in any other cell nucleus, a surprising finding has been the short chromatin repeat length in adult cortical neurones which appears to arise early in development (8,9). The reason for this structural change, which appears to be unique among mammalian cells, is unknown but presumably it represents a fundamental event in the maturation of a cortical neurone.

2. PROCEDURE FOR THE ISOLATION OF THE N^1 (NEURONAL) AND N^2 (NON-ASTROCYTIC GLIAL) NUCLEAR POPULATIONS FROM THE CEREBRAL CORTEX AND FOR THE ISOLATION OF NEURONAL NUCLEI FROM THE CEREBELLUM

2.1 **Materials**

The animals used are 3 × 2–2½ kg adult New Zealand white rabbits.

0.32, 1.8, 2.0, 2.2 and 2.4 M sucrose solutions are required, which must all contain 1 mM $MgCl_2$. The sucrose and $MgCl_2$ used must be of Analar grade. It is convenient to make up 100 ml of the 0.32 M, 25 or 50 ml of the 1.8 M, 200 ml of the 2.0 M, 100 ml of the 2.2 M and 25 or 50 ml of the 2.4 M. The sucrose solutions should be made up fresh on the day of the experiment in volumetric flasks. $MgCl_2$ (100 mM) dissolved in glass-distilled water is stored in aliquots at −20°C, thawed and added to a final concentration of 1 mM $MgCl_2$. It is important (especially with the 2.4 M sucrose solution) that the sucrose solutions are made up accurately (hence volumetric flasks), the more concentrated sucrose solutions will eventually dissolve if repeatedly inverted at room temperature. The 25 or 50 ml volumes of 1.8 and 2.4 M sucrose are in excess of the volumes actually required for the isolation procedure. This is because concentrated sucrose solutions are difficult to pipette accurately and the best way of doing this during the isolation procedure is with a 5 ml graduated pipette (preferably with a relatively wide-bore tip). Depending on the particular manufacturer, some 25 ml volumetric flasks do not have a wide enough stem to admit a 5 ml pipette, whereas most varieties of 50 ml volumetric flasks are sufficiently wide. See *Table 1* for other materials.

The original purification procedure was devised using an Aldridge-type glass homogenizer with a Teflon pestle (radial clearance 0.075 mm) driven by an overhead motor at 840 r.p.m. The overhead motor was an advantage since homogenizing in the presence of hypertonic sucrose solutions is mechanically difficult and the use of a motor ensures reproducibility between experiments. The clearance of the pestle is probably important, for example use of a pestle with twice this radial clearance was abandoned since nuclei were produced with tags of adherent cytoplasm visible at the light microscope level. Equivalent centrifuge facilities can obviously be substituted for the Beckman instrument used here.

Table 1. Glassware and other materials.

Aldridge-type homogenizer (0.075 mm radial clearance) and homogenizer motor
Beckman SW.27 centrifuge rotor and tubes
Beckman SW.50 centrifuge rotor and tubes
Two bench centrifuge tubes
Filter paper circles (hardened, Whatman No. 50)
Scissors, secateurs, scalpel, bone forceps
Muslin
Two Pasteur pipettes
100 ml graduated cylinder
Four 10 (or 15) ml graduated conical centrifuge tubes
Ice bucket
Four 100 ml beakers

2.2 Methods

(i) Sacrifice the rabbits by a blow on the base of the skull, decapitate and expose the skull.

(ii) The easiest way to remove the brain is by 'springing' the two lateral halves of the skull apart with the secateurs with one blade inserted into the foramen magnum. Remove further pieces of skull carefully using bone forceps, and remove the whole brain as quickly as possible.

(iii) Place each brain immediately into a beaker containing 0.32 M sucrose and 1 mM $MgCl_2$ pre-cooled in an ice-bucket. Transfer operations immediately to a 4°C cold-room.

(iv) Remove each brain separately from the ice-bucket, place on Whatman No. 50 hardened filter paper and divide into cerebrum and cerebellum. It is important that the filter-paper used is the hardened variety, since brain tissue sticks too avidly to the qualitative type.

(v) Remove the meninges as far as possible from each brain using fine forceps, which is necessary to stop erythrocytes sedimenting with the nuclei.

(vi) Section the corpus callosum, reflect each half of the cerebral cortex and scrape away white matter with a scalpel.

(vii) Pool all cerebral cortices (total wet weight should be ~10 g) and cerebella (should be ~4 g) separately, in ice-cold empty 100 ml beakers.

2.2.1 *Homogenizing the cerebral cortex*

(i) Chop the cerebral cortices finely in the bottom of the beaker using a scalpel or scissors and add enough 2.0 M sucrose containing 1 mM $MgCl_2$ to give a 20% w/v homogenate (i.e. for 10 g of cortex the final volume should be 50 ml). Homogenize with four complete up-and-down strokes of the homogenizer.

(ii) Pour the homogenate into a pre-cooled 100 ml measuring cylinder and add an equal volume of 2.0 M sucrose containing 1 mM $MgCl_2$, mix thoroughly by pouring back into the homogenizer and then into the measuring cylinder 4–6 times.

(iii) Filter through two layers of muslin, changing the muslin half-way through. (This is simply done by poking muslin into the top of a second 100 ml measuring cylinder and pouring in the diluted homogenate.)

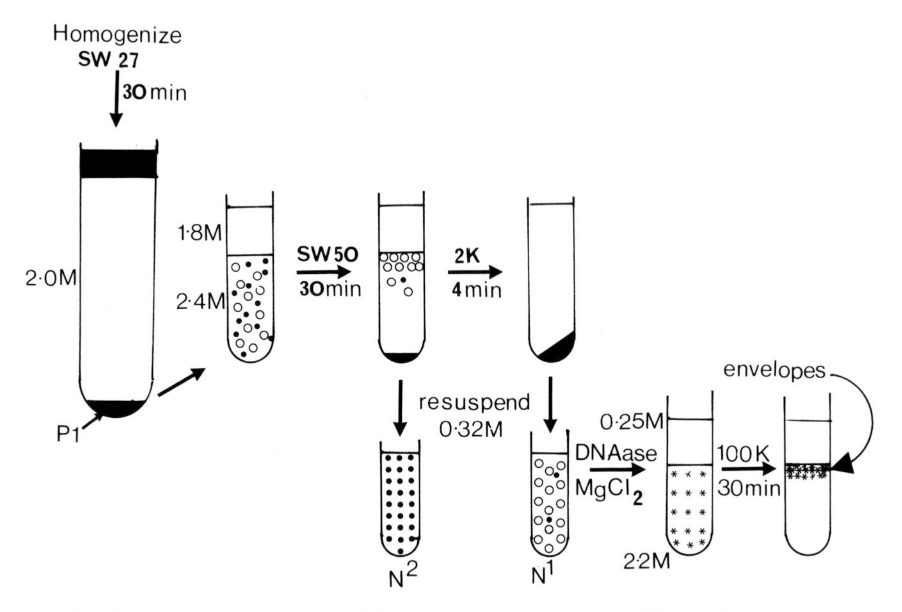

Figure 1. Diagrammatic representation of the isolation procedures for the N^1 and N^2 nuclear populations from the cerebral cortex and for the isolation of nuclear envelopes from neuronal nuclei.

(iv) Distribute the homogenate equally between two SW.27 centrifuge tubes and place on ice, while the cerebella are being homogenized.

2.2.2 *Homogenizing the cerebellum*

Chop the cerebella finely, add sufficient 2.0 M sucrose containing 1 mM $MgCl_2$ to form a 10% w/v homogenate (i.e. if the starting weight of cerebellar tissue is 4 g, the final volume should be 40 ml), homogenize with four complete up-and-down strokes of the pestle (wash the homogenizer out with ice-cold glass-distilled water between homogenizing cerebra and cerebella), dilute with an equal volume of 2.2 M sucrose containing 1 mM $MgCl_2$, mix and filter through muslin as before. Distribute the cerebellar homogenate equally between two SW.27 centrifuge tubes.

2.2.3 *Isolating cerebellar nuclei and the P1 pellet from the cerebral cortex (see Figure 1)*

(i) Centrifuge at 64 000 g_{av} for 30 min at 4°C in a Beckman SW.27 rotor or equivalent. The tubes show a 'pellicle' on top of the hypertonic sucrose, which is much more prominent with the cerebral cortex homogenate than with the cerebellar homogenate. The simplest way to remove it is to detach it from the walls of the tube with the back of the scalpel blade and then decant the pellicle and hypertonic sucrose supernatant from the nuclear pellets at the bottom of the centrifuge tubes.

(ii) Invert the tubes on top of the ice in an ice-bucket to allow the final drops of hypertonic sucrose to decant and then wipe the inside of the tubes with Kleenex tissue to remove the remains of sucrose solution from the walls of the tube, taking care not to disturb the nuclear pellets.

(iii) Resuspend the cerebellar nuclear pellets directly in approximately 3 ml of 0.32 M sucrose containing 1 mM $MgCl_2$ and use immediately; optionally they can be washed by re-centrifugation in 0.32 M sucrose containing 1 mM $MgCl_2$.

2.2.4 *Isolation of the N^1 and N^2 nuclear populations from the P1 pellet from the cerebral cortex*

(i) Resuspend each P1 pellet in 4 ml of 2.4 M sucrose/1 mM $MgCl_2$, and add to each SW.27 tube with a 5 ml graduated pipette with a wide-bore tip. The best way to achieve resuspension of the nuclei is with a 'Whirlimixer', which should have as vigorous an action as possible. Resuspension using stirring with glass rods or Pasteur pipettes is not as satisfactory and leaves the nuclei in sticky clumps which do not separate well in the next stage of the procedure.
(ii) Tip the resuspended P1 nuclei into two SW.50 (or equivalent) centrifuge tubes, leaving the SW.27 centrifuge tubes inverted over the SW.50 tubes as it takes several minutes for the viscous 2.4 M sucrose suspension of nuclei to drain out.
(iii) Overlay each nuclear suspension with 1.5 ml of 1.8 M sucrose + 1 mM $MgCl_2$, again using a 5 ml graduated pipette with a wide-bore tip, and centrifuge for 30 min at 85 000 g_{av} at 4°C in a Beckman SW.50 rotor (or equivalent). At the end of the centrifugation there is a thin pellet at the bottom of each tube (the N^2 nuclear population), and a diffuse suspension of N^1 nuclei above this, which concentrates into a layer at the top of the 2.4 M sucrose.
(iv) Remove the 1.8 M sucrose overlay with a Pasteur pipette, decant the 2.4 M sucrose suspension into two conical or other graduated test-tubes, and invert the tubes with pellets of N^2 nuclei over ice to allow the remains of the 2.4 M sucrose suspension to drain off.
(v) Add two volumes of 0.32 M sucrose + 1 mM $MgCl_2$ to the 2.4 M sucrose + 1 mM $MgCl_2$ suspensions in each graduated test-tube and mix by inversion with Parafilm, or similar material, over the end of each tube.
(vi) Centrifuge at 2000 g_{av} for 3–5 min in a refrigerated bench-centrifuge or equivalent (e.g. at low speed in the SW.50 rotor).
(vii) Decant off the 0.32 M sucrose supernatants and resuspend the N^1 nuclear pellets in about 1.5–2.0 ml of 0.32 M sucrose + 1 mM $MgCl_2$.
(viii) After the centrifuge tubes containing the N^2 nuclear pellets have drained, wipe any remaining 2.4 M sucrose from the inside of the tubes with Kleenex (being careful not to disturb the N^2 nuclear pellets), and resuspend the nuclei in 1 ml of 0.32 M sucrose + 1 mM $MgCl_2$.

2.3 Comments on the procedure and the nuclear populations produced

(i) The concentration of sucrose (2.4 M) used to separate the N^1 and N^2 nuclear populations is critical. The use of 2.3 M sucrose leads to greater contamination of N^2 nuclei with N^1 nuclei, whereas use of 2.5 M sucrose leads to greater contamination of N^1 nuclei with N^2 nuclei.
(ii) The purpose of the 1.8 M overlay is to remove damaged nuclei and residual capillaries from the N^1 population, which float on top of the 1.8 M sucrose. It has been claimed, using bovine brain, that insertion of a layer of 2.1 M sucrose

between the 1.8 and 2.4 M sucrose layers improves the purity of the N^1 and N^2 nuclear populations (10).

(iii) The three nuclear populations can be examined morphologically under phase contrast and should appear as in *Figure 2*. Alternatively they can be stained by mixing a drop of 0.5% crystal violet with the nuclear suspension and examining under normal optics. The N^1 nuclear population consists of 75–80% large pale nuclei; of these, 87% have a single prominent nucleolus, 12% have two nucleoli,

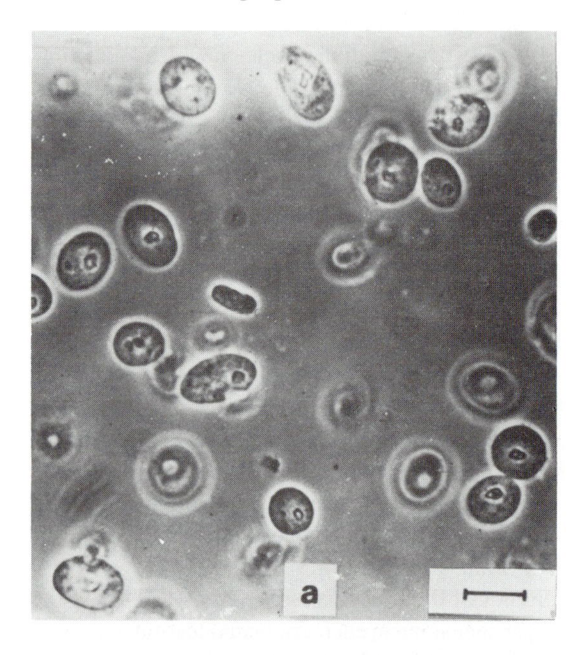

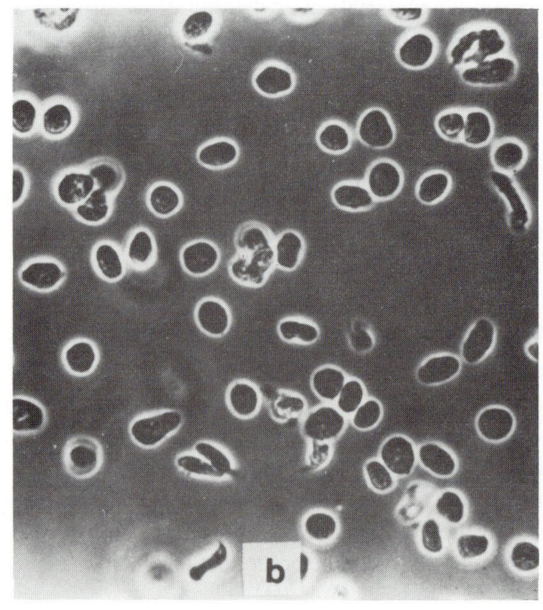

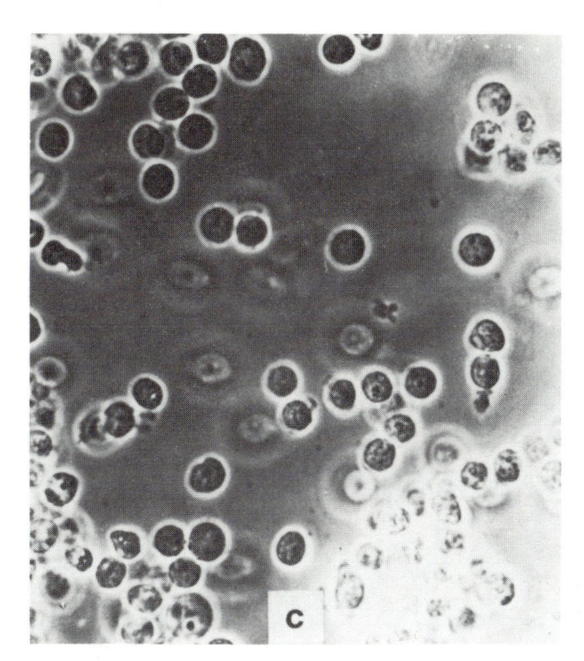

Figure 2. The appearance of **(a)** the N^1 nuclear population, **(b)** the N^2 nuclear population and **(c)** cerebellar nuclei under phase contrast microscopy. The horizontal bar in **(a)** represents 10 μm and all three populations were photographed at the same magnification.

and very few ($\sim 1\%$) have three nucleoli. The remainder of the N^1 nuclear population consists of small dark N^2 nuclei. The separation procedure, as originally used on rabbit brain, produces very few nuclei characteristic of astrocytes: the reasons for this are not clear. The N^2 nuclear population consists of 90–95% small dark nuclei with no obvious nucleolus, the rest of the population consists of large pale nuclei as above. Attempts to improve on this separation, for example by the use of sucrose gradients or by resuspending the isolated N^1 nuclear population in 2.4 M sucrose + 1 mM $MgCl_2$ and re-centrifuging, have not been successful even though flow cytofluorometry has indicated that the small dark nuclei in the N^1 nuclear population appear morphologically identical to those in the N^2 population (11). The cerebellar nuclear population appears as a rather homogeneous collection of small spherical nuclei, usually showing a single nucleolus (*Figure 2*) with a small number ($<1\%$) of very large pale nuclei which are considered to originate from Purkinje cells.

(iv) The nuclear populations can be prepared for electron microscopy by pelleting the nuclear suspensions by low-speed centrifugation, fixing in 2% glutaraldehyde for 2 h, washing in sodium phosphate buffer (pH 7.4, 500 mM), and treating for 3 h in 2% osmium tetroxide in the same buffer. Dehydrate, embed, and section in the usual way. Stain the sections with 10% uranyl acetate in methanol and Reynolds lead citrate. Under the electron microscope N^1 and N^2 nuclei should appear as in *Figure 3*; note the virtual absence of heterochromatin in an N^1 nucleus.

(v) Total numbers of nuclei can most simply be determined by counting in a

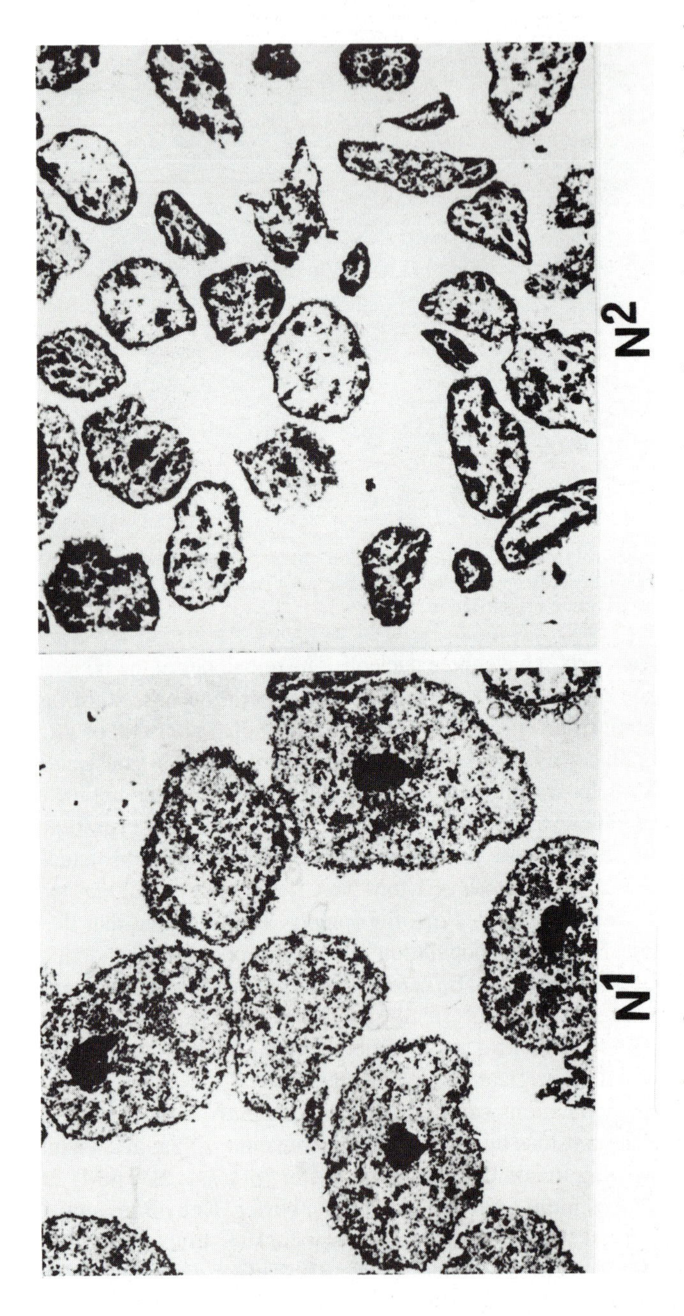

Figure 3. The appearance of the N^1 and N^2 nuclear populations under the electron microscope. Note the prominent, well-structured nucleolus and dispersed nucleoplasm in the N^1 nucleus. Magnification ×4800.

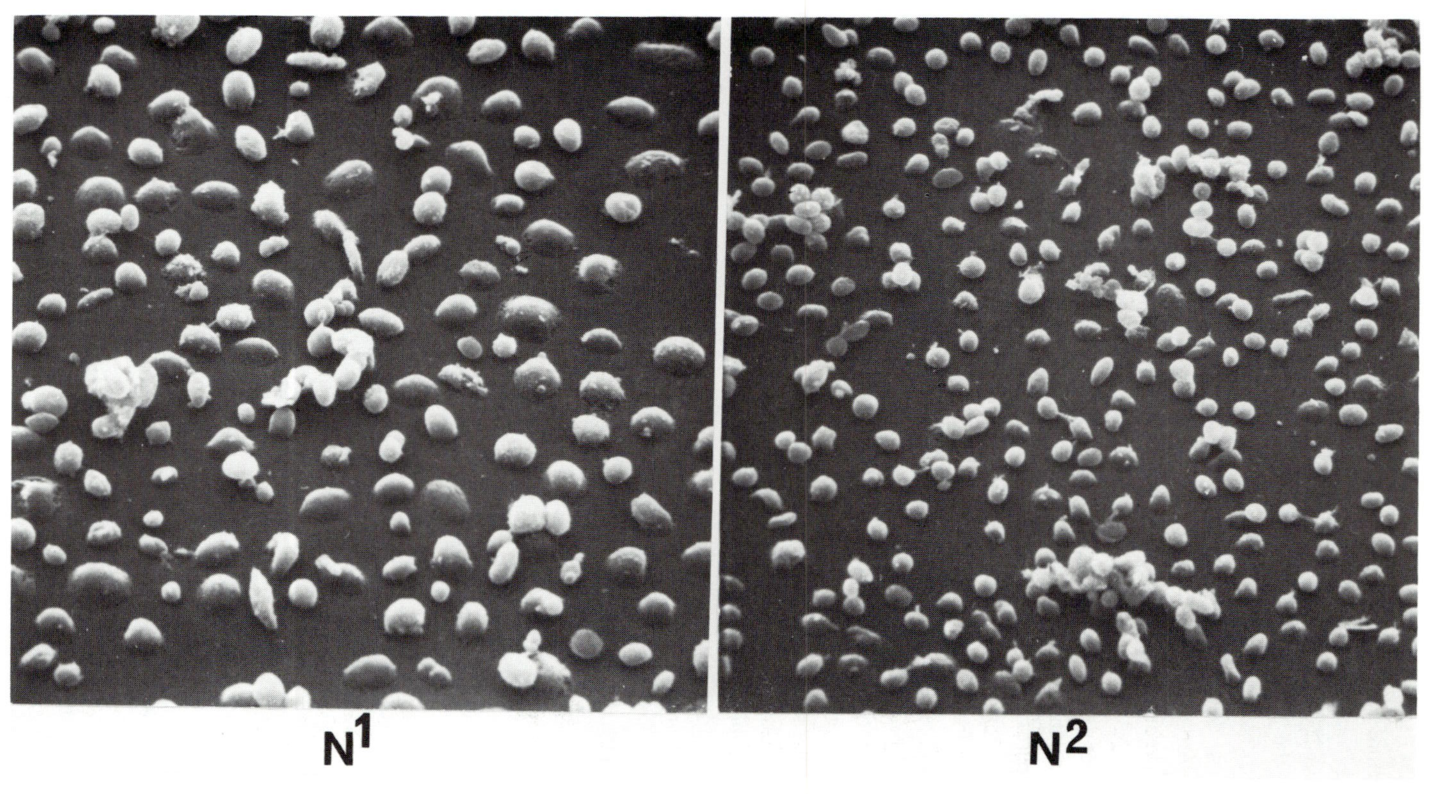

Figure 4. Scanning electron micrographs of the N^1 and N^2 nuclear populations. Magnification ×500 (courtesy of Professor L.Herman).

haemocytometer. Dilute the cerebellar nuclear population 1 in 20, the N^1 population 1 in 10, and the N^2 population 1 in 5 (all in 0.32 M sucrose + 1 mM $MgCl_2$) before being counted under the microscope. The total numbers of nuclei obtained from three rabbit brains are typically 100×10^6 for N^1, 50×10^6 for N^2, and 400×10^6 for cerebellar nuclei. Coulter counter size distribution analysis of the N^1 and N^2 nuclear populations (2) has shown that N^1 nuclei range from 4 to 14 μm in diameter (with 69% between 6 and 14 μm) while N^2 nuclei range from 3 to 9 μm with 85% between 3 and 6.5 μm in diameter. The more heterogeneous size distribution of the former nuclear population can perhaps be appreciated from *Figure 4*, which shows scanning electron micrographs of N^1 and N^2 nuclei. It should be emphasized that the above size distribution ranges were determined in 0.32 M sucrose + 1 mM $MgCl_2$, higher divalent cation concentrations lead to chromatin condensation and a shift to smaller size ranges.

(vi) Apart from morphological inspection, the high purity of the three nuclear preparations can be checked by enzyme marker studies. These indicate that the nuclei are virtually free from other cell constituents, the enzyme markers and methods used can be found in the original publication, as can methods for DNA, RNA and protein estimations (2).

(vii) An optional extra, especially if nuclear proteins (histones or otherwise) are being examined, is to include 0.25 mM phenylmethylsulphonyl fluoride (PMSF) in the sucrose solutions used in the nuclear isolation procedures. This should be stored as a 50 mM stock solution in isopropanol at 4°C in a tightly stoppered glass bottle and added (with immediate vigorous mixing) to the above final concentration just before the dissolved sucrose solutions are made up to an accurate final volume.

3. ISOLATION OF NUCLEAR ENVELOPES FROM BRAIN NUCLEI

The nuclear envelope consists of two membranes, each 70–80 Å thick and separated by the perinuclear space of 100–700 Å. The inner and outer membranes fuse at the

Table 2. Isolation of nuclear envelopes.

Materials

Sucrose solutions and materials as for the preparation of N^1 and N^2 as above;

0.25 M sucrose, 5 mM $MgCl_2$
2.2 M sucrose, 500 mM $MgCl_2$, Tris-HCl pH 7.5, 50 mM
Electrophoretically pure DNase (Sigma) 1 mg/ml in glass-distilled water

Method

1. Prepare N^1 nuclei as in Section 2 and resuspend in 1.5–2.0 ml of 0.32 M sucrose + 1 mM $MgCl_2$.
2. Take 1.0 ml of the N^1 nuclear suspension on ice and add 0.1 ml of the DNase solution, mix thoroughly and allow to reach room temperature for 1–2 min. Replace on ice, add 3.5 ml of 2.2 M sucrose + 500 mM $MgCl_2$, mix thoroughly and transfer to a pre-cooled Beckman SW.50 tube (or equivalent).
3. Overlay with 1.5 ml of 0.25 M sucrose + 5 mM $MgCl_2$. Centrifuge at 100 000 g_{av} for 30 min. The nuclear envelopes collect at the 0.25 M/2.2 M interface. Remove the whitish layer with a Pasteur pipette, dilute with 4.5 ml 0.32 M sucrose + 1 mM $MgCl_2$ and sediment the nuclear envelopes at 10 000 g_{av} for 5 min. Resuspend the pellet in 1 ml of 0.32 M sucrose + 1 mM $MgCl_2$.

nuclear pores, which have been estimated to occupy 13–20% of the nuclear envelope in rat brain (12). The method described here for the isolation of nuclear envelopes from the N^1 nuclear population was devised while investigating an active incorporation of CTP into an acid-insoluble product which was much more prominent in N^1 nuclei than in N^2 or cerebellar nuclei. This incorporation was due to the formation of CDP-diglyceride (13), and the enzyme responsible (CTP-phosphatidic acid phosphatidyl transferase) was later found to be entirely localized in the nuclear envelopes of N^1 nuclei (R.J.Thompson, unpublished results). See *Table 2* for the materials required and the details of the method.

The procedure will also work on N^2 nuclei but a much greater starting number of nuclei is required to produce a visible nuclear envelope layer (e.g. 12 rabbit cortices). The appearance of the nuclear envelopes in the electron microscope (processed as for *Figure 3*) is shown in *Figure 5*. Note the preservation of the nuclear pore complexes. The nuclear envelopes prepared from N^1 nuclei contain about 10% of the total nuclear protein and virtually all of the nuclear phospholipid.

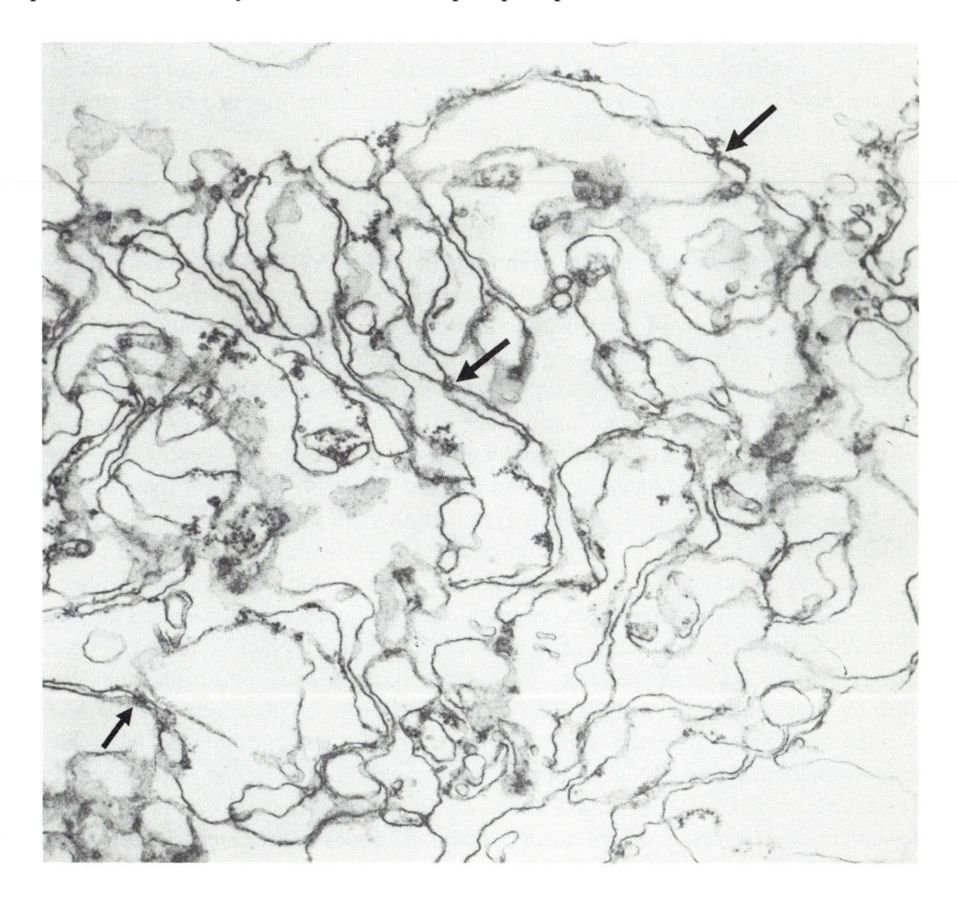

Figure 5. The appearance of the nuclear envelope preparation under the electron microscope. Note the preservation of the nuclear pore structures (arrowed). Magnification ×25 000.

4. IN VITRO APPLICATIONS OF NUCLEI FROM THE MAMMALIAN BRAIN

4.1 Chromatin structure in neuronal and glial nuclei

The DNA within a eukaryotic nucleus is a linear array of linked nucleosomes, a nucleosome consisting of an octamer of two each of the histones H2A, H2B, H3, and H4 with typically 200 base pairs (bp) of DNA wound in two turns around the outside of the histone complex. The DNA passing from one nucleosome to the next is termed the linker DNA and with it is associated a fifth histone, H1 (or H5 in the case of chicken erythrocytes). The nucleosome filament forms the basic chromatin fibre of 10 nm diameter, this filament can be further folded into a filament 30 nm in diameter which can be seen in interphase nuclei. This folding *in vitro* can be induced by high ionic strength and requires the presence of H1, the condensed higher order structure found is a solenoid with six nucleosomes per turn. For a recent review of chromatin structure see ref. (14).

Digestion of the linear array of nucleosomes in the 10 nm filament with micrococcal nuclease produces double-strand cuts in the linker DNA (presumably because DNA in other areas is protected by interaction with histones) and releases mononucleosomes and oligonucleosomes. A released mononucleosome — consisting of a histone octamer, histone H1 and typically 200 bp of DNA — can be further digested by trimming of the ends of the linker DNA to form a 'chromatosome' (a histone octamer, histone H1, and 166 bp of DNA). Finally, further digestion releases H1 and forms the nucleosome 'core particle' consisting of the histone octamer and 146 bp of DNA.

If chromatin digested with micrococcal nuclease is run on a gel, a ladder of DNA bands is seen, since random double strand cuts in the linker DNA release mononucleosomes and multiples of mononucleosomes still held together by undigested linker DNA. The amount of DNA in an individual nucleosome (the chromatin repeat length, since the nucleosome is the basic repeating unit in chromatin) can be determined by measuring the difference in size of successive DNA bands on the gel. This eliminates the effects of 'trimming' of the mononucleosome DNA after release, described above.

Measurement of the chromatin repeat length from many different sources in this way has shown a 'canonical' value of 200 bp (originally established in rat liver) in most cell types examined. Repeat lengths longer than this are known, for example 218 bp for bull sperm and 240 bp for sea urchin sperm (14). Measurement of the repeat length of chromatin from the N^1 nuclear population from rabbit brain shows a value of 160–165 bp (15) which appears to be unique among higher eukaryotic cells although values as short as this have been demonstrated in fungi (14). Nuclei from the N^1 (non-astrocytic glial) nuclear population, from cerebellar neurones, and from rabbit liver show the 'normal' repeat length of 200 bp (15). The repeat length in astrocyte chromatin is unknown. Short repeat lengths of the order of 160–165 bp have now been demonstrated in adult neuronal chromatin from several species (rabbit, rat, guinea-pig, mouse and ox). The establishment of a short chromatin repeat length in cortical neurones appears to be developmentally regulated, that is neurones in the cerebral cortex begin with a 'normal' repeat length of 200 bp which shortens to 160–165 bp over a period of a few days, either shortly after birth in the rat and rabbit, or before birth in the guinea-pig (8,9). The reason for this apparently total rearrangement of nuclear DNA in the developing neurone is unknown. When originally discovered it was tentatively suggested that the short chromatin repeat length somehow enabled high rates of transcription *in vivo* (15). The 'core particle' from 'short repeat' chromatin appears the same as that from other chromatins, and even though linker DNA is virtually absent from short repeat chromatin, folding into a higher-order solenoidal structure *in vitro* is apparently normal (16). The short chromatin repeat length in neuronal chromatin can be explained structurally by less histone H1 per nucleosome in neuronal chromatin, which balances the lack of linker DNA. For discussion of histone stoichiometry, DNA contents and repeat lengths of neuronal and glial nuclei, see ref. 11.

4.2 Measuring chromatin repeat length

See ref. 17 in conjunction with this section. See *Table 3* for materials required and *Table 4* for details of the method.

When the photograph of the gel has been produced (see *Figure 6*), measure the migration distance of the bands on the photograph from the origin to the mid-point of each band. Plot a calibration curve of distance migrated against DNA size in base pairs using

Table 3. Measurement of chromatin repeat length: materials.

Gel apparatus, e.g. BRL Series 1087 Model H5
Agarose (e.g. Sigma Agarose 3768, Type V, high gelling temperature)
Ethidium bromide (10 mg/ml in glass-distilled water, store at room temperature in a tightly stoppered vial, enclosed in aluminium foil)
10 × Tris-acetate running buffer for agarose gels

- 200 ml of 2 M Tris base
- 7.44 g of sodium EDTA
- 13.8 ml of glacial acetic acid

Make this up to 1 litre with distilled water; the pH should be 8.0
Sample buffer

- 2 ml of the above running buffer, 2 g of sucrose, make up to 10 ml with water, add a few drops of 1% (w/v) bromophenol blue in methanol

Micrococcal nuclease (e.g. Worthington)

- Make up to 1 mg/ml in 5 mM sodium phosphate buffer pH 7.0 containing 2.5×10^{-5} M $CaCl_2$. Distribute into 0.2 ml aliquots and store frozen at −20°C

0.1 M $CaCl_2$ solution (store frozen in aliquots at −20°C)
0.1 M Na EDTA (store at 4°C)
0.2 mM Na EDTA (store at 4°C)
10% sodium dodecyl sulphate (SDS) in water (store at room temperature)
5 M NaCl in distilled water (store at room temperature)
Chloroform/pentan-2-ol mixture, 24:1, stored at room temperature in a tightly stoppered glass bottle wrapped in aluminium foil
Buffer A:

- 10 mM Tris-HCl, pH 7.5
- 0.25 mM PMSF
- 0.34 M sucrose
- 1 mM $MgCl_2$
- Store at 4°C

0.1 M NaOH
Ice-cold absolute ethanol and 70% ethanol

Table 4. Measurement of chromatin repeat length: method.

1. Prepare the N^1, N^2, and cerebellar nuclear populations as in the preceding section.
2. Add 10 μl of the N^1 and cerebellar nuclear suspensions to 0.990 ml of 0.1 M NaOH and 50 μl of the N^2 nuclear suspension to 0.950 ml of 0.1 M NaOH, read the A_{260} against the appropriate blank (i.e. 10 μl and 50 μl of 0.32 M sucrose added to the relevant volumes of 0.1 M NaOH). The aim is to digest the nuclei at a concentration of ~ 50 A_{260} units/ml. From the readings of the diluted nuclear suspensions at A_{260} in 0.1 M NaOH, take a volume of nuclear suspension calculated to contain five absorbance units [this will be a smaller volume (e.g. < 100 μl) for the cerebellar nuclei than for the N^1 and N^2 nuclei]. Centrifuge at 4000 *g* for 4 min in glass tubes in a refrigerated Sorvall centrifuge or equivalent.
3. Resuspend the nuclear pellets in 100 μl of Buffer A in glass tubes.
4. Warm to 37°C for 2 min in a waterbath. Add 1 μl of 1 M $CaCl_2$, mix well by tapping the tube. Take an aliquot of the micrococcal nuclease solution (previously thawed) and dilute.
5. Add 3 μl to the 100 μl of nuclear suspension, mix well and incubate at 37°C.
6. Take samples at, e.g. 1, 2, 4 and 8 min withdrawing the equivalent of 1 A_{260} unit (i.e. ~ 20 μl) at each time point.
7. Place the 20 μl sample immediately into an Eppendorf tube on ice containing 2 μl of 0.1 M EDTA.
8. Add 190 μl of 0.2 mM EDTA and bring to room temperature.
9. Add 1/7 vol (i.e. 30 μl) of 10% SDS, mix well.
10. Add 2/7 vol of 5 M NaCl, mix well.
11. Add an equal vol (i.e. 300 μl) of the $CHCl_3$/pentan-2-ol mixture and mix for at least 30 s vigorously on a 'whirlimixer'.
12. Centrifuge the sample (e.g. in an MSE Micro-Centaur bench centrifuge at top speed). Remove the *upper* layer with a Pasteur pipette into a fresh Eppendorf tube and repeat the extraction procedure exactly as above.
13. Remove the upper layer into a glass Corex tube (or similar glass centrifuge tube), add 5 vol of ice-cold absolute ethanol and leave at -20°C overnight.
14. Centrifuge the sample at 10 000 *g* in a Sorvall refrigerated centrifuge (or equivalent) for 25 min.
15. Wash the faintly visible pellets twice with 1 ml of ice-cold 70% ethanol, centrifuging at 10 000 *g* for 5 min each time.
16. Remove the ethanol carefully with a Pasteur pipette and dry the pellets in a vacuum desiccator for 10 min, taking care to release the vacuum steadily at the end of this period. The aim is to load 0.04 A_{260} units/track on a 1% agarose gel.
17. Dissolve the pellets in 100 μl of glass-distilled water. Do this by scraping the pellet away from the bottom of the tube for several minutes after the addition of the water using a Pasteur pipette with a sealed end.
18. Take 4 μl of the dissolved DNA, 3.5 μl of distilled water and 7.5 μl of sample buffer, mix and apply the 15 μl sample to the well in a 1% agarose gel.
19. The gel is prepared in the usual way[a] in 1 × running buffer to which 50 μl of ethidium bromide solution per litre has been added.
20. Run at 10 V/cm and photograph as usual[a]. An example is shown in *Figure 6*.

[a]See ref. 17.

the known sizes of the restriction fragments used, as illustrated in *Figure 7*. Using the calibration curve, determine the DNA repeat length by plotting a graph of band number versus DNA size in base pairs as illustrated in *Figure 8*, and measuring the gradient of the graph. It is usually difficult to measure more than five or six bands in the N^1 nuclear digest lane because the background fluorescence is higher than in other lanes, presumably because of N^2 nuclei (200-bp repeat chromatin) contaminating the N^1 nuclear population (165-bp repeat chromatin).

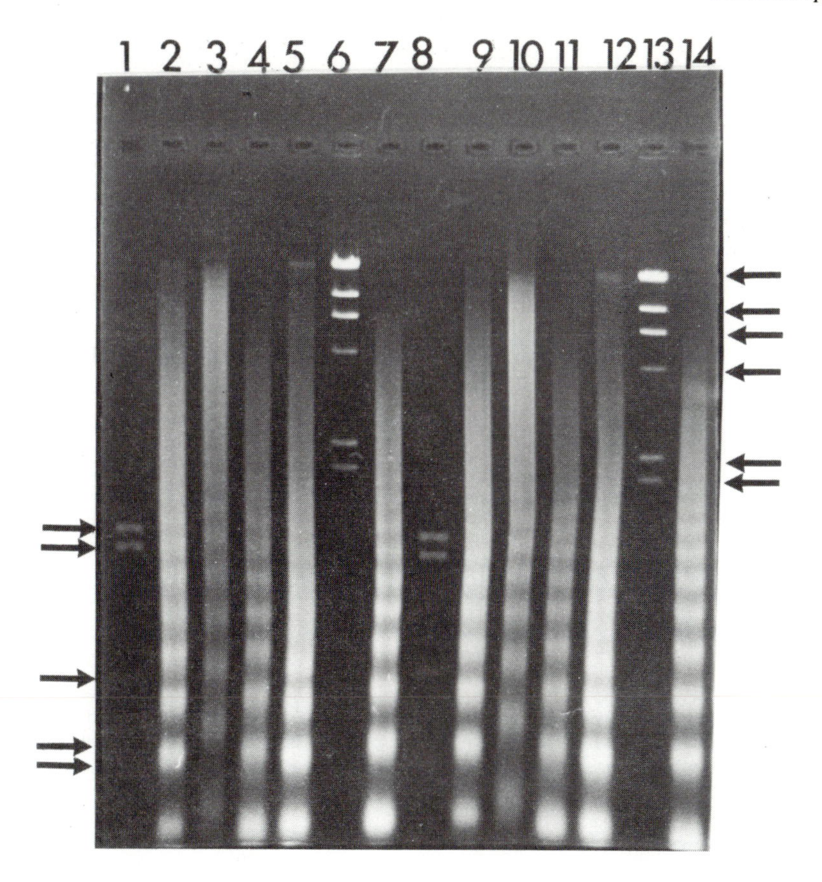

Figure 6. Micrococcal nuclease digests of N[1], N[2], and cerebellar nuclei. Nuclear populations were digested and electrophoresed as described in the text. **Lanes 1** and **8**, *Taq*I pBR322 DNA fragments (arrowed on the left). From top to bottom 1.44, 1.307, 0.616, 0.368, 0.315 kb, respectively; **lanes 2** and **9**, N[2] nuclei micrococcal digest; **lanes 3** and **10**, sea urchin sperm micrococcal nuclease digest; **lanes 4** and **11**, cerebellar nuclei micrococcal nuclease digest; **lanes 5** and **12**, N[1] nuclei micrococcal nuclease digest; **lanes 6** and **13**, *Hin*dIII lambda DNA fragments (arrowed on the right). From top to bottom 23.72, 9.46, 6.67, 4.26, 2.25, 1.96 kb, respectively; **lanes 7** and **14**, rat liver chromatin micrococcal nuclease digest. Note that the bands in N[2] nuclei, cerebellar nuclei, and rat liver nuclei are all 'in-phase', whereas the N[1] nuclei go out-of-phase on the 'short' side while the sea urchin sperm chromatin is progressively out-of-phase on the 'long' side.

4.3 Electrophoresis of histones

Total nuclear protein in brain nuclei can be examined by dissolving pelleted N[1], N[2] or cerebellar nuclei directly in the SDS sample buffer described in *Table 5*. Histones can be electrophoresed alone if first extracted from the nuclear preparations with acid.

4.3.1 *Acid extraction procedure*

(i) Take nuclei equivalent to ~500 A_{260} units and pellet them in glass tubes.
(ii) Add 2 ml of 0.2 M H_2SO_4, leave to extract on ice, stirring frequently with a glass rod.

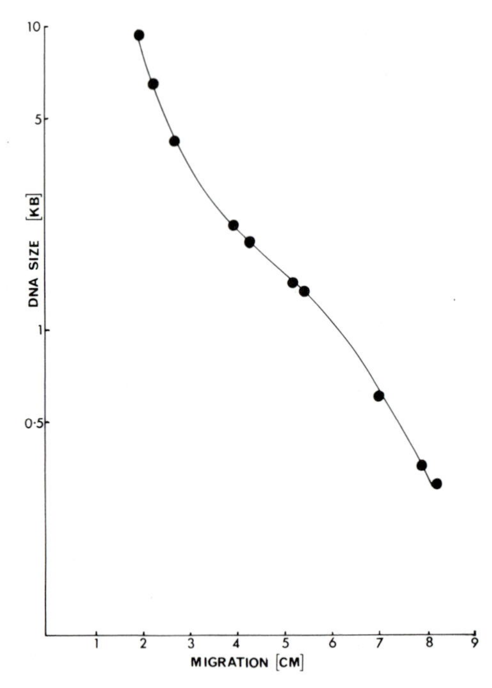

Figure 7. Calibration curve for determining chromatin repeat lengths. The migration distances of the restriction fragments used as standards in *Figure 6* (minus the largest *Hin*dIII lambda fragment) are plotted against the known DNA size in base pairs.

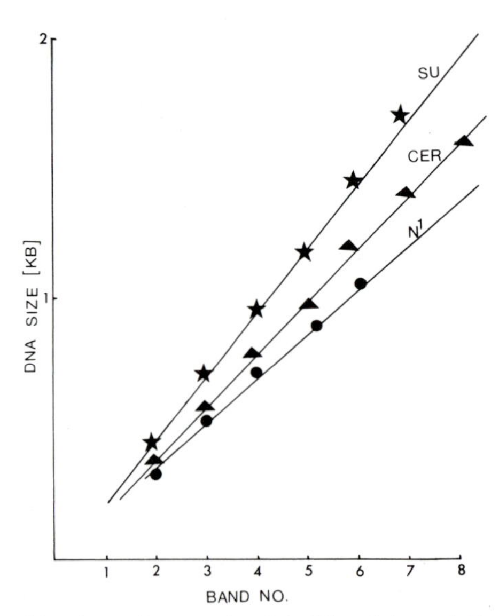

Figure 8. Determination of the chromatin repeat length in N^1 nuclei, cerebellar nuclei (CER), and sea urchin sperm (SU). Using the calibration curve shown in *Figure 7* the DNA band numbers in the micrococcal nuclease digests shown in *Figure 6* have been plotted against DNA size in base pairs. The slopes (i.e. the DNA sizes in base pairs divided by the number of bands) of the lines give the chromatin repeat length. N^1 nuclei show a repeat length of 165 bp, cerebellar and N^2 nuclei (the latter not shown) have a repeat length of 200 bp; for comparison sea urchin sperm chromatin (repeat length 240 bp) is also shown.

Table 5. Gel system for electrophoresis of histones.

1.	Dissolve in SDS sample buffer:	
	Water	6.9 ml
	Glycerol	1.0 ml
	2-mercaptoethanol	0.1 ml
	10% SDS	1.0 ml
	0.5 M Tris pH 6.8	1.0 ml
2.	Gel system for 18% SDS resolving gel:	
	For a volume sufficient for a standard 15 × 15 cm polyacrylamide slab gel	
	Acrylamide/bisacrylamide 30 g/0.15 g in 100 ml	30 ml
	Water	6.25 ml
	Tris-HCl, pH 8.8, 3 M	12.5 ml
	Mix and de-gas, e.g. on a freeze-drier pump	
	Then add:	
	TEMED	10 μl
	10% SDS	0.5 ml
	Ammonium persulphate (10%, made up fresh in distilled water)	0.5 ml
	Pour into slab gel apparatus, overlay with water-saturated butanol	
3.	Stacking gel system:	
	Acrylamide/bisacrylamide 10 g/0.8 g in 100 ml	6 ml
	Tris-HCl pH 6.8 (0.5 M)	4.8 ml
	Water	8.8 ml
	De-gas as above	
	Then add:	
	TEMED	10 μl
	10% SDS	0.2 ml
	Ammonium persulphate (as above)	0.2 ml
	Polymerize on top of the resolving gel after removing the water-saturated butanol and rinsing with distilled water.	
4.	Running buffer:	
	30 g of Tris base	
	140 g of Glycine	
	50 ml of 10% SDS	
	Make up to 1 litre and dilute five times before use.	
5.	Run the gel, typically, at about 30 mA total current. Remove the slab gel and fix in methanol/glacial acetic acid/water (5/1/5 by vol.).	
6.	Stain the gel in the same solution containing 0.1% Coomassie Blue.	
7.	De-stain in methanol/glacial acetic acid/water (5/7.5/87.5 by vol.) at room temperature.	

For the appearance of brain histones and a discussion of histone stoichiometry in brain nuclei, see ref. 11.

(iii) Centrifuge at about 10 000 *g* in a refrigerated Sorvall centrifuge for 5 min.

(iv) Remove the supernatant and keep on ice.

(v) Re-extract as above with 1 ml of 0.2 M H_2SO_4 for 30 min on ice. Pool these two supernatants in a glass tube.

(vi) Add an equal volume of 50% trichloroacetic acid (TCA), leave on ice for 45 min, centrifuge at about 9000 *g* in a Sorvall centrifuge for 15 min. Wash the pellet in 2 ml of acetone containing 1 mM HCl, then in 2 ml of acetone and dry in a vacuum desiccator. See *Table 5* for the gel system used.

4.4 Other *in vitro* applications of brain nuclei

There is extensive literature on studies of cell nuclei *in vitro* from the mammalian brain (see refs 3 and 8). It is generally agreed that the *in vitro* transcriptional activity of neuronal nuclei is higher than that of glial nuclei, however these studies have measured total RNA synthesis under various conditions without examining the production of neuronal- or glial-specific transcripts (3,8). Methods which have been used to study protein synthesis and nuclear RNA species, acetylation and phosphorylation of histones, nuclear enzyme activities and nuclear non-histone proteins can be found in the original publication (see 3,8, for references). As a generalization there is no clear evidence that neuronal nuclei in particular show properties qualitatively different from those of other mammalian nuclei, apart from the distinctive change in chromatin structure described above.

5. REFERENCES

1. McEwen,B.S. and Zigmond,R.E. (1972) In *Research Methods in Neurochemistry.* Marks,N. and Rodnight,R. (eds), Plenum Press, New York, Vol. 1, p. 139.
2. Thompson,R.J. (1973) *J. Neurochem.*, **21**, 19.
3. Takahashi,Y. (1984) In *Handbook of Neurochemistry.* Lajtha,A. (ed.), Plenum Press, New York, Vol. 7, p. 205.
4. Poduslo,S.E. and Norton,W.T. (1975) In *Methods in Enzymology.* Lowenstein,J.M. (ed.) Academic Press, New York, Vol. 35, p. 561.
5. McEwen,B.S., Plapinger,L., Wallach,G. and Magnus,C. (1972) *J. Neurochem.*, **19**, 1159.
6. Banks,S.P. and Johnson,T.C. (1973) *Biochim. Biophys. Acta*, **294**, 450.
7. Takahashi,Y., Araki,K., Ikeda,K. and Oyanagi,S. (1973) *Brain Res.*, **73**, 189.
8. Brown,I.R. and Greenwood,P. (1982) In *Molecular Approaches to Neurobiology.* Brown,I.R. (ed.), Academic Press, New York, p. 41.
9. Whatley,S.A., Hall,C. and Lim,L. (1981) *Biochem. J.*, **96**, 115.
10. Harris,M.R., Harborne,N., Smith,B.J. and Allan,J. (1982) *Biochem. Biophys. Res. Commun.*, **109**, 78.
11. Pearson,E.C., Bates,D.L., Prospero,T.D. and Thomas,J.O. (1984) *Eur. J. Biochem.*, **144**, 353.
12. Feldherr,C.M. (1972) *Adv. Cell Mol. Biol.*, **2**, 273.
13. Thompson,R.J. (1975) *J. Neurochem.*, **25**, 811.
14. Thomas,J.O. (1983) In *Eukaryotic Genes, Their Structure, Activity and Regulation.* MacLean,N., Gregory,S.P. and Flavell,R.A. (eds), Butterworths, London, p. 9.
15. Thomas,J.O. and Thompson,R.J. (1977) *Cell*, **10**, 633.
16. Pearson,E.C., Butler,P.J.G. and Thomas,J.O. (1983) *EMBO J.*, **2**, 1367.
17. Sealey,P.G. and Southern,E. (1982) In *Gel Electrophoresis of Nucleic Acids – A Practical Approach.* Rickwood,D. and Hames,B.D. (eds), IRL Press, Oxford and Washington DC, p. 39.

CHAPTER 9

Functional expression in the *Xenopus* oocyte of mRNAs for receptors and ion channels

ERIC A.BARNARD and GRAEME BILBE

1. INTRODUCTION

We shall consider in this chapter the special requirements for mRNA expression where the product is a cell membrane receptor or ion channel. A receptor may, of course, contain its own ion channel, as in the case of the nicotinic acetylcholine receptor (nAChR), but in most cases it will not. A separate transduction system will in the latter cases be required, which may be either co-expressed or inherent in the expression system, as we shall see.

1.1 **Aims of receptor expression**

A need for the experimental expression of the mRNAs encoding a membrane-located receptor or ion channel of the nervous system (or other tissue) arises in several different contexts and can be summarized as follows.

(i) To identify suitable sources and developmental stages where the mRNA involved is adequately produced, so as to facilitate the preparation of its cDNA for molecular cloning.

(ii) To follow the time-course during development of the appearance of the mRNA concerned, and to study its regulation. Where several genes encode one receptor, the coordination of their expression is a further important subject of study.

(iii) To obtain the receptor or channel in an isolated state but incorporated in a cell membrane (e.g. that of the *Xenopus* oocyte), so as to overcome the difficulties often encountered in its electrophysiological analysis in the native environment. Such difficulties occur especially in many complex and poorly accessible structures in the brain; they include the presence of pre-synaptic release systems, polysynaptic pathways, interacting multiple receptor systems on one target cell, physical difficulty in the impalement of very small neurones or of dendrites, ionic imbalance or damage due to the pipette insertion in the latter type of case, restricted access of ligands, imperfect voltage clamping of some structures *in situ* and parallel problems in applying patch-clamping to them. To implant a receptor or ion channel in a foreign membrane in a controlled and accessible environment can often overcome these problems and permit a wealth of intracellular or patch-clamp recording data to be accumulated rapidly. This implantation can be most readily achieved by the translation of the mRNAs (naturally-occurring or pure) in a cellular system such as the *Xenopus* oocyte.

(iv) To identify the set of cloned cDNAs encoding the full structure of the receptor or channel. One cannot be sure that all of the genes involved have been cloned, until the full channel or transduction function has been shown to be generated by the expression or co-expression of the mRNAs proposed to encode the structure. The *Xenopus* oocyte is at present the best system available for such testing.

(v) To test for the requirement for a particular subunit in a functional property of the entire protein structure. This includes the important recent development of constructing hybrid receptor molecules in the oocyte membrane, in which subunits derived from more than one source of the receptor are co-expressed. A property or quantitative index characteristic of one source (or of an embryonic versus an adult form) can hence be assigned to a particular subunit [as exemplified in the study of Mishina *et al.* (1) and discussed further below].

(vi) To interpret fully the amino acid sequence of the receptor or channel protein, by expressing mutagenized or partly-deleted versions of its mRNAs. By this means, identified segments or amino acids can be shown to be associated with particular functional or structural properties.

We can note that two different types of expression are represented in the above-listed approaches, that is the expression of naturally-occurring mRNA [aims (i), (ii) and (iii)], or alternatively the expression of pure mRNAs obtained by transcription from cloned cDNAs [aims (iv), (v) and (vi)]. Both of these modes of expression are needed at different stages of investigation of a receptor or channel structure.

1.2 **Vehicles of receptor expression**

1.2.1 *Cell-free translation*

Translation in the reticulocyte lysate or in the wheat germ system is the conventional approach in molecular biology to the expression of natural or cDNA-transcribed mRNAs (2). However, this is generally not applicable to the mRNA(s) encoding a receptor or an ion channel. Firstly, channel or other transduction activity requires a membrane and a specific organization of protein(s) within or around the channel. Secondly, post-translational processing of several kinds is commonly required for the expression of the mRNAs of these intrinsic membrane proteins, and this can only be adequately provided in a cellular translation system. Thus, attempts to recognize the four subunits of the nAChR after cell-free translation of *Torpedo* electric organ mRNA failed (3) due to lack of glycosylation, signal peptide removal and other (as yet unspecified) natural post-translational events (4). Even the addition of rough microsomal membranes to the cell-free system, although it introduces signal peptidase action and membrane insertion of the nascent polypeptides as well as some N-glycosylation, did not produce the authentic mature forms of all of the subunits (5,6). In particular, the subunits do not assemble into the nAChR structure after any such translation, and the binding properties characteristic of the functional receptor are not attained (5,7). In fact, a functional multi-subunit receptor of any kind has never been produced in a cell-free translational system.

1.2.2 *Cellular translation*

The only system in use at present for the translation in a living cell of exogenous mRNA is the *Xenopus laevis* oocyte. Microinjection of mRNA into this cell was introduced by John Gurdon and colleagues (8) and has been used in many studies of secreted proteins.

Vertebrate somatic cells are likely in general to be far less efficient for such expression, since the translational machinery will be much more occupied by endogenous mRNAs and the degradation of mRNA and protein is high, compared with the quiescent state of the pre-fertilization oocyte (9). Considering also the exceptional size of the *Xenopus* immature oocyte ($\sim$ 1 mm diameter), which greatly facilitates the microinjection and much increases the yield of translation product per injection, the advantages of this cell type are great.

Possibly, however, other such exceptional cells might be found in the future. Large invertebrate cells have not been explored for this purpose. Other amphibian oocytes may also be applicable: the oocyte of a newt, *Cynops pyrrhogaster*, has recently been shown (10) to translate well electric organ poly(A) mRNA to form the functional nAChR channel, as in *Xenopus*. These oocytes are of the order of twice the diameter of the *Xenopus* oocyte and are reported to be more viable *in vitro* than the latter. However, the surrounding follicular layer was more difficult to remove; the use of some other such species where the converse is true and where the microvilli which convolute the *Xenopus* oocyte surface are sparse or absent could be an improvement even on the case of *X. laevis*.

Parenthetically, it should be noted that there has been a continuing confusion on the taxonomy of *X. laevis* in the literature on the practical applications of its oocytes. It is referred to by some authors as a frog and by others as a toad. In fact, taxonomic authorities state that the genus *Xenopus* is midway between the toads and the frogs; since there is no common name for this specialized group, it is acceptable to use either 'toad' or 'frog' as a common name. We prefer the usage 'toad'.

2. RELEVANT PROPERTIES OF THE *XENOPUS* OOCYTE

2.1 Translation of foreign mRNAs

Since the initial work of John Gurdon and his colleagues in 1971 (8) demonstrating that *X. laevis* oocytes synthesized rabbit globins after the microinjection of rabbit reticulocyte RNA, these oocytes have been increasingly used to study protein synthesis and post-translational processing. The procedures used have generally involved micro-injection of the exogenous mRNA of interest, subsequent incubation, where necessary, in the presence of a radioactive precursor (e.g. [^{35}S]methionine, [^{3}H]lysine or [^{3}H]hex-ose), and extraction and analysis by immunoprecipitation, SDS–gel electrophoresis, enzymic activity or bioassay, as appropriate. This system was used in the first phase of such studies for the study of the translation of mRNAs for viral proteins, storage proteins, immunoglobulins and other secreted proteins (11).

The amphibian oocyte is specialized for production and storage of proteins for later use in embryogenesis and includes those components required for a post-fertilization translational burst, for example ribosomes and tRNAs (12,13). Since the endogenous protein biosynthesis is largely quiescent in the 'prophase arrest' period prior to hor-monal stimulation to produce the egg, this cell type at this stage is far more available for the translation of exogenous mRNA than any somatic or developing cell would be.

Injected poly(A) mRNA is, therefore, translated very efficiently in the immature *Xenopus* oocyte: the intrinsic efficiency for globin mRNA was estimated to be about 30-fold that of the reticulocyte lysate, for a given incubation period (8). Further, many injected vertebrate mRNAs are very stable in the oocyte, although no general conclusions on this can yet be drawn (14,15); the oocyte then reaches a steady translational and processing state by 24 h (16) which can last up to 7 days in the culture conditions (14). The oocytes in culture are, therefore, available for the biosynthesis for far longer periods than the relatively unstable cell-free systems, and this can raise the advantage of the former over the latter to well over 1000-fold. Even a low abundance mRNA species will, therefore, be translated demonstrably in the oocyte (17–19).

The *Xenopus* oocyte has been shown to be inherently capable of the post-translational modification of many proteins as in their native cells (20). These events are known to include signal peptide removal, polyprotein cleavage [but not of some secretory protein precursors such as proinsulin or promelittin (21,22)], N-glycosylation, proline hydroxylation, amino-acetylation and phosphorylation (11,23). The capabilities of the oocyte in this respect far exceed those of the cell-free systems, even when the latter are fortified with microsomal membranes (see Section 1.2.1 above). Other such processes which may be important for some oocyte receptors are O-glycosylation and the trimming of N-glycosyl core oligosaccharides (both of which occur in the native situation in the Golgi apparatus), fatty acid acylation and inositol lipid addition. The oocyte has been shown to be capable of at least some of these steps, for example some trimming of the sugar cores (24,25) and the addition of fucose (26), but terminal sialylation may be deficient. How far the secondary stages of post-translational modifications proceed appropriately on receptor proteins as translated in the oocyte has not yet been determined.

2.2 Assembly of oligomeric proteins

Most receptors, on present information, probably exist in their native cell membrane as oligomers, even when they contain only one type of polypeptide. These oligomers contain subunits either bound together entirely non-covalently (as in the nAChR) or with the aid of disulphide (or other) bridges, as in the insulin receptor (27). The *Xenopus* oocyte has been shown (28) to fully assemble and secrete soluble, active immunoglobulins, whose multiple chains become correctly linked by disulphides. For foreign proteins where the oocyte has an abundant endogenous form of its own, assembly can proceed, however, to form the mixed oligomer: with lysosomal β-glucuronidase a mouse–toad hybrid appears to form (29).

The latter complication will generally be absent for receptor and channel proteins which do not occur in the oocyte. In general, however, for receptor proteins (with or without native intrachain disulphides), the assembly is a complex process following upon insertion of the chains into the membrane, for example as with the nAChR (30) or the insulin and epidermal growth factor receptors.

A vitally important feature of the oocyte translation system is that it can execute these assembly processes for membrane proteins and can segregate the products. The protein sequences themselves contain the information for initiating all of the steps concerned, including ensuring residence in the cell membrane as the final destination, and for any vertebrate receptor this information is recognized in the oocyte. Thus, it was initially shown (7,31,32) that the four subunits of the nAChR, after biosynthesis from *Torpedo* or muscle mRNA in the oocyte, are spontaneously processed and completely assembled in the cell membrane to the fully functional receptor. In all subsequent cases where other receptor mRNAs have been successfully translated in the *Xenopus* oocyte (see Section 3, and reviews in refs. 33 and 34) the same phenomena of adequate processing and spontaneous assembly have been found, and functional receptors implanted in the oocyte cell membrane have been formed.

The study of post-translational control of receptor or channel function is another important application, for which the oocyte is at present the best system. Thus, co-injection of the mRNA-injected oocytes with tunicamycin, swainsonine or other inhibitors of specific post-translational modifications is an effective procedure. Various requirements in the subunit complement for assembly can be investigated, as exemplified by the study of Sakmann *et al.* (35).

2.3 Native oocyte receptors

One of the advantages of the oocyte over other types of cell which might in principle express, at least to some extent, injected heterologous mRNAs, is that the immature oocyte, as a resting germ line cell, contains very few active receptors or ion channels of its own. Nevertheless, some of these do occur in its cell membrane, mostly attributable to its own maturation or housekeeping requirements. Obviously these can interfere with the expression of one of the same components from another source. Even then, their expression may still be recognized by virtue of large quantitative differences between endogenous and exogenous receptors or because the *Xenopus* form may be distinguishable by some characteristic. It is desirable, therefore, to list these native components of the oocyte.

2.3.1 *Native acetylcholine receptors*

The cholinergic response of non-injected *Xenopus* oocytes is muscarinic, being completely atropine-sensitive (36). This component has been characterized in detail by Dascal and Landau (37–39): the main response to acetylcholine (ACh) or other muscarinic agonists is a fluctuating series of depolarizations, although the response is complex and comprises four sequential phases. Mainly Cl^-, but also K^+, channels are opened. This response is variable between individuals and seasonally, and is also often removed by collagenase defolliculation (see Section 5.3). Sugiyama *et al.* (40), injecting rat brain poly(A) mRNA and employing collagenase inactivation, and Kubo *et al.* (41) injecting a muscarinic receptor mRNA (derived from a cloned cDNA encoding a pig brain muscarinic receptor) into oocytes in which the native muscarinic response was naturally minimal, have been able, in fact, to record implanted muscarinic responses.

There are no native nAChRs in the oocyte, so that all types of nicotinic receptors can, with advantage, be sought to be expressed there.

2.3.2 *Native adenosine and catecholamine receptors*

There is a strong and reproducible response in non-injected *Xenopus* oocytes to adenosine or ATP, and this is blocked by low levels of theophylline (42,43). Both the P1 and P2 purine receptor types appear to contribute to the complex response (44).

β-Adrenergic agonists and dopamine produce smaller and variable responses in the native oocyte (43,45). These responses disappear when the follicular cells are removed manually or by collagenase treatment. This suggests that these native receptors are located on the follicular cells, the response seen in the impaled oocyte being due to gap junctions between the two cell types or to incorporation by impalement of follicular cell material into an oocyte. However, it cannot as yet be excluded that, instead, defolliculation itself damages the native receptor or channel.

2.3.3 *Other oocyte receptors*

Serotonin has been described as evoking a membrane hyperpolarization in *Xenopus* oocytes, albeit variably (46). In our and collaborating laboratories, this is, in fact, found to be negligible in defolliculated oocytes and rare and small in folliculated oocytes. It can be ignored in comparison with the large effect seen after appropriate mRNA injection (Section 3).

No significant native sensitivity to neuropeptides has been described, with one interesting exception. Vasoactive intestinal polypeptide (VIP), which occurs in the brain and spinal cord as well as in the intestinal tract, has been shown recently here to evoke a significant and very frequent response in non-injected defolliculated oocytes (V.Reale and E.A.Barnard, unpublished observations).

Some steroid receptors appear to be located on the oocyte plasma membrane (47,48). Other receptor types, beyond those noted here, have not been reliably detected in the native oocyte. This paucity of the native receptor complement emphasizes the value of this cell in screening for a wide range of receptor mRNAs.

2.4 **Native oocyte ion channels**

Voltage-dependent Ca^{2+} channels have been detected in the native *Xenopus* oocyte by intracellular recording in the presence of Ba^{2+} (49). As a true Ca^{2+} current it would be very small ($\sim$1 nA, in Ringer). A transient Ca^{2+}-dependent chloride current of the oocyte can also be detected (50), which is larger (30$-$100 nA, in Ringer), although of the same voltage-dependence. The former is likely to evoke the latter. Removal of the external Ca^{2+} by EGTA gives rise to a large increase in the membrane conductance, apparently due to a Na^{+} flux (51), of uncertain mechanism. In such a medium the oocytes quickly deteriorate, however. This can be overcome by adding 20 mM Mg^{2+} or Mn^{2+} (52).

A slow K^{+} current is activated by depolarization when the Ca^{2+} channel is blocked (50). At +10 mV in Ringer it is approximately 30 nA. A slow voltage-dependent Na^{+} channel appears, curiously, after long depolarization (53); this is largely tetrodotoxin (TTX)-insensitive. This type of channel has so far only been found in amphibian oocytes. The absence in the *Xenopus* oocyte of the classical highly TTX-sensitive fast voltage-dependent Na^{+} channel of excitable tissues has made possible the expression therein of rat brain pure mRNA encoding that channel (54). The native channels of the oocyte, as catalogued here, should in general interfere little with the expression of exogenous channels.

Cl^{-} currents are readily evoked in the oocyte by receptor activation, for example of the oocyte's own muscarinic receptors (55) and of some of the mRNA-induced receptors (see Section 3). Some of these have a fluctuating phase, which is also Cl^--mediated, this being a characteristic endogenous oocyte system. Hence, care must be taken in interpreting an agonist-evoked Cl^- current after a receptor mRNA is translated in the oocyte, since the product in some cases may become coupled to an oocyte Cl^- channel.

2.5 The resting membrane potential (RMP) and input resistance

The normal RMP of the native or manually defolliculated *Xenopus* oocyte in the undamaged state is about −60 mV to −80 mV. The equilibrium potentials there (56) are:

K^+ −100 mV
Na^+ + 80 mV
Cl^- − 25 mV

A strong electrogenic Na^+/K^+ pump maintains equilibrium.

After the insertion of a microelectrode or micropipette (for injection of mRNA) the RMP generally falls, and likewise the input membrane resistance. The membrane usually re-seals around it. These effects have been studied in detail by Dascal *et al.* (39). After manual defolliculation the RMP is restored in most, but not all, healthy oocytes to near its normal level (say to −50 mV or better), over a period of several hours, and after penetration it is restored after 15−30 min. The input resistance should be in the range 0.3−1 MOhm. After impalement or injection, these values, also, should be restored.

2.6 Second messenger systems of the oocyte

The *Xenopus* oocyte has an active adenylate cyclase and protein kinase A system (43, 44). Upon intracellular injection of cAMP or an active analogue thereof, or of forskolin, a large outward current appears, mostly due to K^+. Forskolin strongly potentiates the response to external adenosine (57). Defolliculated oocytes still respond to forskolin by a rise in cAMP (58), suggesting that this system is in the oocyte proper. The basal level of cAMP has been quoted as, for example, 1.4 pmol per oocyte and adenosine (external) increases this (59). GTP-binding regulatory proteins have been shown to be intermediates in the membrane current response to, for example, glutamate, in brain RNA-injected oocytes, which is being suppressed by incubation with pertussis toxin (84).

There is also present the phosphoinositide/protein kinase C system. Injected inositol 1,4,5-trisphosphate ($InsP_3$) gives rise to large periodic membrane currents which mimic the muscarinic effect (60). Phorbol esters evoke no currents themselves but, at nanomolar levels, inhibit the responses to adenosine or to intracellularly-injected cAMP (52).

3. EXAMPLES OF RECEPTOR mRNA EXPRESSION

3.1 Nicotinic acetylcholine receptors

The first cellular expression of a receptor was obtained with *Torpedo* electric organ poly(A) mRNA, which produced the α-bungarotoxin-binding ACh receptor in the *Xenopus* oocyte membrane: this contained the four native subunits (7) and the receptor Na^+/K^+ channel opened by ACh (31). The receptors produced are located primarily in the oocyte membrane and are assembled to the native oligomeric size (9S) (*Figure*

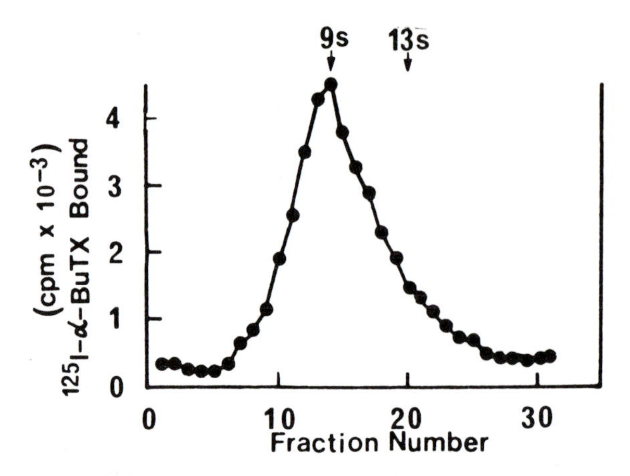

Figure 1. Analysis of the nAChR produced in the oocyte membrane after injection of *Torpedo* electric organ mRNA, to show the molecular size of the active product. After injection of sucrose gradient-selected active fractions of the poly(A) RNA and 48 h incubation, 30 injected oocytes were extracted in 1% Triton X-100 medium and the extract placed on top of a 5−20% sucrose density gradient. After centrifugation and fractionation, every fraction was assayed for the receptor as measured by the specific binding of [^{125}I]α-bungarotoxin. The arrows show the positions of markers of sedimentation constant 9S and 13S in a parallel tube. The receptor is seen to be formed in the oocyte as a 9S species: the same profile is given by the monomeric native nAChR when extracted and analysed similarly. The 9S form is the $\alpha_2\beta\gamma\delta$ active pentamer. From ref. 61.

1). The latter was shown by fractionation of the oocyte membrane detergent extract on a sucrose density gradient and measuring [^{125}I]α-bungarotoxin binding activity across the gradient (7,61). This response is specific to electric organ or skeletal muscle mRNAs and is proportional to the total poly(A) mRNA from a given source. The electrophysiological response of the electric organ and muscle nAChRs induced in the oocyte membrane exhibits the predicted nicotinic pharmacology and they are correctly orientated since intracellular agonist administration has no effect (4,31).

Tunicamycin (2 μg/ml, present in the medium for 48 h after its co-injection, at 2 ng per oocyte, with the mRNA) abolishes the ACh response and the toxin binding, although the subunits of the receptor are still produced and deglycosylation of the native receptor protein does not remove its activity. This illustrates the approach to a study of the role of N-glycosylation and other processes in receptor assembly, which is possible in the oocyte system.

The nAChR subunits produced in the oocyte membrane via *Torpedo* mRNA were shown at an early stage to be the authentic exogenous subunits, since a one-dimensional peptide map of, for example, the α-subunit was identical between the native *Torpedo* nAChR and the oocyte-produced version (32). Subsequently the possibility of a stimulation of the host cell genome to produce induced receptor or channel has been totally excluded by the recent elegant studies by two laboratories (1,62,63) of totally synthetic subunit-specific mRNAs. The cloning of the cDNAs encoding all the four subunit types of the *Torpedo* nAChR made possible the injection into the oocyte of relatively large amounts of the corresponding pure specific mRNAs, showing that these four are both necessary and sufficient for the full binding and ACh channel-activation functions. This system has facilitated quantitative studies on the channel function, and the probing of

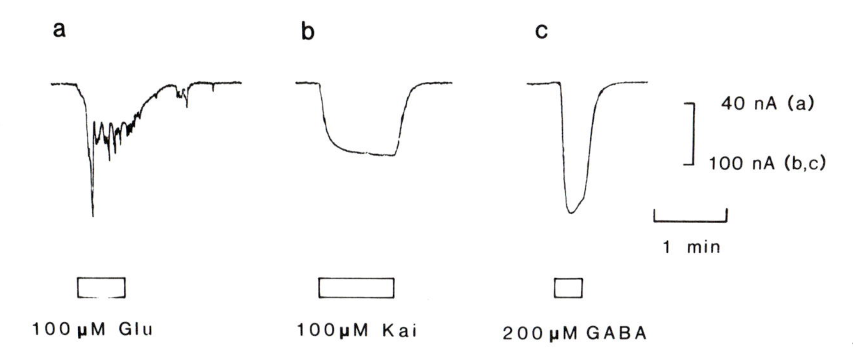

Figure 2. Recording of the response of receptors, induced in the oocyte membrane by translation of chick brain mRNA, to a specific (bath-applied) agonist. Inward membrane currents are shown, voltage-clamped at −70 mV. (From ref. 33.) **(a)** Response to L-glutamate. Note the typical oscillatory response. **(b)** Response to kainic acid. Note that there is no significant desensitization. **(c)** Response to GABA. The GABA response desensitizes very readily at concentrations above 4 μM, by a biphasic process (69).

the structural requirements for the receptor functions, by the use of (i) hybrid subunit mRNA mixtures, from different species or developmental stages (1,35,64); (ii) *in vitro* mutagenesis, either by point mutations or by large deletions, in regions hypothesized to be functionally significant (1).

3.2 **GABA and glycine receptors**

When, at an early stage in the history of this approach, a brain poly(A) mRNA sample was prepared in this laboratory by K.Sumikawa, in the course of our studies on the chick optic lobe receptors, for collaborative testing in the *Xenopus* oocyte for γ-amino-butyric acid (GABA) receptor induction, a response to iontophoretically-applied GABA was indeed seen on the cell membrane of the injected oocytes (65). This response was examined in more detail and more conveniently in a bath-application system (66,67), using both the embryonic chick brain and the young rat brain as sources. The results showed that a receptor is induced in the oocyte membrane which responds to GABA agonists with a Cl^- conductance (*Figure 2c*). The allosteric responses of the channel activity to benzodiazepines and anti-convulsive barbiturates were shown to be as in the native $GABA_A$ receptor.

Expressed GABA receptors are present in the *Xenopus* membrane for at least a week, and can persist for up to 3 weeks. The quantitative potencies of a series of GABA agonists are also found to be as on native neuronal $GABA_A$ receptors, and the dose−response relationship (*Figure 3*, triangles) gives a log−log plot limiting slope of 2, showing that two GABA molecules must be bound for channel opening (68). A Schild plot could be obtained for the antagonist bicuculline, with a slope of unity and pA_2 value of 5.9 (as found on neurones). The bicyclophosphate cage-convulsant channel blocker, TBPS, gave a 'mixed-type' of inhibition and properties indicating a closed channel block (69). It is of interest that the Cl^- channel opened in the implanted GABA receptor was shown to differ from the Cl^- channels opened in spontaneous fluctuations or in muscarinic or serotonin responses native to the oocyte, the latter three types being TBPS-insensitive.

Glycine receptors were also expressed from rat brain poly(A) mRNA. These gate

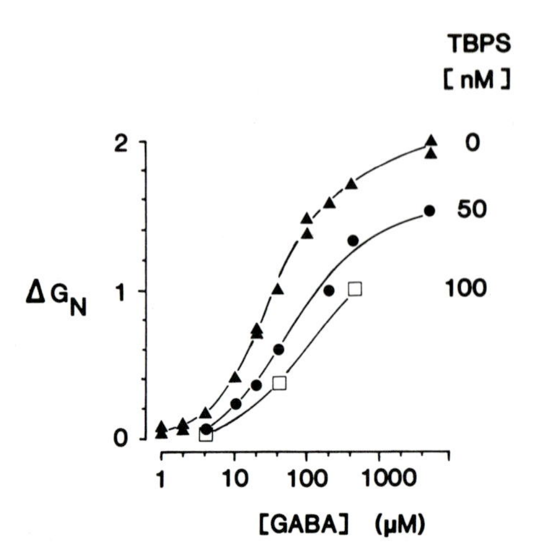

Figure 3. The dose–response relationship for GABA (▲) bath-applied to oocytes pre-injected with chick brain mRNA, and the blockade by low concentrations of the channel-blocking cage convulsant drug, TBPS (●, 50 nM; □, 100 nM). (From ref. 69.) The conductance under voltage-clamp conditions (at −70 mV) is normalized by showing its ratio to the peak current (see *Figure 2c*) evoked by 40 μM GABA in TBPS-free conditions. (Mean values for 15 oocytes.)

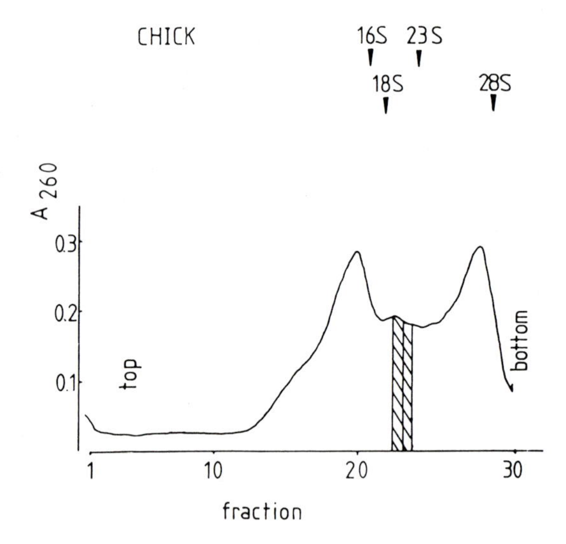

Figure 4. Separation and identification of the $GABA_A$ receptor mRNA fractions from total brain poly(A) RNA. The day-19 embryonic chick brain was used for the preparation of poly(A) RNA. This was applied to a 10–31% isokinetic sucrose gradient and centrifuged, as described in Section 7.2. The RNA concentration in the fractions (measured by A_{260}) is shown by the continuous line. The positions of RNA size markers in parallel gradients are shown by arrows. The RNA of each fraction was injected into oocytes, whose subsequent response to GABA was recorded. The hatched zone shows the fractions which produced an unequivocal response to bath-applied GABA. (From ref. 34.)

a Cl^- channel and are strongly antagonized by 1 μM strychnine (67). The voltage dependence of these implanted glycine receptors was shown to differ significantly from the implanted $GABA_A$ receptors (68). Fractionation of the poly(A) mRNA on a sucrose density gradient separated the GABA and glycine sets of mRNAs, and showed that the former (from rat brain) is all about 19S in size (*Figure 4*).

3.3 **Excitatory amino acid receptors**

Rat brain poly(A) RNA-injected oocytes acquired a sensitivity to glutamate, kainic acid and quisqualate (67,70). The glutamate response is an oscillatory Cl^- conductance plus a smooth response, while the kainic acid response (the most sensitive) is purely a smooth one, with a reversal potential of -11 mV, and is sensitive to the external Na^+ concentration, suggesting a Na^+/K^+ channel (33). Examples recorded in the voltage-clamp mode are shown in *Figure 2a,b*.

3.4 **Ion channels**

The most informative induction in the oocyte of the voltage-dependent Na^+ channel has come from the injection of a cDNA clone-derived mRNA encoding this. Only one species of mRNA, corresponding to the 250-kd α-subunit of rat brain, is needed for the expression of the voltage-dependent, TTX-sensitive channel function (71,83). This, as before, has permitted a site-directed mutagenesis approach to the structural requirements for the channel. Using mRNA from rat brain and muscle, new voltage-dependent Ca^{2+} channels have been induced in the oocyte membrane (49), of two types, fast (lifetime <0.5 sec) and slow (predominant and dihydropyridine-sensitive).

3.5 **Transport systems**

The successful expression in the oocyte of the ‘band 3’ anion transporter of mouse erythrocytes, in a functional state, provides an example here, from the work of Morgan *et al.* (72). This protein does not contribute an electrical conductance when active, but is an anion exchanger. Poly(A) RNA from anaemic mouse spleen was fractionated on a sucrose gradient and injected. After 16 h of incubation, ^{35}S-labelled collagenase-pretreated oocytes, upon SDS extraction, showed a specifically immunoprecipitated protein of the correct 96 kd size and a marked Cl^- flux which was sensitive to the specific band 3 blocker, diisothiocyano-dihydrostilbene disulphonate. It was estimated that approximately 20×10^6 band 3 proteins were produced in 16 h in one oocyte.

4. EXTRACTION AND PURIFICATION OF mRNA

4.1 **Precautions**

Expression of mRNA, whether by cell-free translation or in the *X. laevis* oocyte, is dependent on several factors, the most important of which is the use of undegraded mRNA.

Ribonuclease contamination of samples is the major cause of RNA degradation. Ribonucleases are very robust enzymes (boiling for 15 min at 100°C will not inactivate them) and minute amounts are sufficient to destroy mRNA activity. Potential sources which may be contaminated with ribonucleases are treated as follows.

(i) Wash all glassware in detergent and acid, then bake in an oven (180°C, 3 h) or autoclave.

(ii) Prepare all solutions from the purest possible sources using autoclaved, double glass-distilled water.
(iii) Ideally, add 0.05–0.1% (v/v) diethyl pyrocarbonate (DEPC) to the buffers (this being a strong inactivator of ribonucleases), and then autoclave or filter the solutions through a sterile Millipore filter (0.22 μm).
(iv) Heat to 100°C for 15 min those solutions treated with DEPC but not autoclaved to remove any remaining traces of DEPC. Note that DEPC is unstable with Tris buffers and should not be used to treat these solutions.

Table 1. Guanidinium thiocyanate (GTC) extraction of RNA.

1. (a) Homogenize fresh tissue (1 g in 10 ml) in GTC solution[a] containing 2% sodium N-lauroyl sarcosinate using, for example, a Polytron homogenizer for 15 sec at half-maximum speed.
   (b) If stored tissue is to be used, this is quick-frozen originally in liquid nitrogen for storage; grind to a fine powder in liquid nitrogen before adding the GTC solution.
2. Centrifuge the homogenate in sterile, siliconized 30 ml Corex tubes[b] for 10 min at, for example, 10 000 r.p.m. in a Sorvall HB4 rotor[c], to remove insoluble debris.
3. Remove each supernatant and layer it onto a 3 ml cushion of sterile 5.7 M caesium chloride containing 100 mM EDTA (pH 7.5), in swing-out bucket cellulose nitrate (e.g. Ultraclear) 12 ml tubes. Rinse these tubes, before use, firstly in 10% (v/v) H_2O_2 and secondly six times with double-distilled water.
4. Centrifuge, for example, in a Beckman SW40Ti rotor, 18 h, 32 000 r.p.m., 15–20°C.
5. Remove the tubes, aspirate off the GTC solution down to within 1 cm of the bottom of the tube, then invert rapidly and leave to drain. Cut the tubes 1.5 cm above the pellet, dry the walls thoroughly using 95% ethanol (stored at −20°C) and resuspend the pellets in up to 2 ml (total) solubilizing buffer[d].
6. Transfer the suspension to sterile capped Universal tubes. Heat these in a 65°C bath for 2 min, vortex mix for 20 sec, and repeat the cycle until fully solubilized.
7. Extract in a sterile capped 50 ml Falcon tube with an equal volume of phenol/chloroform/isoamyl alcohol (25:24:1, by vol)[e] pre-heated to 60°C. Shake vigorously for 1–2 min or mix on a rotary wheel for 5 min (at room temperature). Centrifuge (5 min, 10 000 r.p.m., HB4 rotor, room temperature). Re-extract the upper aqueous layer again with the organic mixture, until the interface is completely clear. Extract finally with an equal volume of chloroform/isoamyl alcohol (24:1, v/v).
8. Add 5 M NaCl to a final concentration of 100 mM. Mix thoroughly. Add 2.5 volumes of 95% ethanol, pre-cooled to −20°C. Allow the RNA to precipitate overnight at −20°C, or at least 2 h at −70°C.
9. Recover the RNA by centrifugation for 30 min at 10 000 r.p.m. with a HB4 rotor at 4°C. Wash the pellet twice with 80% ethanol (pre-cooled to −20°C), vortex-mix three times for 10 sec each, resuspending and centrifuging each time at 10 000 r.p.m. for 20 min at 4°C. Dry the pellet finally in 95% ethanol (−20°C), without vortex-mixing.
10. Store under 70% ethanol at −20°C, *or* re-dissolve the pellet in sterile distilled water (to the maximum concentration usable) and store in 5 μl aliquots at −70°C.
11. Determine the RNA concentration of a suitable dilution of the stock solution by measurement of A_{260} (see Section 4.3).

[a]Make up 4 M GTC solution by dissolving 50 g of solid GTC (Eastman Kodak or Fluka) in 50 ml of water. Stir for 10 min at 40°C to dissolve the GTC. Add 2.5 ml of 1 M sodium citrate or Tris-HCl (pH 7.4). Spin and/or filter (Millipore/Nalgene) to clear. Add 1 ml of 2-mercaptoethanol. Finally make the volume up to 100 ml (pH 7.4) including EDTA to 10 mM final concentration, and store in a dark bottle at 4°C. Before use, warm gently to re-dissolve totally. This solution is stable for up to 1 month.
[b]Siliconized Corex (or similar type) tubes are prepared by filling them with dimethyldichlorosilane solution, leaving them for 10 min, then rinsing with sterile water and autoclaving or leaving overnight in a 180°C oven. These Corex tubes are used for all centrifugation steps except where stated.
[c]Any swing-out bucket rotor reaching an equivalent *g* value will suffice.
[d]Solubilizing buffer is 10 mM Tris-HCl (pH 7.5), 5 mM EDTA, 1% SDS.
[e]Analytical grade phenol is made up with 0.1% hydroxyquinoline to act as a stability indicator, and saturated with a solution of 100 mM Tris-HCl (pH 7.4), 10 mM EDTA, 50 mM NaCl, 0.2% 2-mercaptoethanol.

(v) Thoroughly wash large items of glassware which cannot be autoclaved with the extraction buffer before use.
(vi) Pre-packaged sterile disposable plasticware is essentially free of ribonucleases.
(vii) Finally, it is advisable to protect solutions from skin ribonucleases by wearing gloves during all manipulations involving mRNA.

4.2 RNA extraction by the guanidinium thiocyanate (GTC) method

This method has been successful in routinely producing undegraded mRNA suitable for oocyte microinjection. The procedure, modified from Chirgwin *et al.* (73), is a quick, simple, one-step purification of RNA from protein and DNA, following homogenization of the tissue sample in GTC solution, by centrifugation through a 5.7 M CsCl cushion (*Table 1*). The buoyant density of the RNA is such that it passes freely through the cushion to form a translucent pellet, while the DNA forms a band within the CsCl cushion and the protein remains in the GTC layer. It is important to remove all traces of GTC after centrifugation, so as not to contaminate the RNA. (Some investigators superficially wash the pellet twice with ice-cold ethanol.) It is important also to remove excess protein and other contaminants of the RNA by a subsequent extraction with phenol/chloroform (step 7 in *Table 1*), before collection of the RNA by ethanol precipitation.

4.3 Purification of mRNA

mRNA can be purified using techniques which bind the poly(A)$^+$ tail of the molecule, via hydrogen bonds, to a matrix of cellulose and oligodeoxythymidine [oligo(dT)] or polyuridylic−Sepharose at a high salt concentration (*Table 2*). Oligo(dT)−cellulose is readily available commercially and can bind up to several milligrams of mRNA per

Table 2. Oligo(dT)−cellulose chromatography of RNA.

1. Pre-soak 0.5 g of oligo(dT)−cellulose in 5 ml of 10 mM Tris-HCl (pH 7.5), 5 mM EDTA, 0.1% SDS, for 10 min. Apply to a sterile syringe or Pasteur pipette and wash with 20−30 volumes of elution buffer[a] containing 1 mM EDTA. Then wash sequentially with five column volumes each of 0.1 M NaOH and binding buffer. Equilibrate to pH 7.5 with binding buffer[b].
2. Heat the RNA solution to 65°C for 2 min and rapidly chill on ice. Add an equal volume of twice-concentrated binding buffer and apply to the column. Re-pass the flow-through twice through the column.
3. Wash the column with 10 column volumes of binding buffer or until the A_{260} of the effluent is close to zero. Repeat with binding buffer containing 0.1 M NaCl.
4. Elute the bound mRNA with elution buffer and collect in 0.5 ml fractions until all of the RNA is eluted (as monitored by A_{260}). Re-adjust the mRNA-containing fractions to 0.5 M NaCl (final concentration) and then repeat steps 2−4.
5. Add NaCl to a final concentration of 100 mM, then add 2.5 volumes of ethanol. Allow to precipitate at −20°C overnight.
6. Pellet the RNA by centrifugation in sterile, siliconized Corex tubes (see *Table 1*) for 30 min; 10 000 r.p.m. (Sorvall HB4 rotor), 4°C. Rinse and dry the pellet with ethanol and store as in *Table 1* (steps 9 and 10).
7. Regenerate the oligo(dT)−cellulose by successive washings with water, 0.1 M NaOH, water and finally binding buffer as in step 1.

[a]Elution buffer is 10 mM Tris-HCl (pH 7.5), 1 mM EDTA, 0.5% SDS.
[b]Binding buffer is 10 mM Tris-HCl (pH 7.5), 1 mM EDTA, 0.5 M NaCl, 0.5% SDS.

gram of cellulose. Up to 10 mg of total RNA can be loaded onto a 1–2 ml column of oligo(dT)–cellulose, which is then extensively washed in a buffer containing 0.5 M NaCl. This wash removes primarily rRNA and tRNA. Poly(A)$^+$ RNA is subsequently eluted by lowering the ionic strength of the wash buffer. rRNA is a major contaminant in the first mRNA elution, but is reduced by a second passage over the column. For *in vitro* translation, it is advisable to precipitate the mRNA as a potassium salt by addition of one-tenth of the volume of 20% potassium acetate (pH 6.5) and 2.5 volumes of ethanol. When possible, the mRNA is stored as an ethanol precipitate at –20°C, which avoids repeated freeze–thaw cycles of aqueous solutions which will inactivate some mRNAs. The yield and purity of a mRNA preparation is assessed by measuring the absorbance at 260 nm and 280 nm. One A_{260} unit is equivalent to an RNA concentration of 50 μg/ml. Ideally, pure RNA has a A_{260}:A_{280} ratio of 2.0. In practice it is often found to be about 1.8. A value below this indicates too much contamination by protein or phenol, or degradation, and the preparation should be discarded.

4.4 Assessing mRNA quality

The following two methods are used as routine assays to determine, firstly, that the mRNA prepared is undegraded and, secondly, that this mRNA can direct protein synthesis.

Table 3. Electrophoresis of mRNA following glyoxal denaturation.

1. Make up glyoxal (Sigma) as a 6 M solution and de-ionize by continuous passages down a mixed-bed ion-exchange column, e.g. of MB1 resin (from BDH Chemicals Ltd) pre-washed in ethanol, ethanol/ether (1:1), dried, and packed into a sterile Pasteur pipette. Check the pH of the glyoxal solution after each pass and continue until it reaches pH 6. Use immediately or store in 50 μl aliquots at –20°C.
2. Mix in a sterile, siliconized Eppendorf tube 5 μl of the RNA solution (3–6 μg RNA) or suitable marker RNA[a], 3 μl of 6 M glyoxal at pH 6, 2 μl of 100 mM sodium phosphate buffer (pH 7) and 10 μl of dimethyl sulphoxide (Sigma).
3. Incubate the RNA/glyoxal solution at 50°C for 1 h. Add 5 μl of gel loading buffer[b] and use immediately or store up to 24 h at 4°C.
4. Make up a 1–1.5% agarose gel in 10 mM sodium phosphate buffer (pH 7). Electrophorese in 10 mM phosphate buffer (pH 7) at <25 mA for 3–4 h, with continuous buffer recirculation. Alternatively, change the buffer every 30 min.
5. Dissociate glyoxal from the RNA by shaking for 30 min in 50 mM NaOH. Transfer to 100 mM sodium phosphate buffer (pH 6.5) with shaking and several changes of buffer until the pH of the gel returns to a value of 7.
6. In the dark, stain the gel using a 0.5 μg/ml ethidium bromide solution in water for 5 min, followed by at least 10 min de-staining in 100 mM sodium phosphate buffer (pH 7). Visualize under u.v. light. Photograph on a u.v. transilluminator with type 57 Polaroid film, e.g. at f4.5 for 1 sec, using red and yellow filters.

   Destaining for 2 h or even overnight can reduce the background substantially: the optimum intermediate time for contrast should ideally be found[c].

[a]One of the several commercially available sets of RNA standards or 'RNA ladder' kits is satisfactory.
[b]Loading buffer is 10 mM phosphate buffer (pH 7), 50% glyoxal, 0.1% bromophenol blue.
[c]If necessary, sharper staining is obtained if formaldehyde/formamide replaces the glyoxal (see Section 4.4.1). The method is equivalent except that the agarose, too, contains formaldehyde (6%). One version of the latter method is given in *Focus,* **8:2**, 14 (1986), obtainable from Bethesda Research Laboratories (Gaithesburg, MD 20760, USA).

4.4.1 *Agarose gel electrophoresis of mRNA*

A convenient system to measure the molecular weight range and detect degraded mRNA is denaturation with de-ionized glyoxal and agarose gel electrophoresis (as modified from ref. 74). Glyoxal may react reversibly with all the bases of RNA, but forms a stable adduct with guanosine bases at pH 7; this sterically hinders the formation of G−C base pairs (74). To maintain RNA in this denatured state during electrophoresis, the buffer pH must be maintained below pH 8. After electrophoresis, glyoxal is removed by incubation in 50 mM NaOH and the pH of the gel re-equilibrated to pH 7 before staining with 0.5 μg/ml ethidium bromide to visualize the mRNA under u.v. light. The method is detailed in *Table 3*.

Denaturation with glyoxal constitutes less of a hazard than using methylmercuric hydroxide or formaldehyde (75). However, the fluorescence of the stained RNA bands and the gel background are better in a formaldehyde, rather than a glyoxal, gel.

4.4.2 *Rabbit reticulocyte lysate assay of mRNA activity*

The reticulocyte lysate assay is used as a routine screen to assess the activity of the prepared mRNA against a standard such as globin mRNA (*Table 4* and ref. 2). The lysate has a high translational capacity with exogenous mRNAs, each active mRNA molecule being translated 40−70 times in a 90 min incubation. The mRNA activity

Table 4. Rabbit reticulocyte lysate assay of mRNA activity and size range.

1. The micrococcal nuclease-treated reticulocyte lysate mixture from Amersham International Ltd. (N90) or Bethesda Research Laboratories or other standard source is suitable. To 40 μl of this lysate, in a sterile, siliconized Eppendorf tube, add 2 μl of mRNA (0.5 mg/ml), and 50 μCi of L-[^{35}S]methionine. Mix gently by vortex and incubate at 30°C.
2. Remove duplicate 1 μl samples at intervals of 0, 5, 10, 20, 30, 40 and 60 min. Add to 0.5 ml of 0.5 M NaOH (to hydrolyse aminoacyl tRNA) in 5% (v/v) H_2O_2 (to bleach the samples) in a 3 ml LP3 tube (Luckham).
3. Place the tubes in a water bath at 37°C for 10 min.
4. Remove the tubes to an ice bath and add 2 ml of ice-cold 25% (w/v) trichloroacetic acid (TCA) plus 100 μl of 1% bovine serum albumin. Leave on ice for 30 min to precipitate the protein.
5. Filter the tube contents through a 2.5 cm Whatman GF/C glass fibre filter, wash twice with 5 ml of ice-cold 5% (w/v) TCA, then with 5 ml of cold ethanol.
6. Dry the filters under an infra-red lamp and determine the radioactivity by liquid scintillation counting.
7. From the final incubation mix at 60 min, transfer 5 μl into an Eppendorf tube containing gel sample buffer[a] (pre-heated to 80°C). Vortex gently and maintain at 80°C for 3 min. Before loading the samples onto a 12.5% SDS−PAGE gel[b], centrifuge for 2 min in a benchtop microcentrifuge. Run[c] with controls in parallel lanes (mRNA omitted) and a lane of protein molecular weight markers (preferably a radio-labelled set, e.g. from Amersham International Ltd.). The markers should be added to medium obtained from a control in which all the steps were performed except the addition of [^{35}S]methionine, to ensure that the markers run in the same conditions.
8. After electrophoresis, fix the gel with 50% aqueous methanol in 7% acetic acid for 30 min, and rapidly wash with 15% aqueous methanol. Immerse in 100 ml of Amplify (Amersham International Ltd.) fluorographic reagent for 30 min and prepare the gel for fluorography.

[a]Gel sample buffer is 20 mM Tris-HCl (pH 7.5), 2% (w/v) SDS, 10% (w/v) glycerol, 0.1 M dithiothreitol (or 70 mM 2-mercaptoethanol), 0.007% bromophenol blue.
[b]Prepared, for example, according to Laemmli (ref. 82).
[c]For some further notes and precautions see Section 5.2 of ref. 2.

is judged by the incorporation of [^{35}S]methionine into synthesized polypeptides, as shown by trichloroacetic acid (TCA) precipitation or SDS–polyacrylamide gel electrophoresis. Commercial kits (e.g. from Amersham) are readily available and easy to use. In this laboratory, after a 60 min incubation at 37°C, active, intact mRNA (prepared as in *Table 1*) characteristically displays a greater than 10-fold stimulation of [^{35}S]methionine incorporation into the precipitated protein, over endogenous lysate protein synthesis in parallel assays, and the labelled proteins are produced in sizes ranging up to 100 kd.

5. MICROINJECTION OF mRNA

5.1 Maintenance of *Xenopus*

Adult female *X. laevis* (100–150 g) are obtainable from Xenopus Ltd (Homesdale Nursery, Mid Street, South Nuffield RH1 4JY, UK). Over a period of several years use, we have found that the wild animals (i.e. obtained from South Africa) are usually larger and provide more active oocytes than laboratory-reared toads (both from Xenopus Ltd).

Maintain the toads in large covered fish tanks (100 cm × 30 cm × 50 cm) in at least 10 cm of water, up to 10 to a tank. Feed them on alternate working days with chopped liver and/or trout food pellets (Xenopus Ltd). Water temperature in the tanks should be 19–20°C, clearing is usually carried out several hours post-feeding and the tanks are refilled using tap water which has been standing at room temperature for at least 24 h to free it from chlorine. Although not essential, water filters will keep the tanks clean between feeding times.

Some seasonal variation is often encountered: the period of December to March in the Northern hemisphere is the time when yields of oocytes responsive to exogenous receptor and other mRNAs are lowest. The native muscarinic receptor response is lowest then, as another indicator. Considerable individual variation in responses between oocytes is found at that time. A few individual donor animals give poorly-responsive oocytes at all times and should be eliminated from the pool of donors.

5.2 Oocyte removal

A large, mature toad contains in excess of 40 000 large (>1 mm diameter) oocytes in its multi-lobe ovary. For injection experiments using up to several hundred oocytes at a time, anaesthetize the toad for 10–15 min in a 0.1% aqueous solution of ethyl-*m*-aminobenzoate (Sigma, also known as Tricaine or MS222). Gloves should be worn during this procedure, as the anaesthetic is a potential carcinogen.

(i) To remove the oocytes, place the toad ventral side uppermost and make a small abdominal incision (~1–5 cm long) through the skin and the gut wall 1 cm above the bladder and to one side of the midline.

(ii) Using blunt-ended forceps placed through the incision into the abdominal cavity, remove one or more ovarian lobes.

(iii) Immediately transfer the ovarian lobe to a 90 mm Petri dish containing modified Barth's solution (*Table 5*).

(iv) Stitch the gut wall and skin separately and allow the toad to recover for about 10 h in a container having only 0.5 cm depth of water to avoid drowning during revival.

Table 5. Modified Barth's solution.

1. Composition[a]:
   88 mM NaCl
   1 mM KCl
   2.4 mM $NaHCO_3$
   0.33 mM $CaNO_3 \cdot 4H_2O$
   0.82 mM $MgSO_4 \cdot 7H_2O$
   0.41 mM $CaCl_2 \cdot 6H_2O$
   7.5 mM Tris-HCl, pH 7.6.
2. Make as a 5-fold concentrate and autoclave.
3. Before use add sodium penicillin and streptomycin sulphate, each to 10 μg/ml, or gentamycin to 100 μg/ml and nystatin (optional) to 50 U/ml. For post-injection incubations, some nutrients, e.g. fetal calf serum to 0.1%, pyruvate to 1 mM etc., may be added but their effects have not been systematically evaluated.

[a]All reagents are Analytical grade.

Using this procedure, one female can be used for several donations, allowing 1–2 months recovery time between each operation.

5.3 Oocyte stages and layers

In the ovary of *Xenopus*, the female germ line cells initially enter meiosis; after the first prophase they enter the first meiotic arrest and then develop only slowly. These form the 'immature oocytes', which develop in six stages (76). The most developed of these oocytes are 1.2–1.3 mm in diameter. Each has a dark brown ('animal') hemisphere and a yellow–green ('vegetal') hemisphere. The surface is covered at the ultrastructural level with microvilli, which are denser in the animal hemisphere. Most of the oocyte RNA is concentrated in the subcortical vegetal region (77). Further maturation to the eggs, which are laid into the water, occurs when triggered by progesterone or gonadotrophin.

Oocytes to be used for exogenous mRNA translation should be of stages 5–6. These are distinguished by their bright equatorial band and very distinct polar hemispheres. (In some stage 6 oocytes the bright band is hard to discern and their animal hemisphere has turned a greenish hue, but they are also recognizable by their large size.) Oocytes with white speckles in the animal pole or dark speckles in the vegetal pole are in the process of maturation to eggs and must be avoided. A furrow or wrinkles on the surface is also a negative indicator.

The oocytes are surrounded first by the vitelline membrane, a non-cellular fibrous, highly porous layer (for a picture, see ref. 23), and then by the follicle cells, which form many gap junctions to the oocyte cell membrane (78). Outside these are a connective tissue ('theca') layer and epithelial attachments to the ovary wall. The follicular and thecal layers are often removed, either by collagenase or mechanically, in Ca^{2+}-free medium. The vitelline layer then remains.

Both mechanical and enzymic defolliculation can damage the oocyte cell membrane (79). Collagenase treatment abolishes the native muscarinic response to ACh (37) and the native β-adrenergic or adenosine responses (43), in at least some of the oocytes. Responses induced by subsequently injected mRNA in the oocyte are often preserved after defolliculation, if a recovery period (e.g. 3 h in modified Barth's at 21°C) is given.

In general, mechanical defolliculation is less damaging in this context than is collagenase treatment.

The vitelline layer proves no barrier, but it interferes with the sealing of the membrane to a micropipette tip for patch-clamping: for such purposes it is elevated mechanically after pre-incubation in a hyperosmotic medium (54,64).

5.4 **Preparation of oocytes**

Rinse the Petri dish containing the ovarian lobe several times with modified Barth's solution (*Table 5*) to remove blood and tissue debris, and then break the ovarian lobe into clumps of about 50 oocytes using fine forceps. These small clumps can be treated in several ways. If individual oocytes are not needed, then divide each clump of 50 oocytes into clusters of 4–5 oocytes which are easily handled for microinjection. However, in most cases single oocytes are required and these are prepared as follows.

(i) Hold a 50-oocyte clump using a fine pair of forceps; strip off individual oocytes by repeatedly passing another pair of forceps across the first.
(ii) Rinse the resulting mixture several times with modified Barth's solution and transfer single oocytes to a clean Petri dish using a wide-mouthed plastic Pasteur pipette.

Depending on the oocyte batch used, a varying amount of mechanical damage occurs during this procedure; however, undamaged oocytes are easily recognized after an overnight incubation at 21°C. Alternatively, 50-oocyte clumps can be incubated (at 19– 21°C) in 1 mg/ml collagenase (Sigma Type II) in modified Barth's solution for 1–3 h (or overnight at lower concentration) until swirling of the preparation releases the individual oocytes.

Each of these methods gives a preparation containing various sizes of oocytes but for convenience only large ones are used for microinjection. The oocytes should be collected and stored at 21°C. It is very useful to have an incubator accurately thermostatted to this temperature in the dissection/injection room, which itself should be temperature-controlled.

5.5 **Microinjection equipment**

Essentially one needs a good stereomicroscope which contains a lens with adjustable magnification (to 25× or 100×; see below) and a good depth of field. For good lighting, use a fibre optic system (e.g. Schott LK1500) to provide a high intensity, cold light illumination of the oocyte and microinjection needle. A vertical electrode puller is also desirable but not essential (see Section 5.6). A coarse-action micromanipulator is used: the Prior instrument is very satisfactory. Finally polyethylene tubing connects the microinjection needle to a suitable syringe for the injection of mRNA solutions into the oocytes. This can be an 'Agla' micrometer syringe (from Wellcome Reagents Ltd, Beckenham, Kent, UK) with the dead space filled by paraffin oil, or an equivalent construction (80) or (very effective) a syringe of the type used in patch-clamping work (54).

A pressure microinjection system is more time-consuming to set up and use but has the advantage that the volumes are very accurate and reproducible and that smaller pipette tips (10 μm) and higher concentrations of RNA (i.e. of higher viscosity) can be employed, in the much smaller volumes readily controlled in this equipment. The recovery of the oocyte from the injection is very variable if more than 50 nl is injected into a large (~1 mm) oocyte and the re-sealing of the membrane is definitely better if a smaller pipette tip is used. It is in general recommended to use smaller volumes (say ~10 nl) if mRNA of sufficient concentration is available, and to use a pipette tip of 10 μm, with a pressure injector. (A suitable instrument is marketed by WP Instruments Inc., New Haven, CT 06515, USA.)

5.6 Making micropipettes

(i) For the first stage of preparation of the microinjection needle, heat a hard glass capillary 10 cm tube (BDH Chemicals Ltd.) to melting point in a flame and then rapidly stretch it to about twice its original length, to give an external diameter of about 200 μm.
(ii) Use a vertical electrode-puller apparatus to draw out the central section to the required tip diameter. (A home-made system can also be used: see ref. 11.)
(iii) After breaking obliquely across in the centre, trim the tip, either using fine forceps, or in a microforge.

The needle tip diameter should be 20−30 μm for a manual system and about 10 μm for a pressure system. For the latter size, a microforge is essential. Ideally the barrel of the needle should be parallel-sided for at least 30 mm from the tip to allow accurate volume measurement (in a manual system). Note that some practice may be needed to obtain a good needle.

5.6.1 *Calibration*

This is needed only if a manual system is used. The easiest empirical method is to draw up 1 μl of sterile distilled water, then mark off the extent of travel using a fine marker pen. Expel the water from the needle, then, using a graduated eyepiece, mark off 10 equal divisions between the tip and the maximum mark. These divisions are approximately equal to 100 nl and are a convenient volume to work with. At those times when a more accurate calibration is needed, a small volume of radioactive tracer in water is drawn into the pipette and equal volumes are spotted onto filter discs and counted. Microinjection needles ready for use are stored tip-uppermost in sterile capped tubes.

5.7 Microinjection of mRNA

Purified mRNA for microinjection is dissolved in sterile distilled water and stored as 5 μl aliquots at −70°C.

(i) Before use, rinse the microinjection needle with sterile distilled water; cover a glass slide with Parafilm and spot on 1 μl of water and rinse the pipette by suction.
(ii) Next, place 2 μl of a mRNA solution onto the Parafilm. The minimum concentration of total poly(A) mRNA usable is of the order of 1 mg/ml, as it may be necessary to deliver up to 50 ng of mRNA per oocyte and 50 nl is the maximum volume (see Section 5.5). Less mRNA can be injected and provide satisfactory

results when combined with a suitably lengthened incubation time. With RNA concentrations over 2 mg/ml the risk of particulate matter blocking the micropipette must be guarded against, and (as noted above), these concentrations are best used with a pressure injector.

(iii) Keep the unused RNA and the slide on ice.

(iv) Fill the microinjection needle to the maximum mark or (for manual injection) until the meniscus and oocyte are comfortably within the fields of view. With pressure injection the need to see the meniscus disappears, and a higher magnification (up to 100×) can be maintained on the oocyte dish. This is a distinct advantage for accurate pipette insertion and for recognizing when damage inadvertently occurs.

(v) Transfer 4–5 oocytes singly onto a clean glass microscope slide in a small drop of modified Barth's solution, then place under the microscope. We immobilize the oocyte for microinjection by moving it into a hemispherical well (diameter a little greater than the oocyte size) in a plastic holder constructed to contain a set of these wells.

(vi) Insert the needle into the yellow–green vegetal side of the oocyte (i.e. avoiding the nucleus) and inject the measured volume of the mRNA solution. Do not allow the oocyte to dry out during this procedure as this would greatly increase the mortality rate.

(vii) Transfer the injected oocytes to a fresh Petri dish containing modified Barth's solution and incubate overnight at 21°C. With practice up to 200 oocytes per hour can be injected.

6. ANALYSIS OF PROTEINS PRODUCED IN THE OOCYTE

6.1 **Preparation for analysis**

Table 6 details methods for preparing oocyte material for analysis. These are basic outlines which should be modified to the investigator's system of interest. For example, in the study of expressed neurotransmitter receptors in this laboratory, the detergents and buffers used are designed to maximize extraction of the particular receptor in a form which can be precipitated by antibodies or used in binding assays.

When working with oocyte homogenates, the yellow, floating lipid layer which is present after centrifugation should be carefully avoided when removing supernatants, as it will greatly increase the background in immunoprecipitations. Homogenization is usually carried out on relatively few oocytes in a correspondingly small volume and a simple method for accomplishing this is to modify a Pasteur pipette. Break back a glass Pasteur pipette until the broken end is slightly larger than the bottom of an Eppendorf tube. Flame the broken end until the closed form resembles the end of an Eppendorf tube. With a little practice a good, tight-fitting Pasteur pipette homogenizer can be made. For larger batches of oocytes, a glass Potter homogenizer can be used.

6.2 **Modes of analysis**

6.2.1 *Biochemical analysis*

The membrane proteins produced in the oocyte by translation of mRNA from an ap-

Table 6. Oocyte homogenization, immunoprecipitation and fractionation.

A. Homogenization and Extraction

1. Separate oocytes from the incubation medium by centrifugation in Eppendorf tubes[a] for 5 min in a microcentrifuge.
2. Transfer the pelleted oocytes to an Eppendorf tube, add up to 300 μl of homogenization buffer[b] and homogenize using a modified Pasteur pipette, or a small glass Potter homogenizer for larger volumes.
3. Add detergent to 1–2% and shake the homogenate for 30–60 min at 4°C.
4. Centrifuge at 10 000 *g* for 20 min at 4°C.
5. Remove the supernatant, avoiding the yellow floating lipid phase.
6. Assay, or add directly to gel sample buffer[c] for analysis by gel electrophoresis (as in Table 4).

B. Immunoprecipitation

1. Using oocytes injected and labelled with, e.g. [^{35}S]methionine, divide the supernatant after solubilization or homogenization (at step 5 in A) into two equal fractions (~100–150 μl). Add antibody buffer[d] and an appropriate dilution of antibody. Incubate for 2 h at room temperature.
2. Precipitate the immune complex using the appropriate second antibody or *Staphylococcus aureus* membranes. Incubate on ice for 30 min.
3. Centrifuge the immune complex in a microcentrifuge for 10 min, and wash it three times with antibody buffer[d].
4. Finally resuspend the pellet in a gel sample buffer and at once heat at 100°C for 3 min. Analyse by gel electrophoresis (as in Table 4, steps 7 and 8).
5. Run negative controls with pre-immune serum or an irrelevant antibody[e] (for monoclonal antibodies). A very good additional control is obtained by pre-treating the relevant antibody with an excess of its own protein antigen: the specific band, alone, will be suppressed or greatly weakened. It is common for irrelevant cross-reaction or entrapment in the precipitate to give several other bands, which can be recognized as such in these controls.

C. Fractionation[f]

1. Homogenize 20–30 oocytes in 0.5 ml of 10% sucrose in sucrose buffer[g] and layer onto a step gradient of 1 ml of 20% sucrose on 1 ml of 50% sucrose. Centrifuge in a swing-out rotor at 15 000 *g* for 30 min at 4°C.
2. Remove fractions using a sterile plastic pipette. The 10% sucrose layer contains cytosolic components, the 10–20% interface mainly mitochondria and the 20–50% interface contains membrane and secretory proteins.
3. To prepare membranes alone, layer the homogenate onto a 20% sucrose cushion and spin as in step 1.
4. Resuspend the pellet (containing membranes and yolk protein) in sucrose buffer and centrifuge for 3 min in a microcentrifuge. Remove the supernatant containing the membranes using a sterile Pasteur pipette.

[a]Eppendorf tubes should be treated with dimethyldichlorosilane solution (e.g. from BDH Chemicals Ltd.), rinsed twice with distilled water and autoclaved.

[b]Homogenization buffer is 20 mM Tris-HCl (pH 7.6), 1 mM EDTA, 1 mM EGTA and (for example, as protease inhibitors) 1 mM phenylmethylsulphonyl fluoride (PMSF), 10 μg/ml soybean trypsin inhibitor or aprotinin (Sigma), 1 mM benzamidine and 100 μg/ml bacitracin.

[c]Sample buffer is 20 mM Tris-HCl (pH 7.5), 2% (w/v) SDS, 10% (w/v) glycerol, 0.1 M dithiothreitol, 0.007% bromophenol blue. Heat immediately at 80°C, 3 min. Alkylation by iodoacetamide should be performed at this stage if there is a suspicion of cross-linking occurring: details are given by Colman (ref. 11, Section 5.6).

[d]Antibody buffer is 20 mM Tris-HCl (pH 7.5), 150 mM NaCl plus detergent(s) used in the solubilization.

[e]Any irrelevant antibody which recognizes other products present which are very abundant should be avoided, due to the trapping effects. Thus, for brain poly(A) mRNA, anti-tubulin antibody and, for muscle poly(A) mRNA, anti-contractile protein antibodies, can give false positives for the products of rare messengers present.

[f]Adapted from ref. 11.

[g]Sucrose buffer is 20 mM Tris-HCl (pH 7.6), 50 mM NaCl, 10 mM magnesium acetate, 1 mM PMSF.

propriate source can be recognized by one of the following methods.

(i) *In vivo* labelling with co-injected [^{35}S]methionine (Section 7.1), homogenization and detergent solubilization [either from whole oocytes or from the membrane fraction (*Table 6C*)]. This is followed by immunoprecipitation when a specific antibody is available, then SDS–PAGE and fluorography or slicing of the gel.

(ii) Similar use of unlabelled oocytes, with the detergent extract being subjected to SDS–PAGE and Western blotting (assuming that an antibody recognizing a known denatured subunit is available).

(iii) The detergent extract can be applied to a receptor-specific affinity gel, if one is available. Examples are binding of the translated nAChR to α-bungarotoxin–Sepharose 6B and of the translated $GABA_A$ receptor to a benzodiazepine affinity gel (34). Elution is then either with a suitable specific ligand or with SDS solution. This method also requires the prior use of [^{35}S]methionine labelling, or the subsequent use of a receptor-specific antibody, either in Western blotting or with a ^{125}I-labelled second antibody.

(iv) A specific receptor ligand, in radiolabelled form, can be used in binding studies after oocyte translation. When a rare message is involved, ^{3}H-labelling would require a large number of oocytes. When pure, specific mRNAs (Section 7.3) are used, however, ^{3}H-labelling on about 20 injected oocytes can give a sufficient signal, using a high-affinity ligand. Thus, Kubo *et al.* (41) detected muscarinic receptors produced from their specific *in vitro*-transcribed mRNA using only 0.27 nM [^{3}H]quinuclidinyl benzilate (42 Ci/mmol), when 3–7 fmol per oocyte were bound. ^{125}I, ^{35}S or ^{32}P are much better for natural mRNA mixtures, and often essential. The labelling can be performed on whole oocytes, which can take up label only on the plasma membrane if the reagent is impermeable (e.g. [^{125}I]α-bungarotoxin), or it can be performed on a homogenate or membrane fraction or detergent extract. Irreversibly bound ligands, such as radioactive photo-affinity reagents, offer the additional possibility of recognizing receptor subunits. An example is the [^{3}H]flunitrazepam labelling of a 53 000-dalton subunit seen after natural brain $GABA_A$ receptor mRNA is translated. The oocytes are treated with this ligand (10 nM) under u.v. light, the membrane fraction is prepared and deoxycholate-extracted and the labelled subunit of the $GABA_A$ receptor is visualized in SDS–PAGE by fluorography or by gel-slicing and counting (34).

(v) Bulk measurement of ion flux into the oocyte can be used to estimate the channel activity where time-resolution is not required. The oocytes, after foreign mRNA-directed channel synthesis is complete, are suspended in the isotope in modified Barth's medium. Examples are with $^{36}Cl^-$ or $^{86}Rb^+$ (both from Amersham International Ltd.). Incubation at 21°C of 10–20 oocytes per experimental point is followed by washing in modified Barth's solution until the supernatant is non-radioactive and then counting the radioactivity taken up by the oocytes. Thus, Morgan *et al.* (72) found, after implantation into the oocyte membrane of the red cell band 3 anion transporter, a linear uptake of $^{36}Cl^-$ over a 5-h total period, and Breer and Benke (81) made similar observations with $^{86}Rb^+$ entry after nAChR implantation.

6.2.2 *Electrophysiological analysis*

As noted above, intracellular recording is readily applied to the very large *Xenopus* oocyte. The expression of a channel *per se* must always rely upon this mode of analysis for its demonstration. The full functional properties of a receptor are also demonstrated thus after translation.

For simple testing, potential recording can be used. For further analysis, current-clamping (66) or voltage-clamping (*Figure 2* and ref. 69) or patch-clamping (54) of the oocyte membrane is now employed. The electrophysiological techniques used are beyond the scope of this chapter, and are described in the references cited on such results in the oocyte. It should be noted that temperature control at around 20°C is desirable during the recording, and that a small-volume (0.2−0.5 ml), very rapid superfusion chamber with dual pump control is desirable for rapid bath-application and wash-out.

For most purposes, folliculated but detached or manually defolliculated single oocytes are used. The latter should be used if possible, but often the response is just as good with the follicle present. (See Section 5 for a discussion of the effects of defolliculation upon the response.) Some workers (e.g. ref. 54) use collagenase treatment after mRNA injection for all such studies. The oocytes are selected at the time of the injection. By electrode impalement of some parallel oocytes from the same batch (which are not then used further) the resting membrane potential (RMP) of the batch is checked to be about −50 mV or more negative, and the input membrane resistance to be 1 MOhm or greater. After injection the oocytes are left in modified Barth's medium at 21°C overnight, or for longer periods. The optimum period is 48 h for many mRNA responses. Some types appear within 3 h while in a few cases it may take up to 5 days to become maximal. If the injected oocytes are stored at 4°C for up to a few weeks and then allowed to recover at 21°C, for some mRNAs the response, although weaker, still persists. Shaking of the oocytes (e.g. if in transit elsewhere) after injection can much reduce their viability.

When the recording electrode is inserted, then after a further recovery period (at least 15 min) the RMP should be found to be restored. Oocytes with RMP of negative potential smaller than −50 mV or input resistance less than 0.3 MOhm should be discarded.

Control injections with medium alone or an irrelevant message should from time to time be added.

(i) *Patch-clamping.* For patch-clamping, defolliculation by collagenase and removal of the vitelline membrane by osmotic treatment are needed, following the procedures of Methfessel *et al* (54). These workers recommend a multi-stage procedure.

(1) mRNA injection into folliculated oocytes.
(2) On the second day, collagenase (1 mg/ml, 1 h) defolliculation, and storage at 19°C.
(3) 'Stripping' in hypertonic medium: 200 mM potassium aspartate, 20 mM KCl, 1 mM $MgCl_2$, 10 mM EGTA, 10 mM Hepes (pH 7.4), at 0.475 mOsmolar. After 5 min the vitelline membrane becomes elevated and (under 2200 magnification, dark-field) is removed by forceps (Dumont no.5).

The stripped oocytes adhere tightly to a Falcon 30001F plastic chamber, which is perfused. High-resistance seals are then readily made.

It was noted that spontaneous stretch-cultivated channels are common in these patches, equally from injected and non-injected cells, which interfere with implanted channel recording. This was overcome by the use of smaller (~ 0.5 μm) tip pipettes with thick rims. Fluoride (80 mM) is advantageous in the pipette solution. For details of these and other points of oocyte patch-clamping, see ref. 54.

(ii) *Surface area and channel density.* The surface area of a denuded oocyte, of 1–1.2 mm diameter, was found by capacitance measurements to be about 2×10^7 μm^2 (54). This shows that the surface is increased 4–8 times by the microvilli present (76). These do not, however, interfere with patch-clamping. The density of channels when the specific mRNAs are co-injected reaches the high value of 40 per μm^2 or above for the nAChR, and presumably will tend towards this value with other specific mRNAs, assuming that the mRNA is equally long-lived. This is more than ample for cell-attached patch-clamp recording.

7. ADDITIONAL BIOCHEMICAL PROCEDURES

7.1 **Radioactive labelling of oocytes**

The simplest and easiest way to label the protein produced by the oocyte is to incubate the injected oocytes in a minimum volume of Barth's solution (10–20 oocytes/100 μl) with the appropriate radioactive amino acid or sugar. Microtitre plates or siliconized sterile Eppendorf tubes serve equally well in this respect. Incorporation of the radioactive compound depends on the oocyte pool for each particular compound and may be affected by non-uniform uptake processes for each radiolabel. Amino acid pools are listed in Table 6 of ref. 11. The oocyte pool for phosphate is particularly large (13) and the precursor pool for methionine is the smallest of all the amino acid pools (11).

A more efficient method of labelling is to co-inject the radiolabel. The advantages in this approach are that the label is initially more concentrated near the injected mRNA than elsewhere in the oocyte where endogenous mRNAs may be present and less free radioactive label is present in the incubation medium when a secreted protein is synthesized. For co-injection of a radiolabelled precursor, fill a second micropipette with the latter in water and inject immediately following the mRNA injection. Alternatively, if possible, freeze-dry a suitable amount of the radiolabel, for example 50 μCi/10–20 oocytes, then re-dissolve this in the mRNA solution. This has the advantage of not requiring a second injection into the oocyte, which would increase the mortality rate.

7.2 **Fractionation of mRNA**

When a response is obtained in the oocyte to a given crude poly(A) mRNA preparation, it is sometimes of value to fractionate the latter in active form. This can be done by, for example, sucrose density gradient centrifugation. This can give an enrichment, and also an estimate of the size(s) of the mRNA(s) involved. If more than one mRNA is involved in the receptor or channel and these have quite different sizes, combination of different fractions in turn will be needed to identify these. Further, the size-selection and enrichment of the particular mRNA which can occur in favourable cases can be of marked value in cDNA library construction, when these size-selected active fractions are used.

A suitable sucrose gradient is an isokinetic one of 10–31% sucrose, containing 10 mM Hepes (pH 7.5) buffer, 1 mM EDTA and 0.1% lithium dodecyl sulphate, run at 2°C (e.g. in a Beckman SW41 rotor, 120 000 g_{av}, 18 h, with non-braking deceleration). A parallel gradient contains ribosomal and other RNA size markers. The sample RNA is layered on, in the same medium (0.2 ml). Aliquots of every 0.4 ml fraction are taken for reading the A_{260} and for oocyte expression (after collection by ethanol/NaCl precipitation in the presence of 2 μg of tRNA as carrier). In this way, Sumikawa *et al.* (61) showed that the four mRNAs of the *Torpedo* nAChR are in a fairly narrow size range, centred at 18S, and likewise (32) for the mRNAs of chick muscle; several such applications using oocyte monitoring of sized mRNA have been described since. An example with the brain $GABA_A$ receptor mRNAs is shown in *Figure 4*.

7.3 Production of a specific single mRNA

When the cloned cDNA(s) for the protein in question are available, the specific mRNA is at present best obtained for oocyte injection by applying *in vitro* transcription in the phage SP6 or T7 system, by techniques which will be obvious to the laboratories producing the cDNA. Capping of the mRNA, for example with diguanosine triphosphate, is needed in general for the oocyte translation; exogenous poly-adenylation is not. This approach can be expected to produce a much greater number of the receptor or channel proteins in the oocyte membrane. This was realised in practice in estimates by Methfessel *et al.* (54): *Torpedo* natural poly(A) RNA gave about 50 times less of the α-bungarotoxin binding sites than its specific mRNA (the mixture of four types for all the subunits) and about 10 000 times less of the estimated channel density.

The specific, pure mRNA (8–16 ng per oocyte) encoding the muscarinic response, which has only one type of polypeptide, produced a channel, as well as a ligand-binding response, in the injected oocytes (41). This shows that the channel recorded need not be provided by the translated protein, but can be supplied to the latter from the accessory systems of the oocyte.

A further dimension in the use of a specific mRNA is the production of the specific anti-sense mRNA, on the complementary duplex strand of the cDNA. When it is not the case, or not known, that all of the specific mRNAs needed can be produced from the cDNA(s) cloned, the anti-sense mRNA for any of the receptor subunits should, if pre-injected in excess with the natural mRNA mixture, block (at least in part) the response to the injection of the latter. This 'hybrid-arrest' test can be of much value during the stages of cloning the cDNAs for a hetero-oligomeric receptor or channel protein.

8. ACKNOWLEDGEMENTS

Contributions to the experimental work reviewed here were made by (in chronological order): K.Sumikawa, B.M.Richards (at Searle Research Laboratories, High Wycombe),

R.Miledi (at University College, London), K.Houamed, T.G.Smart and A.Constanti (at the Department of Pharmacology, School of Pharmacy, London), Catherine Van Renterghem, S.J.Moss, F.A.Lai, M.G.Darlison, D.Carpenter-Ferguson, Barbara Demeneix and Vincenza Reale. Citations to some of their published work are given in the References. We thank Mrs Mary Wynn for invaluable secretarial assistance. The methods described in this chapter were developed since 1980 in the authors' laboratory at the Department of Biochemistry, Imperial College of Science and Technology, London, and more recently after moving that laboratory to the MRC Molecular Neurobiology Unit, Cambridge. It has been supported throughout by the Medical Research Council (UK). G.B. was supported by the Science and Engineering Research Council (UK) when engaged in this work.

9. REFERENCES

1. Mishina,M., Takai,T., Imoto,K., Noda,M., Takahashi,T., Numa,S., Methfessel,C. and Sakmann,B. (1986) *Nature,* **321**, 406.
2. Clemens,M.J. (1984) In *Transcription and Translation – A Practical Approach.* Hames,B.D. and Higgins,S.J. (eds), IRL Press Ltd., Oxford and Washington, DC, p. 231.
3. Mendez,B., Valenzuela,P., Martial,J.A. and Baxter,J.D. (1980) *Sceince,* **209**, 695.
4. Sumikawa,K., Miledi,R., Houghton,M. and Barnard,E.A. (1983) In *Cell Surface Receptors.* Strange,P.G. (ed.), Macmillan, London, p. 249.
5. Anderson,D.J. and Blobel,G. (1981) *Proc. Natl. Acad. Sci. USA,* **78**, 5598.
6. Anderson,D.J. and Blobel,G. (1983) *Cold Spring Harbor Symp. Quant. Biol.,* **48**, 125.
7. Sumikawa,K., Houghton,M., Emtage,J.S., Richards,B.M. and Barnard,E.A. (1981) *Nature,* **292**, 862.
8. Gurdon,J.B., Lane,C.D., Woodland,H.R. and Marbaix,G. (1971) *Nature,* **233**, 177.
9. Taylor,M.A., Johnson,A.D. and Smith,L.D. (1985) *Proc. Natl. Acad. Sci. USA,* **82**, 6586.
10. Kobayashi,S., Ho,H. and Aoshima,H. (1986) *Mol. Brain Res.,* **1**, 93.
11. Colman,A. (1984) In *Transcription and Translation – A Practical Approach.* Hames,B.D. and Higgins, S.J. (eds), IRL Press Ltd., Oxford and Washington, DC, p. 271.
12. Dumont,N.J. (1972) *J. Morphol.,* **136**, 153.
13. Woodland,H. (1982) *Biosci. Rep.,* **2**, 471.
14. Gurdon,H.B., Lingrel,J.B. and Marbaix,G. (1973) *J. Mol. Biol.,* **80**, 539.
15. Lane,C.D. (1983) In *Current Topics in Developmental Biology.* Moscona,A.A. and Monroy,A. (eds), Academic Press, Inc., New York and London, Vol. **18**, p. 89.
16. Asselbergs,F.M. (1979) *Mol. Biol. Rep.,* **5**, 199.
17. Marbaix,G. and Huez,G. (1980) In *The Transfer of Cell Constituents into Eukaryotic Cells.* Celis,J.E., Graessman,A. and Loyter,A. (eds), Plenum Press, New York, p. 347.
18. Nudel,U., Soreq,H., Littauer,U.Z., Marbaix,G., Huez,G., Leclercq,M., Hubert,E. and Chantrenne,H. (1976) *Eur. J. Biochem.,* **64**, 115.
19. Richter,J. and Smith,L. (1981) *Cell,* **27**, 183.
20. Marbaix,G. and Lane,C.D. (1972) *J. Mol. Biol.,* **67**, 517.
21. Rapoport,R.A. (1981) *Eur. J. Biochem.,* **115**, 665.
22. Lane,C.D., Champion,J., Haiml,Z. and Kreil,G. (1981) *Eur. J. Biochem.,* **113**, 273.
23. Soreq,H. (1985) *CRC Crit. Rev. Biochem.,* **18**, 199.
24. Mous,J., Peeters,B. and Rombauts,W. (1980) *FEBS Lett.,* **122**, 105.
25. Cutler,D., Lane,D.C. and Colman,A. (1981) *J. Mol. Biol.,* **153**, 917.
26. Deacon,N.J. and Ebringer,A. (1979) *FEBS Lett.,* **79**, 191.
27. Ullrich,A., Bell,J.R., Chen,E.Y., Herrera,R., Petruzzelli,L.M., Dull,T.J., Gray,A., Coussens,L., Liao,Y.-C., Mason,A., Seeburg,P.H., Grunfeld,C., Rosen,O.M. and Ramachandran,J. (1985) *Nature,* **313**, 756.
28. Valle,G., Besley,J. and Colman,A. (1981) *Nature,* **291**, 5813.
29. Labarca,C. and Paigen,K. (1977) *Proc. Natl. Acad. Sci. USA,* **74**, 4466.
30. Olson,E.N., Glaser,L., Merlie,J.P., Sebbane,R. and Lindstrom,J. (1983) *J. Biol. Chem.,* **258**, 13946.
31. Barnard,E.A., Miledi,R. and Sumikawa,K. (1982) *Proc. R. Soc. Lond. B,* **215**, 241.
32. Barnard,E.A., Beeson,D., Bilbe,G., Brown,D.A., Conti-Tronconi,B.M., Dolly,J.O., Dunn,S.M.J., Mehraban,F., Richards,B.M. and Smart,T.G. (1983) *Cold Spring Harbor Symp. Quant. Biol.,* **48**, 109.

33. Smart,T.G., Constanti,A., Houamed,K., Bilbe,G., Brown,D.A., Barnard,E.A. and Van Renterghem,C. (1986) *Adv. Exp. Med. Biol.*, **203**, 252.
34. Barnard,E.A., Bilbe,G., Houamed,K., Smart,T.G. and Van Renterghem,C. (1987) *Neuropharmacology*, in press.
35. Sakmann,B., Methfessel,C., Mishina,M., Takahashi,T., Takai,T., Kurasaki,M., Fukuda,K. and Numa,S. (1985) *Nature*, **318**, 538.
36. Kusano,K., Miledi,R. and Stinnakre,J. (1977) *Nature*, **270**, 739.
37. Dascal,N. and Landau,E.M. (1980) *Life Sci.*, **27**, 1423.
38. Dascal,N. and Landau,E.M. (1982) *Proc. Natl. Acad. Sci. USA*, **79**, 3052.
39. Dascal,N., Landau,E.M. and Lass,Y. (1984) *J. Physiol.*, **352**, 551.
40. Sugiyama,H., Hisanaga,Y. and Hirono,C. (1985) *Brain Res.*, **338**, 346.
41. Kubo,T., Fukuda,K., Mikami,A., Maeda,A., Takahashi,H., Mishina,M., Haga,T., Haga,K., Ichiyama, A., Kangawa,K., Kojima,M., Matsuo,K., Kirose,T. and Numa,S. (1986) *Nature*, **323**, 411.
42. Lotan,I., Dascal,N., Cohen,S. and Lass,Y. (1982) *Nature*, **298**, 572.
43. Van Renterghem,C., Penit-Soria,J. and Stinnakre,J. (1985) *Proc. R. Soc. Lond. B*, **223**, 389.
44. Lotan,I., Dascal,N., Cohen,S. and Lass,Y. (1986) *Pflugers Arch.*, **406**, 158.
45. Kusano,K., Miledi,R. and Stinnakre,J. (1982) *J. Physiol.*, **328**, 143.
46. Gundersen,C.B., Miledi,R. and Parker,I. (1983) *Proc. R. Soc. Lond. B*, **219**, 103.
47. Sadler,S.E. and Maller,J.L. (1982) *J. Biol. Chem.*, **257**, 127.
48. Morrill,G.A., Ziegler,D., Kunar,J., Weinstein,S.P. and Kostellaw,A.B. (1984) *J. Membr. Biol.*, **77**, 201.
49. Dascal,N., Snutch,T.P., Lubbert,H., Davidson,N.R. and Lester,H.A. (1986) *Science*, **231**, 1147.
50. Barish,M.E. (1983) *J. Physiol.*, **342**, 309.
51. Belle,R., Ozon,R. and Stinnakre,J. (1977) *Mol. Cell. Endocrinol.*, **8**, 65.
52. Dascal,N., Lotan,I., Gillo,B., Lester,H.A. and Lass,Y. (1985) *Proc. Natl. Acad. Sci. USA*, **82**, 6001.
53. Baud,C. and Kado,R.T. (1984) *J. Physiol.*, **356**, 275.
54. Methfessel,C., Witzemann,V., Takahashi,T., Mishina,M., Numa,S. and Sakmann,B. (1986) *Pflugers Arch.*, **407**, 577.
55. Dascal,N. and Landau,E.M. (1984) *J. Exp. Zool.*, **230**, 131.
56. Kado,R.T. (1983) In *Physiology of Excitable Cells.* Grinnell,A.D. and Moody,W.J. (eds), Alan R.Liss Inc., New York, p. 247.
57. Stinnakre,J. and Van Renterghem,C. (1987) *J. Physiol.*, in press.
58. Schorderet-Slatkine,S., Schorderet,M. and Baulieu,E.E. (1982) *Proc. Natl. Acad. Sci. USA*, **79**, 850.
59. Lotan,I., Dascal,N., Oron,Y., Cohen,S. and Lass,Y. (1985) *Mol. Pharmacol.*, **28**, 170.
60. Oron,Y., Dascal,N., Nadler,E. and Lupu,M. (1985) *Nature*, **313**, 141.
61. Sumikawa,K., Houghton,M., Smith,J.C., Bell,L., Richards,B.M. and Barnard,E.A. (1982) *Nucleic Acids Res.*, **10**, 5809.
62. Mishina,M., Kurosaki,T., Tobimatsu,T., Morimoto,M., Noda,T., Yamamoto,M., Terao,M., Lindstrom,J., Takahashi,T., Kuno,M. and Numa,S. (1984) *Nature*, **307**, 604.
63. White,M.M., Mixter-Mayne,K., Lester,H.A. and Davidson,N. (1985) *Proc. Natl. Acad. Sci. USA*, **82**, 4852.
64. Leonard,J., Snutch,T., Lubbert,H., Davidson,N. and Lester,H.A. (1986) *Biophys. J.*, **49**, 386a.
65. Miledi,R., Parker,I. and Sumikawa,K. (1982) *Proc. R. Soc. Lond. B*, **216**, 509.
66. Smart,T.G., Constanti,A., Bilbe,G., Brown,D.A. and Barnard,E.A. (1983) *Neurosci. Lett.*, **40**, 55.
67. Houamed,K.M., Bilbe,G., Smart,T.G., Constanti,A., Brown,D.A., Barnard,E.A. and Richards,B.M. (1984) *Nature*, **310**, 318.
68. Smart,T.G., Houamed,K., Van Renterghem,C. and Constanti,A. (1987) *Biochem. Soc. Trans.*, **15**, 117.
69. Van Renterghem,C., Bilbe,G., Moss,S., Smart,T.G., Brown,D.A. and Barnard,E.A. (1987) *Mol. Brain Res.*, **2**, 21.
70. Gundersen,C.B., Miledi,R. and Parker,J. (1984) *Nature*, **308**, 421.
71. Goldin,A., Snutch,T., Hubbert,H., Dowsett,A., Marshall,J., Auld,V., Downey,W., Fritz,L.C., Lester, H.A., Dunn,R., Catterall,W.A. and Davidson,N. (1986) *Proc. Natl. Acad. Sci. USA*, **83**, 7503.
72. Morgan,M., Hanke,P., Grzynczyk,R., Tintschl,A., Fasold,H. and Passow,H. (1985) *EMBO J.*, **4**, 1927.
73. Chirgwin,J.M., Przybyla,A.E., MacDonald,R.J. and Rutter,W.J. (1979) *Biochemistry*, **18**, 5294.
74. Carmichael,G.G. and McMaster,G.K. (1980) In *Methods in Enzymology.* Grossmann,L. and Moldave,K. (eds), Academic Press, New York, Vol. 65, p. 381.
75. Maniatis,T., Fritsch,E.F. and Sambrook,J. (1982) *Molecular Cloning. A Laboratory Manual.* Cold Spring Harbor Laboratory Press, New York.
76. Dumont,J.N. and Brumett,A.R. (1978) *J. Morphol.*, **155**, 73.
77. Capco,D.J. and Jeffery,W.R. (1982) *Dev. Biol.*, **89**, 1.
78. Browne,C.L., Wiley,H.S. and Dumont,J.N. (1978) *Science*, **203**, 182.

79. Frank,M. and Horowitz,S.B. (1980) *Am. J. Physiol.*, **238**, C133.
80. Contreras,R., Cheroutre,H., Degrave,W. and Fiers,W. (1982) *Nucleic Acids Res.*, **10**, 6353.
81. Breer,H. and Benke,D. (1986) *Mol. Brain Res.*, **1**, 111.
82. Laemmli,U.K. (1970) *Nature*, **277**, 680.
83. Noda,M., Ikedu,T., Suzuki,W., Takeshima,H., Takahashi,T., Kuno,M. and Numa,S. (1986) *Nature*, **322**, 826.
84. Sugiyama,W., Ito,I. and Hirono,C. (1987) *Nature*, **325**, 531.

INDEX